90 0418957 2

Conservation of Brick

Butterworth-Heinemann Series in Conservation and Museology

Published titles:

Artists' Pigments c.1600–1835, 2nd Edition (Harley)
Care and Conservation of Geological Material (Howie)
Care and Conservation of Palaeontological Material (Collins)
Chemical Principles of Textile Conservation (Tímár-Balázsy, Eastop)
Conservation and Exhibitions (Stolow)
Conservation and Restoration of Ceramics (Buys, Oakley)
Conservation and Restoration of Works of Art and Antiquities (Kühn)
Conservation of Building and Decorative Stone, Volumes 1 and 2 (Ashurst, Dimes)
Conservation of Glass (Newton, Davison)
Conservation of Historic Buildings (Feilden)
Conservation of Library and Archive Materials and the Graphic Arts (Petherbridge)
Conservation of Manuscripts and Painting of South-east Asia (Agrawal)
Conservation of Marine Archaeological Objects (Pearson)
Conservation of Wall Paintings (Mora, Mora, Philippot)
Historic Floors: Their History and Conservation (Fawcett)
The Museum Environment, 2nd Edition (Thomson)
The Organic Chemistry of Museum Objects, 2nd Edition (Mills, White)
Radiography of Cultural Material (Lang, Middleton)
The Textile Conservator's Manual, 2nd Edition (Landi)

Related titles:

Lighting Historic Buildings (Phillips)
Manual of Curatorship, 2nd edition (Thompson)
Manual of Heritage Management (Harrison)
Materials for Conservation (Horie)
Metal Plating and Patination (Niece, Craddock)
Museum Documentation Systems (Light)
Touring Exhibitions (Sixsmith)

Conservation of Brick

John Warren RIBA, FRTPI, FSA

With a contribution by Pat Ryan and David Andrews of Essex County Council

BUTTERWORTH HEINEMANN

OXFORD AUCKLAND BOSTON JOHANNESBURG MELBOURNE NEW DELHI

Butterworth-Heinemann
Linacre House, Jordan Hill, Oxford OX2 8DP
225 Wildwood Avenue, Woburn, MA 01801-2041
A division of Reed Educational and Professional Publishing Ltd

A member of the Reed Elsevier plc group

First published 1999

British Library Cataloguing in Publication Data
A catalogue record for this book is available from the British Library

ISBN 0 7506 3091 4

Library of Congress Cataloguing in Publication Data
A catalogue record for this book is available from the Library of Congress

Composition by Scribe Design, Gillingham, Kent
Printed and bound in Great Britain by The Bath Press, Bath

Contents

Pug! Why, that's the plastic clay, bor, the brick earth. We had to grind it to make it wukkable, loike, an' git all the stone out. When we chucked the water on it went all sloppy, an' as the mill turned the stones wukked outer the soides. Most o' the water drained away when we filled the bins with pug. That had to lay there a day or two afore we could move it to the mouldin' shed, an' that's where owd Captain done the mouldin'. Ain't you never seen anybody a-mouldin' bricks, then? Well, you hev a mould what's lined with metal, jest the size of a brick, an' the bottom on't is raised a bit for the frog. You picks up a lump o' pug, bang it down hard into the mould an' fill it. Then you pick up a mould with the pug in it an' jounces it a time or two on the bench jest to make sure she's full. When she's nice an' full you take a strike, thasser bit o' wood, an' cut the pug off clean at the sides. Then you pick up a wood pallet what's a mite bigger'n the mould, an' lays it on the mould. You turn the whole lot upsides down an' let it mould off. An' there's yer brick, moulded, frog an' all.

from *Five Miles from Bunkum*
– an account of nineteenth century life near Saffron Walden

Series Editors' Preface

The conservation of artefacts and buildings has a long history, but the positive emergence of conservation as a profession can be said to date from the foundation of the International Institute for the Conservation of Museum Objects (IIC) in 1950 (the last two words of the title being later changed to Historic and Artistic Works) and the appearance soon after in 1952 of its journal *Studies in Conservation*. The role of the conservator as distinct from those of the restorer and the scientist had been emerging during the 1930s with a focal point in the Fogg Art Museum, Harvard University, which published the precursor to *Studies in Conservation, Technical Studies in the Field of the Fine Arts* (1932–42).

UNESCO, through its Cultural Heritage Division and its publications, had always taken a positive role in conservation and the foundation, under its auspices, of the International Centre for the Study of the Preservation and the Restoration of Cultural Property, in Rome (ICCROM), was a further advance. The Centre was established in 1959 with the aims of advising internationally on conservation problems, co-ordinating conservation activators and establishing standards of training courses.

A significant confirmation of professional progress was the transformation at New York in 1966 of the two committees of the International Council of Museums (ICOM), one curatorial on the Care of Paintings (founded in 1949) and the other mainly scientific (founded in the mid-1950s) into the ICOM Committee for Conservation.

Following the Second International Congress of Architects and Technicians of Historic Monuments, which met in Venice in 1964, the Venice Charter was promulgated as a declaration on the fundamental principles of architectural conservation. In the following year, the International Council of Monuments and Sites (ICOMOS) was founded to 'encourage the adoption and implementation of international recommendations concerning monuments, groups of buildings and sites'. Since then, the national committees and international scientific committees of ICOMOS have contributed to the promulgation of several international charters and guidelines on such subjects as historic gardens, historic towns and urban areas, archaeological heritage management, and training. ICOMOS is also one of the bodies delegated to advise the World Heritage Committee of UNESCO on the inclusion of properties on the World Heritage List. A major concern of the various organizations, IIC, ICOM, ICOMOS and ICCROM, is not only to advance the various technical specializations, but also to emphasize the cohesion of conservators and their subjects as an interdisciplinary profession.

The use of the term *Conservation* in the title of this series refers to the whole subject of the care and treatment of valuable artefacts both movable and immovable, but within the discipline conservation has a meaning which is distinct from that of restoration. *Conservation* used in this specialized sense has two aspects: first, the control of the environment to minimize the decay of artefacts and materials; and, second, their treatment to arrest decay and to stabilize them where possible against further deterioration. Restoration is the continuation of the latter process, when conservation treatment is thought to be insufficient, to the extent of reinstating an object, without falsification, to a condition in which it can be exhibited.

In the field of conservation conflicts of values on aesthetic, historical, or technical grounds are often inevitable. Rival attitudes and methods inevitably arise in a subject which is still developing and at the core of these differences there is often a deficiency of technical knowledge. That is one of the principal *raisons d'être* of this series. In most of these matters ethical principles are the subject of much discussion, and generalizations cannot easily cover (say) buildings, furniture, easel paintings and waterlogged wooden objects.

A rigid, universally agreed principle is that all treatment should be adequately documented. There is also general agreement that structural and decorative falsification should be avoided. In addition there are three other principles which, unless there are overriding objections, it is generally agreed should be followed.

The first is the principle of the reversibility of processes, which states that a treatment should normally be such that the artefact can, if desired, be returned to its pre-treatment condition even after a long lapse of time. This principle is impossible to apply in some cases, for example where the survival of an artefact may depend upon an irreversible process. The second, intrinsic to the whole subject, is that as far as possible decayed parts of an artefact should be conserved and not replaced. The third is that the consequences of the ageing of the original materials (for example 'patina') should not normally be disguised or removed. This includes a secondary proviso that later accretions should not be retained under the false guise of natural patina.

The authors of the volumes in this series give their views on these matters, where relevant, with reference to the types of material within their scope. They take into account the differences in approach to artefacts of essentially artistic significance and to those in which the interest is primarily historical, archaeological or scientific.

The volumes are unified by a systematic and balanced presentation of theoretical and practical material with, where necessary, an objective comparison of different methods and approaches. A balance has also been maintained between the fine (and decorative) arts, archaeology and architecture in those cases where the respective branches of the subject have common ground, for example in the treatment of stone and glass and in the control of the museum environment. Since the publication of the first volume it has been decided to include within the series related monographs and technical studies. To reflect this enlargement of its scope the series has been renamed the Butterworth-Heinemann Series in Conservation and Museology.

Though necessarily different in details of organization and treatment (to fit the particular requirements of the subject) each volume has the same general standard. The aim of the series is to provide authoritative texts for university level courses in conservation and, at the same time, be standard reference material for practising conservators.

The authors have been chosen from among the acknowledged experts in each field, but as a result of the wide areas of knowledge and technique covered even by the specialized volumes in this series, in many instances multi-authorship has been necessary.

With the existence of IIC, ICOM, ICOMOS and ICCROM, the principles and practice of conservation have become as internationalized as the problems. The collaboration of Consultant Editors will help to ensure that the practices discussed in this series will be applicable throughout the world.

Preface

The purpose of this volume is to put the conservator and student in possession of a method of approach to brickwork and associated materials which reflects current practice and philosophies in building conservation. It does not seek to offer short cuts or immediate answers but aims to set down fundamental methods of procedure or thinking.

The approach to the problem is analytical. The intention is that each aspect is examined sufficiently to enable the reader to visualize the nature of the problem, to deduce from first principles the available remedies and to foresee dangers. Historic buildings are complex entities in which many simultaneous conditions, circumstances and criteria present problems of interacting complexity. While simple remedies can be helpful their application must be seen within a context of history, aesthetics, evolution and changing circumstance that demands a clear understanding of the principles involved.

I am glad to record with gratitude the help of many friends and colleagues, particularly in my own practice and the Conservation Practice. To colleagues at The Institute of Advanced Architectural Studies in the University of York I owe the stimulus of preparing lectures which have given rise to this volume. As architect to The Open Air Museum at Singleton, and in the Wealden Buildings Study Group I have had the opportunity to investigate processes of building that a professional life does not normally provide. I owe a special debt to Margaret Holt of the Group, ever an enthusiast for the historic brickwork of Sussex. To colleagues in these organizations and in the Amberley Museum of Industrial History I express my gratitude for support and discussion. To Sir John Smith, who founded the Landmark Trust, I record thanks for generous opportunity and shared enthusiasm.

Direct help in producing this volume has been given very generously by Dr Ralph Heeley, late of Redbank Ltd, brick- and tilemakers. Dr Heeley has been a very important mentor. Part of the text of Chapter 3 is taken from an Essex County Council publication on Cressing Temple with the much appreciated cooperation of the County Archaeologist. The Department of Archaeology at the University of Southampton and Dr Maurice Howard of the University of Sussex share my thanks for their technical help and I am glad to acknowledge the specific assistance of Dr Ettore Vio, Alan Robson, Dr A. Walls, Jean-Marie Teutonico, John Fidler, Nick Maclean, Ray Harrison, Nicholas Wood, Anthony Capo-Bianco, John Cannon and those authors and publishers who have kindly allowed their work to be quoted. Inevitably I have drawn on the writings of many workers in the field, including Dr Kathleen Watt and Kim Leslie in particular. The librarian of the Society of Antiquaries has been consistently helpful and to a lesser extent I have exploited the generosity of the RIBA. Ultimately theoreticians rely upon those who work in manufacture and on the buildings. Martin Welch of Mirror Construction and Crispin Fenwick stand out in the memory among many who have been thoughtful collaborators and, in the practical sense, tutors. To them all and many others unnamed I express gratitude for help and happily, in many cases, friendship.

To a very patient publisher, Neil Warnock-Smith, and to a perceptive reader and old friend, Dr Derek Linstrum, I owe a particular debt of thanks in that, but for them, this present volume would still be a compound of brick and mud-brick. To my wife, Judith, who has tirelessly put the manuscript into shape I owe my profoundest thanks and deepest gratitude.

J.W.

1

An outline of the evolution of bricks and brickwork

The oldest solid rock so far recovered is dated to some 3860 million years ago; that is to say that it has been a piece of solid material since the time when a substantial part of the earth's crust had cooled sufficiently to solidify but before the condensation of water. At that point weathering began. The material which had cooled into solids was broken down, moved, impacted upon, twisted, contorted, fractured, compressed, chemically attacked, solidified, eroded, remelted and metamorphosed; and then, with liquefaction of water, dissolved, eroded, transported, deposited, congealed, cemented and hardened under pressure. Some particles were moved by wind and most were transported by water. The processes were repeated again and again to leave after some 3000 million years a planetary surface composed very largely of small particles of that same rock weathered into small pieces, many times recemented or remelted to be broken down again into a residue of small particles covering the majority of the earth's surface. Man has used these particles as soil for his livelihood and structure for his habitations.

The particles resolidified by nature become sedimentary rocks which man may grind again into particles to process them into materials of his own choosing. Others he cements together using the natural formation of crystals. Yet others he compacts and coheres into building elements by bringing them to a temperature at which they partially melt and are melded into a solid. In a very small proportion of cases the adhesion between the particles is created by the introduction of a matrix of organic material almost always deriving initially from the hydrocarbon deposits of the earth's crust – natural bitumens and oils or their more sophisticated manufactured derivatives such as resins. Man has in his history of building created the wealth of artefacts that now draw his energy and ingenuity in resisting their decay under those forces which reduced the original rocks to earths, containing principally silica, carbonates and clays. While silica is unchanged from the original solidification, carbonates are the product of solution and biological action while clays derive from complex chemistries of the decomposition of rock, being formed by chemical reaction from minerals such as felspars in igneous rocks. The reactant is thought to have been water containing carbonic and sulphuric acids. The clays were concentrated and separated by water or wind. These secondary deposits may have been modified by further reaction (e.g. organic acids from the decay of vegetation) on the secondary deposition or by physical action. Slate is a rock formed from a sedimentary clay deposit rendered hard and non-plastic by heat and pressure. Chalk, limestone and marble are the rocks of carbonates, and sandstones are the cemented and compressed rocks of silica.

The Neolithic period has left us with baked terracotta figurines and some evidence of the use of earths for providing shelter, but the hunter gatherer was less likely to be a builder in the significant sense than was his successor, the cultivator. A more settled life style, a more permanent habitation and greater productivity were conducive to the organization of surplus labour into construction and this seems to have been achieved by about 13 000 BC; but the significant events in the early history of brick and earth construction can hardly be given a date more than 10 000 years later. By this time there were flourishing agricultural communities in the great river valleys of the Nile, the Tigris and the Euphrates, the Indus and flood plains of Northern China. Archaeology will continue to unearth earlier and more definitive evidence but present conclusions must be that the fundamental skills in the handling of earths, plasters and burned brick had by then been discovered and were widely exploited. Terracotta now provided him also with a material for the permanent record of his writings. The mark made on soft earth, no doubt long a method of communication, now became a permanent record when fired and the technique was a great deal easier than carving upon stone. It was about this period that the earliest baked bricks are apparent, made again in response to a special need which was simply durability under wet conditions. Both Egypt and Mesopotamia were passing through their early dynastic phases. Greater reserves of surplus manpower became available to the rulers. It is an almost universal human trait that a stable dynasty seeks to reflect its power and to secure itself with monumental and defensive building.

Further to the east in the Indus Valley the same development is apparent. This period, known as the Jamdat Nasser period in Iran and Mesopotamia, may be pinpointed as the emergence of literate and monument-building civilization and is the crucial point at which the beginnings of the significant history of building in earths and baked brick can be seen. The trade routes by land and sea linking the Nile, Mesopotamia and the Indus Valley must have allowed an interchange of technology sufficient to ensure that solutions to practical problems were known across the region. Knowledge of early events in Mesopotamia is greater than that of the other regions because more of its records have been recovered, and unlike those of the Indus valley they have been deciphered and can be read. Nevertheless the buildings of the Indus region at Harappa and Mohenjodaro command special attention because of the immense amount of construction in burned brick laid in flat masses in gypsum mortar.

The one further significant invention yet denied to many early builders of monuments was lime for mortar. While stone was laid dry in earth or perhaps in gypsum mortar, selected earth, as mud, was used for the bedding of mud brick. Rarely, bitumen was also used. But where brick could be baked limestone could be burned and carried as lump-lime to its place of use. So the two components of brickwork emerged together.

The first of the great Mesopotamian ziggurats was erected at Warka (Buruk) in the Sumerian archaic period. In the Dynastic period, the third millennium BC, a series of these great stepped pyramids was built across southern Mesopotamia, each with a massive mud-brick core faced with a skin of burned brick elaborated into decorative devices whose form is now by no means certain.

At Ur about 2100 BC this external skin was laid in a bituminous mortar on transverse beds of reed. At this period plano-convex dried mud brick was laid with its long axes vertically and the brick inclined. The inclination would be reversed every second or third course followed by a series of flat courses, as it was in the great temple complex at Khafaje. By this time Sumerian builders were already using the true arch and a form of pendentive with buttresses and vaulting. The handling of brickwork had entered into a new and more sophisticated phase which was to continue in a pattern of increasingly complex development. Within a millennium the first of the attached columns of spiral decoration had appeared, the earliest survivor being from the second millennium in northern Iraq at Tel Rimah. The tradition of timber framing with brickwork was then and remains now fundamental in the region with the result that even large brick structures included columnar decoration recessed into the faces and the spiral columns of Tel Rima are set back and set forward in exactly the same way as would have been timber members, from which they appear to have derived.

The focus of architectural development moved northwards. Tel Rima lies west of present day Mosul, Nimrud and Nineveh, Khorsabad further north and Bathargadi, Persepolis and Ecbatana on the Persian highlands. Dome building on squinches, vaulting and arching were not the only discoveries to presage the complexity and quality of future brick construction. Polychrome decoration using glazed brick also appears at this point.

These buildings provide the first evidence of bricks being made to specific shape and colour and brick modelling being introduced. These brickmakers had now dealt with the problems of making their bricks to the specific sizes required allowing for shrinkage. Bricks for vaulting were thinner and smaller than those used for the walling possibly to allow for their use in constructing inclined ring arch vaults where the adhesion face to face of the brickwork was important. Carved brick had been introduced with complex spiral patterns and surface ornamentation.

By the beginning of the first millennium at Nimrud raised decoration was used representationally. There carved marble slabs had traditionally protected from damp and physical damage the lower levels of unfired brick in important rooms. The technique now extended to representational wall panels carried out in fired, moulded or carved glazed brickwork. In nearby Khorsabad (722–705 BC, built by Sargon II) surviving glazed brickwork achieves a new brilliance while terracotta drains and brick sewers demonstrate the use of fired brick in practical applications. The earth bricks in contemporary structures were laid while still plastic allowing them to take up the shapes dictated by loading.

The city of Babylon was rebuilt from 605 to 561 BC and such was the productivity of the workforce that virtually the whole of the inner town was built with fired brick. Glazed, multicoloured surfaces faced gates and the great rooms which were modelled with figurative devices, predominantly dragons and bulls on a blue ground. The topmost storey of the ziggurat was faced in brilliantly blue glazed brick. The story continues into Iran through the sixth, fifth, fourth and third centuries BC in the greater sites including Ectabana, Susa and Persepolis. In these cities the glazed fired brick was in competition with painted carved stone or alabaster and despite the interruption to the pattern caused by the brick jointing and the limitations of the palette the intensity, glow and permanence of the fired surfaces gave the material a brilliance appropriate to royal status. The tradition was to continue through the succeeding Seleucid era with its interaction with Greece, through the Parthian and Sasanian periods which ultimately gave way to Islam, providing the continuity of tradition to the present day.

Sumerian builders in southern Mesopotamia had the gift of colour at their command as they moved from mere building to conscious architecture. Excavations at Buruk (Warka) and superficial finds elsewhere have shown that the interiors of the more important buildings were decorated with a form of mosaic patterning based on the use of conical pieces of terracotta given coloured glaze and embedded in masonry. Copper sheathing was also employed and gypsum surfaces were carved or enlivened with frescoes. Earth and brick had become the stuff of architecture and, with its exposure by the trowel, arises the first problem of the architectural conservator – the conservation of archaeological remains.

In the history of brick building the next significant step was the introduction of a mortar with greater strength than mud and greater durability than gypsum (Juss). With the smelting and working of metals, with the discovery of glass and glazing and with brickmaking itself, man ventured into the upper range of temperatures obtainable in simple kilns – up to about 1000°C – and at some point mastered the techniques of breaking down limestone or marble into quicklime to reconstitute it as a durable mortar.

The great and fundamental difference between the architecture of the Ancient Greeks and that of Rome is in the extensive use of lime mortar: indeed it may be the direct consequence of the availability of that material to the Romans that their architecture took on its totally new structural dimension. Perhaps in part it was also the technical advance offered by pozzuolana – aggregates of volcanic ash discovered near Naples – which conferred upon lime mortar the ability to set into a true concrete. Brick dust was soon discovered to be an artificial pozzuolan, for which purpose it has been used ever since.

While lime mortars and burned brick were known in Ancient Greece, monumental building relied upon superbly cut stone accurately fitted together to form trabeated structures in which intimate contact between stone surfaces was an essential mechanism. The resultant forces were balanced by the careful poise of mass and loading. Adhesion between these stones to carry loads into the matrix into which they were keyed was no part of the concept. Stone was handled like timber. It was shaped true to the joint and the components were interlocked so that forces were transmitted by friction. In monumental building throughout the Hellenic period (sixth to fourth centuries BC) this philosophy seems to have applied both in the great established cities of the mainland and in their colonies around the Mediterranean, persisting well into the succeeding Hellenistic period. Even in the Hellenic period, however, most buildings were humble utilitarian structures in which the techniques of monumental masonry had no part. Greek vernacular continued the traditions of earth structures perhaps faced with stone and more specifically stone structures with earth mortars. In these the principle of a bonding matrix was clear, but apparently it was regarded as an inferior form of construction which, in view of the immense achievements of the monumental designers, is understandable. This gulf between great building and the humbler house or fortification persisted until the introduction of stronger lime mortars allowed the construction of monumental buildings no longer dependent upon the immensely laborious shaping of large stone blocks.

As with most fundamental inventions the origins of the first setting mortars are uncertain and it may not have been among the Hellenistic Greeks that the potential of the powder or ash resulting from the heating of limestone became apparent in structural terms. Several centuries before the Romans achieved great engineering structures dependent on strong mortars Greek households were laying mosaic floors with tesserae embedded in cements. Well before the Roman period Etruscan artists were smoothing over gypsum panels as a base for paintwork to decorate the interiors of their more important buildings. However, the structural potential of the stronger mortars made by burning limes was not exploited until the emergence of the power of Rome.

It might be thought that the potential of brickwork for higher structures would not have been achieved until the invention of these more powerful mortars; but this is not the case. The sixth century BC was a period of enlightenment in China and the Indian subcontinent as much as in the Mediterranean. It was a period of intellectual progress in which the Buddha had laid down a philosophy encouraging man to move to a higher metaphysical plane – a level of Enlightenment. Around this philosophy monastic communities were built. The architectural consequences were significant and enduring and it was as a consequence of Buddhism that within a few centuries the highest brick structures ever reared were built using brick bedded in nothing more than earth mortar. Reputedly there were stupas which exceeded even those surviving at Anaradhupura in Sri Lanka, but even those that remain are the largest and tallest individual brick buildings ever achieved. So far as is known they are bedded throughout on mud mortars which contain a high proportion of silt and some gypsum. They are all subsequent to the earliest use of lime mortars in the Mediterranean basin and with the considerable intercourse between the Greek communities and the Indian sub-continent in the Seleucid period it seems less likely that the techniques of lime burning were unknown than that the builders of Sri Lanka were content with the materials available which allowed them to build to the limits of their capability.

As the tallest brick, structures ever built, the three great stupas of Anaradhupura deserve to be very much better known. They were constructed over a long period and the earliest, dating from the third century BC, is known as the Ruwanvelisaya Stupa. It rises approximately 100 metres from its foundation. Two centuries later at the beginning of the Roman era there was built the Abahayagiri Stupa which is a little larger and substantially greater in volume.

Rome was in decline when the fourth was built in the fourth century AD. The Jetavana Stupa is some 120 metres in height, exceeding its predecessors in mass. The volume of this stupa is equivalent to some three hundred and fifty million modern bricks. Typically each ancient brick was approximately 18" × 9" × 2" (500 mm × 250 mm × 50 mm) and the weight has been calculated at 657 000 tons

Figure 1.1 The Ruwanveliseya Dagaba at Anaradhupura, Sri Lanka, second century BC. The Dagaba is built of mass masonry bedded in earth mortar. From a central core radial structural walls are carried out to the massive shell. (Reproduced with permission of the High Commissioner of Sri Lanka, London.)

delivering a load on the base brick and mortar of 72 lb per square inch.

That these stupas were not the mere product of brute force and surplus manpower is clear from an analysis of their profile which demonstrates that their lower sections were ellipsoidal – in this case a part of the solid of revolution of an ellipse about its long axis. The sands and clays used in the manufacture of the bricks and the earth mortar were carefully graded and they differed, the mortar having a slightly higher proportion of sand. As the mortar was placed in a relatively liquid condition this additional sand reduced shrinkage effects and the builders had learned the lesson of homogeneity in the structure. The matrix or mortar being the weaker component but with greater plasticity the immense mass was uniformly stressed without developing weak pockets or voids.

Other comparable brick structures of the time included the City Wall of Anaradhupurna, 6 metres wide, 6 metres high and 4 kilometres in length. But if the overall massing and height are impressive the arching technology is the converse. Spans were largely achieved by corbelling with rare exceptions which show that the radial arch and vault were known though little used.

While the last of these immense structures was being reared the Chinese had achieved equivalent expertise in the use of fired clays and cementicious lime mortars, and it was at this time also that the foundations of modern brick engineering were being laid in the peninsula of Italy. Roman control of structural techniques and their extraordinary expertise in masonry construction lies fundamentally with the combination of available materials and the ingenuity founded on Hellenistic skills. Pumice and pozzuolana were both volcanic materials that could be used in intimate mixture with freshly slaked lime and sands to provide a mortar which set within days into a rigid and strong mass. Initially Roman brick and stone were bonded conventionally with relatively thin mortar joints but the potential of the mortar was developed until its use in massive pockets demands that it is thought of as concrete. The material was contained by permanent brick or stone shuttering, although there are occasional instances of trench-fill techniques having been used, the earth or 'tuf' being subsequently cut away to expose the concrete wall or vault.

It becomes apparent by this point that historians must take account of a factor which powerfully influenced the nature and character of building, but seems never to be recognized. It can be summarized in the one word 'fuel'.

Wood was the fuel of antiquity. Thermally wood is inefficient having only a moderate calorific value in relation to its density. Even in 'dry' condition it contains substantial amounts of water, usually between a fifth and a sixth part by weight. It does not occur in concentrated mass and, therefore, involves substantial transport while the hewing and handling of timber in volume is laborious.

All peoples and all activities are guided by the amount of fuel or energy consumed and few industries are more sensitive in this respect than building. The fundamental reason for the use of enormous stones in antiquity is that weight for weight – or volume for volume – less energy was consumed in the cutting of a large unit than in a small one and even in transport the ratio would often improve as the size increased. Likewise in timber a large girder was easier to prepare from a tree that simply needed squaring than the smaller timbers that required extensive sawing. Standing timber or uncut stone in the quarry did not of itself represent a valuable asset. As the prime cost was of no great moment large timbers were employed where to modern eyes they look uneconomic. Our contemporary observer marvelling at the immense size of ancient stones or timbers is in reality marvelling at the change in ratio between the cost of material and the cost in energy in its preparation and transport.

In terms of bricks and brickwork the phenomenon is the same although the effect is more subtle. To burn marble or limestone into lime, clay into tile and brick earth into brick demanded large volumes of timber. In hot countries these resources might be scarce and in temperate climates there was always the temptation to use the timber directly to create shelter rather than to burn it to prepare building materials. Large supplies of fuel were in either case only obtainable by widespread foraging and considerable labour in felling and transport. It follows that an initial preference was for the smaller and more valuable article – the tile, which almost always preceded the introduction of brick by generations if not centuries. The same truth is to be seen in the preparation of mortar and the lavish Roman use of concrete.

Volume for volume dried clay and broken nodules of limestone or marble demand about the same amount of fuel to bring them to temperatures upwards of 800°C. To make brick or tile satisfactorily, however, the process was necessarily slower. The temperature had to be maintained at its peak for a longer period so there was a significant advantage in firing stone rather than earth. The real advantage, however, arose subsequently when lime was mixed with sand to produce a mortar of four times the volume of the lime. Where volcanic tuf was added the ratio was improved further. Therefore effectively, a given amount of timber produced a far greater volume of built wall if used to burn lime than if used to fire clay. There was, therefore, a great temptation to increase the depth of mortar beds, to use brickwork only as shuttering against a mortar or concrete fill and to construct whole walls and vaults in mortar. It is no coincidence that the use of mass concrete, of high volumes of mortar and the skilful use of fired earths to contain them emerges at the point in history when empire building first developed a powerful sense of logistic analysis as it did among the Romans.

Sustained and consistent methodical thinking in terms of civil organization gave the Romans command over an empire whose frontiers were months of travel apart. Their methodical solutions to problems of organization extended to calculations of efficiency in engineering terms. Amounts of material required and consumed were analysed and effective methods of overcoming their problems were evolved on a sustainable basis, not task by task as in building the pyramids or monument by monument as in the classical temples of Greece but in the administration of parallel industries capable simultaneously of maintaining widespread networks of communication, extended commerce, civil engineering and the military machine, which was also a consistently trained force of builders. This intellectual method has given us also the first consistent and logical accounts of building technology in the writings of the Elder Pliny and Vitruvius among others. Through them we see the practised Roman mind developing an industry in brick- and tile-making in which lessons learned across the wide spread of empire were coordinated and translated into regular practice so that builders in Syria were working in much the same way as their counterparts in Britain. The patterns used by mosaic workers laying the tesserae in floors or setting tegula and imbrex on roofs were for the first time reliably consistent over long periods and wide areas. While this was not a difference in kind it was certainly a difference in scale among the western empires which leaves Roman builders standing out as the first to use bricks and tiles in virtually all the ways possible in building and to use mortars and concretes in conjunction with them. The important parallel achievements of the contemporary Chinese courts must not be ignored, however. The great nations of China were for the first time united with a universal system of law, a coherent culture and language and reliable communications. In the Ban Po culture of Northern China even earlier concrete floors are found. To assess their immense achievements in the use of fired earths one need look no further than the partially exposed armies of terracotta warriors at Xian. Important parallels will be drawn by future historians.

The Romans and their Byzantine successors used brick, tile, mortar and concrete powerfully and inventively and to do this they were obliged to create effective brick- and tile-making industries, the remains of which are now to be found regularly scattered, often on quite a small scale, across their empires. Many would concentrate on meeting local need. Large bricks might be demanded for massive vaulted structures, whereas in a heavily wooded area the prime need would be for roofing tiles to cover structures made of timber framing.

Although the Etruscans, their immediate predecessors, had developed considerable skills with fired clay, the Romans looked back unashamedly to Greek precedents and the invention of tile- and brick-making as they saw it in the Helladic (pre-Hellenic) period. Tilework can now be traced back to 1800 BC at Lerna near Argos in the Peloponnese (Dobson 1850; Brodribb, 1987), while Pliny held that the roof tile was invented in Cyprus. These early tiles from Lerna measured 230 mm in length by 190 mm in breadth and appear to have been laid on a bed with a butt joint. By the sixth century such tiles had upstands along the edges, which clearly indicates that the joints were covered by secondary coved tiles thus introducing the well known tegula and imbrex or flat and coved tile which ultimately was joined into the concave/convex curve structure of the pantile.

This invention can be seen in the later Roman period. Vitruvius cites three types of tile used by the Greeks: the Lydion, pentadoran and tetradoran (Vit 4) indicating the Lydion as the Roman preference. He also indicates that the cover tile or imbrex arose from the practice of plastering or daubing over the joints between the tegulae. Timber and bronze versions were also used by the Greeks. By the sixth century BC great temples were being roofed with combinations of imbrex and tegula virtually 1.2 m in length and 850 mm in

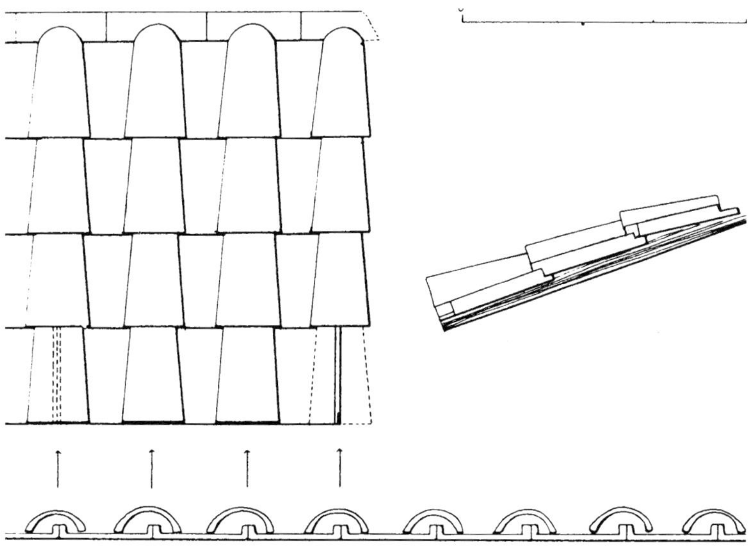

Figure 1.2 The construction of a roof, using tegulae and imbreces. (Courtesy of Gerald Brodribb.)

width. The sophistication of Roman work is apparent in the detailing of upstands with drip channels and the slotting necessary to allow each tile to house neatly into the head of the tile above to maintain weathering. Special items included finials and antefixes which filled the terminal end of the imbrex. These were often stamped or worked with figures and faces.

The Romans used other systems of roofing, particularly imbreces – alternately inverted and normal – a form of roofing known by the Greek term Iconion from its eighth or seventh century BC origins and taken into Byzantine practice. In the north they used a flat plain tile with a rounded or sometimes pointed tail, laid diagonally, probably in imitation of stone slab roofing.

While the material used to make these tiles was not of the greatest refinement it was significantly more clay rich and purified than the materials from which bricks were made, in other words it was a basic terracotta and as such demanded careful manufacturing techniques. This type of material was used for many other specific purposes – particularly drains, gutter members and even cornices. Roman expertise with the material reached heights of invention which demonstrate an extraordinary and virtually modern mastery. Systems of interlocking pipes are a fairly obvious development and the use of

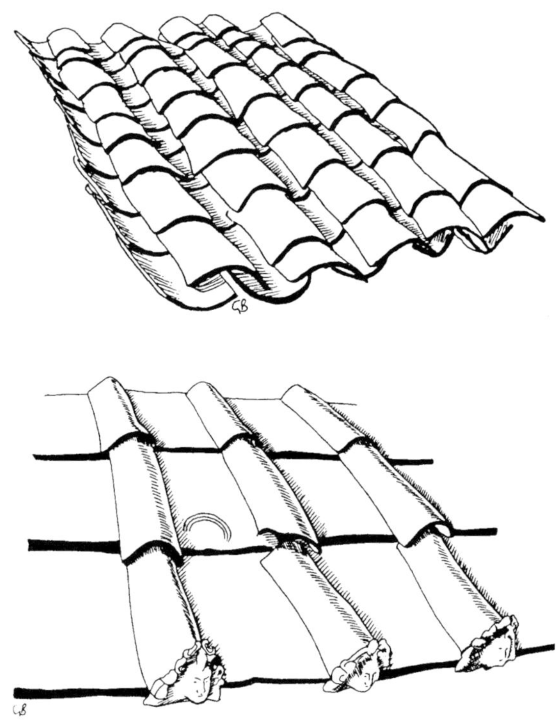

Figure 1.3 Graeco-Roman systems of roofing using imbreces alone and imbreces covering the joints of the tegulae with antefixae covering the terminations. This latter form of roofing was used on a higher calibre of buildings. Both roofs were laid at pitches of less than 30° and often as low as 24°. (Courtesy of Gerald Brodribb.)

amphorae set tail in mouth to form rings of lightweight permanent shuttering in vaults is explicable as an extension of the use of an already available material. Invention moves a stage further with the development of hollow rectangular voussoirs for the formation of arches and further again with the evolution of voussoirs carrying shoulderings to support intermediate tiles so that a series of arches regularly spaced apart would carry intermediate tiles at the level of the intrados and extrados to create lightweight vaulting – a level of sophistication scarcely repeated until the twentieth century.

The system of heating by hypocausts (a word barely known to the Romans) is the best known example of their ingenuity with tiles. Floors were supported on pilae (small pillars) made of specially shaped bricks. A suspended floor was created on these pilae by the positioning of large terracotta slabs on which marble or mosaic might be placed. In the walls batteries of interlocking pipes, often rectangular and sometimes doubled, would provide flues for heated air. Sometimes, however, a cavity construction might be formed by a separate skin of terracotta slabs positioned away from the wall by terracotta spacers retained by metal pins. Terracotta was used for flues, chimney pots, gulleys, drains and their covers, and as surprising as the ingenuity of these many devices is the consistency with which they were used across the Mediterranean and northern Europe.

Wall tiles were formed with protuberances on their back to hold them away from the main wall structure as a separate inner skin. In patterned geometric flooring, *opus sectile*, hexagonal tiles or octagons alternated with triangles laid to a colour pattern. Another flooring in regular use was *opus stacatum* – corn grain – a name as evocative as our own herringbone pattern for the same work. This type of floor would be of bricks laid on edge rubbed to a level and coated with marble dust and lime.

Roman brick was defined by its surface area so that the Lydion, for instance, would be approximately 440 mm by 290 mm and might be as little as 25 mm thick although in some extreme cases it reached six times this thickness. A normal size would be 1.3 Roman feet (400 mm approx.) by 1 foot (280 mm approx.). It was commonly used to form lacing courses in walls, as at Pevensey Castle, Caister and the city wall of Verulamium. The brick one Roman foot square (280 mm by 280 mm) was known as a pedalis and this might typically be 45 mm in thickness. The smaller brick again was known as bessalis, sometimes made circular for building pillars. Examples are known varying from 25 mm in thickness to four times this depth. Among the larger bricks were the bi-pedalis, naturally 2 Roman feet on the sides (580 mm) and typically 50 mm thick forming suspended floors and sometimes being as much as 600 mm by 600 mm by 80 mm. Special bricks included coved skirtings, sometimes designed to receive the inner skin of wall tiling and wedge shaped bricks for use as voussoirs. These might be made solid or hollow and were known as cuneatus up to 450 mm deep and formed, rather than rubbed, to a convergence suiting the span of the arch.

Although Roman brick is predominantly red, often with a black core, the whole range of naturally occurring brick colours is to be found in the Roman repertoire. Coloured bricks were often used deliberately to form patterns. Bricks were scored, embossed, marked and die-stamped and they very often carried the name of a maker or of the authority for which they were made. The letters CLBR stand, for instance, for the many bricks made for the Roman navy in Britain (*Classis Britannica*). The black core typical of so much Roman brickmaking results from over-rapid firing to 600–900°C. The carbonaceous impurity is retained by the vitreous outer skin. Iron oxide is reduced (deoxygenated) by the carbon producing the black pigmentation.

The laying of Roman bricks, tiles and associated vaults and arching displays considerable ingenuity and importantly a very thorough understanding of the integration between brick, mortar and concrete. Brick and tile was used for compartmentation, the most famous example of this, perhaps, being the giant hemispherical vault of the Pantheon in Rome. Bricks were used as permanent shuttering, allowing the core of a wall to be filled with the more economical concrete and aggregate in lifts divided by lacing courses of full brick. The facing brick, however, might be nothing more than triangular units formed by cutting square bricks on the diagonal before they

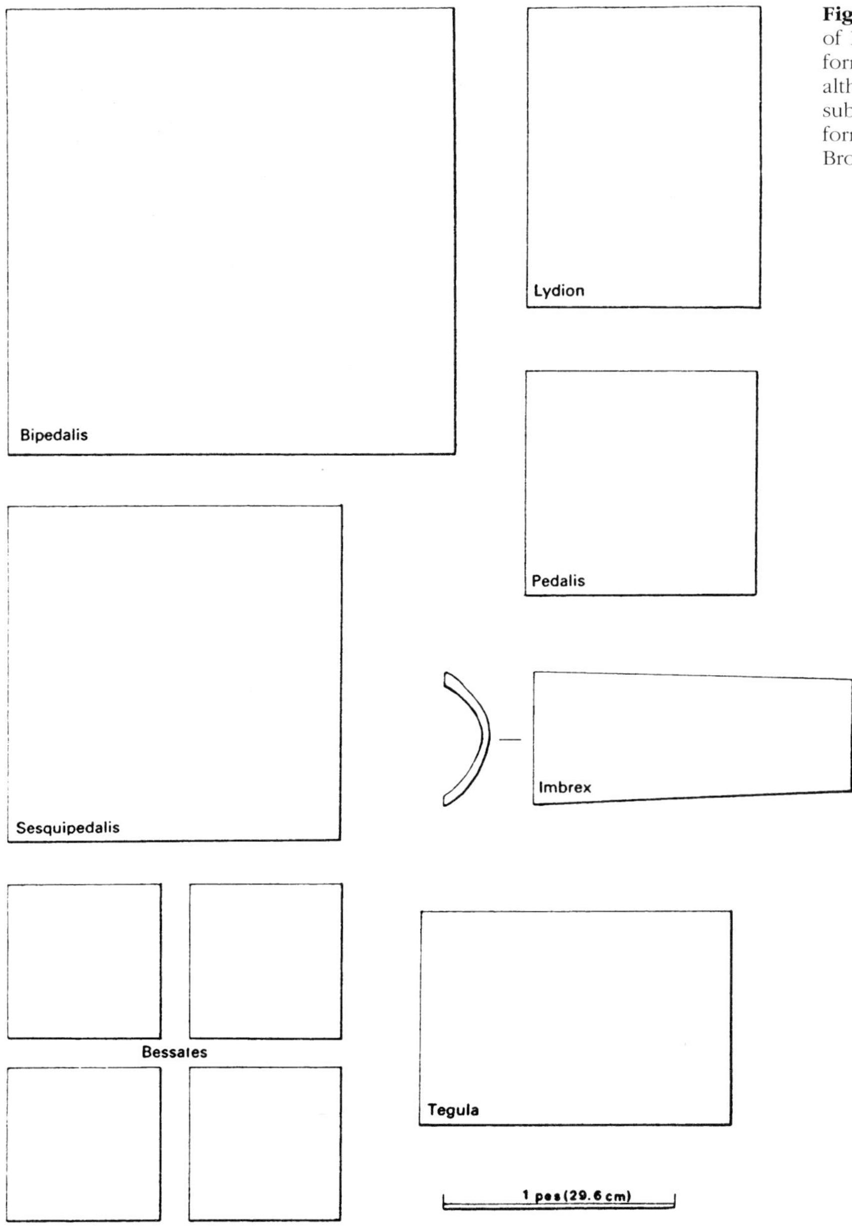

Figure 1.4 Comparative sizes of Roman tile/brick. All are formed of flat clay slab, although the imbrex is subsequently moulded on a former. (Courtesy of Gerald Brodribb.)

were fired. Laid with the diagonal face outwards in broken bond they locked effectively into the wall, giving an appearance of solid brickwork. The mortar was normally finished flush. There is a continuous flow of knowledge, tradition and technique from Imperial Rome into the Byzantine Empire where neither bricklaying nor brickmaking techniques changed, although the structural integration of cross vaulting, arching, dome building, buttressing and fenestration led to the exploitation of structural brick in a manner which has been continuous into modern times.

Across the silk roads of Asia a traffic of commerce left the great civilizations of Rome and China conscious of each other's existence, though there is little evidence of exchange of technical expertise – more perhaps of

Figure 1.5 Roman brickwork, Caister, UK, second to third century AD. Flat Roman brick used in courses to tie in lifts of walling in massive masonry.

embargo. Techniques such as the manufacture of silk were deliberately withheld from the West by Chinese Imperial edict.

By the seventh century Rome had been destroyed, the Sasanian Empire (the power of Persia) was no more and Byzantium was battered and in retreat. If the Dark Ages were not such an eclipse as once was thought they were nevertheless a prolonged and profound period of recession in the building industry of western Europe. Few bricks were burned, few tiles were baked and builders made do with salvaged materials where they could find them. In its heartlands the Byzantine Empire continued in modest prosperity but it is hard to pinpoint areas of major achievement before the year 1000 AD and even the greatest church of the period, the first church of St Marco in Venice, represents no more than a recreation in brick of the earlier church of the Holy Apostles in Constantinople, a single domed cross-in-square. The greatest achievement in brick of the late Roman period was a palace standing on the bank of the Tigris in Mesopotamia in the Sasanian winter capital of Ctesiphon. Now there survives only the great iwan of this structure with one of its flanking side wings, and although the northern wing of its façade has fallen it retains the largest brick vault ever built, rising to almost 37 metres and spanning 25.3 metres (see Fig. 1.7). Its profile is a vertical semi-ellipse and it is constructed in the traditional method of the country, using large square flat bricks in inclined ring arch (or pitched) construction. The arches are built sloping back against an end wall and the profile is maintained by suspended centring positioned through holes still surviving in the vault. A strong lime mortar was used. The building probably dates from the 6th century (Chosroes II), although it may be up to two centuries older. Thus it runs parallel to a brief but important upsurge in building in the Byzantine Empire under Justinian the First (mid-sixth century AD) when brick was widely used throughout his domains, particularly for

Figure 1.6 Byzantine brick and stone coursing, Constantinople (Istanbul), sixth century AD. Cistern: extra-muros. Deep beds of brickwork were alternated with limestone to provide massive structures capable of resisting hydrostatic pressure.

vaulting. This is a period when, for the first time in history there is good evidence for the widespread transportation of brick, tile and pottery. Communications were remarkably good. Roads and shipping were the sinews of the Empire and techniques of building continued to be astonishingly uniform over wide areas. Not only were expensive pieces, such as capitals carved in Constantinople, exported throughout the Mediterranean, but the products of the nearby Thracian brickfields were shipped abroad to form the vaults of palaces, the apses of great churches and the arches of aqueducts thousands of miles away, carried sometimes on camel-back hundreds of miles into the desert. The great palace of a provincial governor at Kasr ibn Wardan (built *c.*640) deep in the Syrian desert is built of materials – brick from Thrace and basalt from the Hauran – carried laboriously to the site far from their place of origin.

By the eighth century Muslim domination of North Africa and the Middle East ushered in a new wave of building based on Sasanian and Byzantine building methods and therefore employing brick extensively. The bricks remained flat slabs for the most part and the mortars which survive are usually lime with a pozzuolanic additive.

The medieval interregnum in Europe marks the change in brick technology, for although the Roman Empire surviving in the east continued to use Roman brick laid in beds of mortar as thick as the brick itself and although similar brickwork was in Muslim use, the re-establishment of a European brick industry north of the Alps seems to have been a sufficiently independent regeneration for new techniques to be introduced, and it is to this period that the brick and clay tile of modern sizes can be traced. This brick – capable of being handled in one hand – brought with it different techniques of laying and a different approach to construction. Likewise the roof-tile. The west European plain tile or the interlocked variant the pantile required a closer setting of the laths which were smaller in section. The rafters, therefore, were more frequent and so roof design evolved afresh.

In the Romanesque period Europe regained social cohesion and prosperity and, in areas lacking stone, brick and tile technology returned as an alternative to timber. The plain tile, an alternative to the stone slab, slate or wooden shingle and depending upon a multi-layer arrangement for its efficiency, was in some ways less demanding of manufacturing technique than the larger brick. It was more akin to pottery. Bricks could be manufactured efficiently only by burning large clamps or kilns which required very substantial amounts of fuel. Tiles could be fired in potters' kilns. If the labour of the gaining and handling of timber is coupled with the labour of preparing, firing and handling the bricks and the stone for the burning of lime it is clear that timber building was a very much more economical process. Nevertheless it was recognizably less permanent and it was not fireproof. Through the later medieval period in

Figure 1.7 Sasanian brickwork, Ctesiphon, Iraq, fourth or sixth century AD. The largest span of unreinforced brickwork known, now surviving as a partially complete parabolic barrel vault and one frontal wing.

Europe the building of fireproof walls, the encasement of fortresses, the building of chimneys, of palaces, of churches, of bridges and other under-water structures increasingly demanded brick with the mortars and renders that went with it and tiles for roofing. It is a curious inversion that the ready availability of structural timber throughout northern Europe inhibited the development of its brick industry; but the economy of labour and the relatively slow rate of decay of timber as a building material gave each generation the material which provided more built volume for an identifiable labour cost. In lowland areas only where permanence, durability or essential fireproofing, as in chimneys, were required did brick dominate initially.

Some of the greatest achievements in brickwork in the Middle Ages are Mediterranean rather than north European – the great cathedral at Albi, huge churches such as Frari in Venice and the Giralda – the tower of the mosque at Seville. Medieval brickmaking under Islam reached new heights with the evolution of the pointed arch and ribbed vault to be seen in the Baghdad Gate at Raqqa (probably eleventh century) and the twelfth century Abassid Palace in Baghdad. However little remains in central Asia and the Middle East from these dates. The Mongol devastations of the mid-thirteenth century and subsequent rebuildings were so total that very few buildings survived in the cities now in Iraq, Iran, Afghanistan, Uzbekistan and Tadjikistan. Three marvellous examples, however, demonstrate that in these eastern regions of the Abbasid Caliphate there had developed a highly sophisticated brick architecture in addition to the immense volume and complexity of earth building. At Bokhara a mausoleum of the early tenth century containing the tomb of Ismail of the Samanid dynasty lay buried under earth and so escaped the destruction of the city at the hands of the Mongols. Its

Figure 1.8 The tenth century tomb of Ismail the Samanid, Bokhara. The translation of stucco carving into the more rigid techniques of brickwork produced overall textures of surface by the use of cut and shaped brick.

interior and exterior surfaces are covered with a complex range of patterns formed in geometric brickwork bonded into the walling. This is no form of mosaic and the intensity of patterning is equivalent to a tapestry or carpet. Restored it survives to demonstrate the great complexity and ingenuity with which brickwork was used with the cutting or formation of countless special bricks for the purpose. In the same city but a couple of centuries later in date stands the enormous cylindrical Kalyan Minaret which survived the Mongol destruction of the remainder of the city only because of its utility as a watch tower. It is over 50 metres in height and elegantly tapered, surface patterned and encircled with decorative inscribed banding. This range of decorative brickwork, monumental brick engineering and delicately cut and carved brick is spread over the whole Islamic world from Cordoba in Andalusia to Central Asia. After the Mongol period examples proliferate and their survivors are legion. The complexities of brick form, of brick carving, of structural interlacement, of delicate and massive polychrome decoration and their encasement in tile and marble demonstrate a range of skills reaching the heights of possible achievement. Inscribed bands of calligraphic origin are carved into brick friezes, modelled in plaster and frozen in bas relief on glazing. The intricacy of muqarnas or stalactite decoration which almost certainly began as a multiple corbelling to form a pendentive is carried from brick into plaster and ceramic. Structural masonry is faced with tile and terracotta mosaic, glazed and modelled in low relief. This work at its most intricate is to be found in Morocco and elsewhere in the Mahgrib, in Islamic Spain and the Muhedjar decoration of Andalusia. The palaces of Granada and Seville carry these techniques to the heights of delicacy, complexity and finish and no more demanding problems of conservation in these materials are likely to be found.

In its long history of the use of fired materials, the Far East and in particular China had produced an architecture in which brick was specific to massive structures and was often used to encase earth. Elaborate forms of roof tiling were in use and ceramic sculpture had been introduced widely into roof and wall decoration.

In all the Eastern Kingdoms ceremonial was immensely important. This was a world hedged about with taboos and restrictions. Certain types of material, special colours and particular forms would be reserved for the Emperor and court officials. Large bricks were used only in Imperial construction. By the end of the medieval period the use of fired brickwork proliferated in paving, facings, engineering of underground services and the core structures of tall buildings carrying decorative timber and painted finishes.

In Europe likewise the medieval period was one of the growing use of fired brick and tile. Italy remained at the forefront of artistic and technical progress, and in the alluvial plains there was good reason for seeking an alternative to stone. The cities of the Po valley used brick intensively in raising the towers and defences of Padua, Verona, Mantua, Bologna, Venice and many a smaller city while churches in the austere Lombardic Gothic rose and were vaulted in brick, perhaps to be faced

Figure 1.9 The Giralda, Seville. Pattern formed by raised and recessed brickwork on the minaret of the Ummayad mosque, now the Cathedral tower.

with marble or dressed with limestone. In Europe this early brickwork was rarely decorative; often it was intended only to be a substructure for more elegant finishes – stone, marble, tile or terracotta – and as a base was correspondingly rough in execution. Its lovable appearance derives from the effects of weathering, age and intrinsic surface qualities. It took on localized characteristics, the clays and methods of firing of each area giving its brickwork distinctive qualities.

An important area of medieval brickmaking lay along the northern shore of Europe from the Baltic to the Low Countries where stone was relatively scarce. At the beginning of the fourteenth century the great monastic castle of Marienburg, now in Poland, was raised in burned brick with decorative coursing, cusping, voussoirs of alternating colour and emphasized mouldings all in the same material, and in the largely brick city of Lubeck in the previous century the church of St Mary,

Figure 1.10 Sixteenth century English terracotta. In the early sixteenth century florescence of terracotta much of the production may be thought of as 'moulded brickwork', as in this example from Sutton Place, Surrey, UK.

enlarged over a century to the size of a cathedral, displays brickwork emulating stone in its detailing. The Low Countries, prosperous and industrial in the late medieval period, built not only churches and cathedrals in brick but town halls, market halls and fortresses such as the castle of Muyden near Amsterdam in the thirteenth century. In these buildings brick and stone were integrated in one masonry system. This was not a matter of facing the brickwork with stone but of using both materials in the one palette.

Prosperity came later to the offshore islands of Europe and the trade across the North Sea may have been responsible for the introduction of brickwork on any substantial scale into eastern England. Ships carrying wool outward returned ballasted with tiles, paving, terracotta and brick although native brick manufacturers were also in production in the later medieval period. Fired medieval floor tiles were relatively rare in lowland England in the Gothic period but by the sixteenth century they had become common and the bricks, though thinner than modern material, are essentially similar and are matched in modern manufacture.

It is doubtful whether the qualities which the conservator values today in these bricks were those which appealed to their original builders and in tracing the development of brickwork over the succeeding centuries there is a discernible trend towards greater precision and perfection of line and surface, culminating in the production of the rubber brick (a brick which is rubbed, usually on another, to produce a very fine surface and accurate bed for jointing) and other bricks which anticipate the accuracy, precision and regularity of industrial manufactures. When ultimately industry was able to achieve and better such precision the irregular and homespun qualities of the earlier brick became valued in their own right. So it is that the conservator turns to the timeworn surfaces of brick and mortar not with an eye to making them look as new but rather of retaining so far as possible their qualities of age and craftsmanship. But the more highly finished and perfect the material, the less likely it is to gain in character with age or perhaps more accurately it may be said that the effects of weathering work against the pristine qualities for which its users originally valued it.

By the eighteenth century European sophistication and skill in architecture and the arts had introduced considerable precision in the manufacture and use of bricks although for the large part neither mechanical methods nor the effects of industry were to be seen upon them. The uniformity was often a result of selection. In particular smooth reds were gauged for size in order to maintain thin uniform joints. The improvements were simply due to a progressive improvement in the nature and quality of manufacture and craftsmanship relying on an increasing availability of trained manpower and its manipulative skills. Until well into the eighteenth century the most notable change in the brickwork of Europe was simply this achievement of higher quality which produced material more regular in colour, more accurate in profile, more consistent, better burned and more highly finished. Inevitably the same effect is observable in mortars. Improved specifications produced mortars with better controlled setting qualities and higher finish. In London the technique of tuck pointing which simulated extreme accuracy in bricklaying is indicative of this trend. But for all this greater skill and capacity to build more accurately with surfaces of greater consistency it remains fundamentally true that the improvements in materials and workmanship were due to higher standards of performance and execution. At the lower end of the market poor quality products proliferated as low-cost manufacturers undercut their rivals. The profound change was the increasing use of coal for firing in the eighteenth century.

To that date wood had been the fuel and the brick clamp was the general method of production in the South of England where the kiln was reserved for tiles, pipes and pottery. Clamps were not a feature of post-medieval brickworks in the north of England. Here kiln firing was favoured because the dense and often carbonaceous clays were not suitable for the relatively uncontrollable firing cycle of the clamp. Sandy clays can better tolerate the wide range of firing temperatures. In the post-medieval period alluvial and estuarine deposits had been dug and manipulated by hand. But with the Industrial Revolution the making of bricks changed, as did the materials and methods of bricklaying. Cements for rendering and mortars for bricklaying had been based on lime, burned in the kiln from chalk or limestone and for the largest part these materials had been manufactured near the point of use. Changes in transport patterns coupled with industrial technology supervened. The techniques of construction likewise evolved with new building methods and there were occasional decorative developments, such as the introduction of tile cladding intended to imitate brickwork.

The colonial era spread European brick technology to much of North and South America, Australia, the fringes of India and Africa and Oceanic Islands but Inner Asia, China and Japan remained unaffected and in possession of their own parallel streams of development.

From 1750 onwards the demands and opportunities afforded by the industrial era produced immense changes in methods of use, available materials and methods of production. The stimulus was competitive improvement, the method was the exchange of information and the rapid growth in the real wealth of society provided the means. This evolution is now itself history and the ageing and decaying brickwork presents the conservator with widely varying problems many of which are to be understood only in the context of the materials and production methods employed. In particular the habit of using coal ash as a component of the mix produced bricks of uncertain, and sometimes very poor quality.

New industrial and scientific methods led to new and enquiring methods of thinking stimulated by engineering and military requirements. The improving gunnery of the eighteenth century demanded greater robustness in fortresses built to resist the improved new weapons. Industrial brickwork was tested analytically for its military effectiveness, perhaps for the first time, when the Navy Board in Great Britain built a series of

Figure 1.11 The spread of brick technology in the Colonial Era. Thin bricks deriving from Roman traditions used in the early sixteenth century in Santo Domingo, Hispaniola.

redoubts at the end of the eighteenth and beginning of the nineteenth century using different tower profiles, different types of mortar and varying bricks to determine, with success, the plan shape, wall thickness, mortar composition and brick which would best withstand naval bombardment. The resulting fortifications, known as 'Martello' Towers, were built round the southern and eastern shores of England during the Napoleonic wars and subsequently in many other parts of the British Empire to say nothing of similar copies built into fortifications elsewhere.

As the century advanced coal, and later coke, supplanted wood as the prime material for firing to be followed ultimately by oil and natural gas. Clays made from shales mined in the carboniferous measures were used for brickmaking, partly because they contained fossilized carbon which reduced the fuel requirements but largely because both clay and fuel were immediately on hand at exactly the point of high demand for the increasing populations of the industrial areas. Mechanical methods of preparation of clays, compaction of brick and tile, forced drying and kiln firing took over from hand working and open clamps. The resultant products were cheaper, more reliably produced, more consistent and, therefore, capable of better performance. Inevitably their character was very different. They could be matched against particular tasks and the increasingly effective transport industry could deliver them to a wider market. Thus it is that bricks originating in one part of the world may have been transported to a diametrically opposite point of use and the conservator will need to be familiar with the problems of identifying the source and manufacturer in order to obtain like material for comparable components.

Mechanical methods of clay extraction, pulverizing, mixing, drying, extruding and pressing were complemented by new methods of firing that were ultimately dependent on new fuels such as natural gas. Almost all brickwork of historical importance, however, will have been fired by wood or coal, perhaps in the form of its derivative coke. The use of wood declined as coal and coke were used increasingly. Both processes, depending on direct combustion, could achieve their final highest temperatures in a reducing atmosphere – that is to say an atmosphere in which the oxygen of the air had been wholly or largely consumed and in consequence the chemistry occurring at highest temperatures was conditioned by the lack of free oxygen. This in turn had significant effects on colour and finish so the variety of bricks was widened.

During the nineteenth century the Hoffman kiln was extensively introduced. This invention consists of a permanent tunnel with a chimney stack and fire door at opposing ends operating on a cyclic (rather than a continuous flow) process. The kiln was essentially an annular tunnel, originally circular but elongated later into two parallel tunnels connected by chambers curved around the ends. Green bricks are loaded stage by stage and the fired product is removed sequentially. At any one time parts of the kiln are being set with green brick while parts are being dried, fired, cooled and drawn. The fire moves progressively and continuously round the annular tunnel in a sequence determined by the burning, its waste heat being used to dry the green bricks ahead of it. This invention provided the most sophisticated technique in general brickmaking until the late twentieth century. It was primarily important in terms of control over the quality of product.

The form of the historical brick was standardized by practicality, its size and weight allowing it to be handled readily by the bricklayer and its shape being determined by the volume and thickness which can readily be moulded and dried. The frog, introduced into the upper surface when moulded, was not, as is often supposed, intended to provide a better key for mortar or even to economize on the use of clay. It was simply a convenient method of forcing the clay to squeeze out into each corner of the mould as the lump of plastic pug was thrown into it. The frog also reduces the weight and eases the processes of drying and firing. With the introduction of mechanical methods this practical device continued, the maker's name often being helpfully embossed on the frog mould itself, providing ready confirmation of the sources of historic brickwork.

The technique of squeezing soft clay through an orifice – i.e. extruding – was in use before the general industrial era for the formation of pipe but the pressures required to squeeze out clay are considerable and it was only with the introduction of heavy machinery and powerful engines that such processes became applicable on a large scale to brickwork. The extruded brick cut by a wire from the emerging column of clay was sometimes reduced in weight to be comparable with its more traditional competitor by the inclusion of holes. Wires or guillotines were used to cut off the bricks. Clearly no frog could be introduced.

Sophisticated products from various adaptations of this technology produced fired hollow blocks some of which, formed in terracotta, were extremely light and were patented to perform inventive functions in flooring, walling and roofing. Extrusion technology was also applied to the formation of roof tiles and each of these processes has produced special and individual characteristics in historic brickwork of the industrial era.

While the term 'face brickwork' is applied to the intended finished surface of brick (facework) there has also arisen a range of techniques for facing the actual bricks themselves with textures and finishes to make them more attractive to the purchaser. These may be designed to make a relatively cheap product more comparable to the expensive types or to produce a wider range of products from one base material. The practice has been in use for about a century and is therefore incorporated in historic structures. Many ancient bricks also have been 'faced' in the sense of being glazed or having had sand applied to their surfaces in the mould. In more recent times the surface characteristics of extruded and pressed bricks and tiles have been frequently modified by the imposition of patterns, scraped textures, blown grit, sand and minerals and by deliberate embossing or contouring. These amendments to the base material could readily be achieved at the production stage. Sometimes a slip of semi-

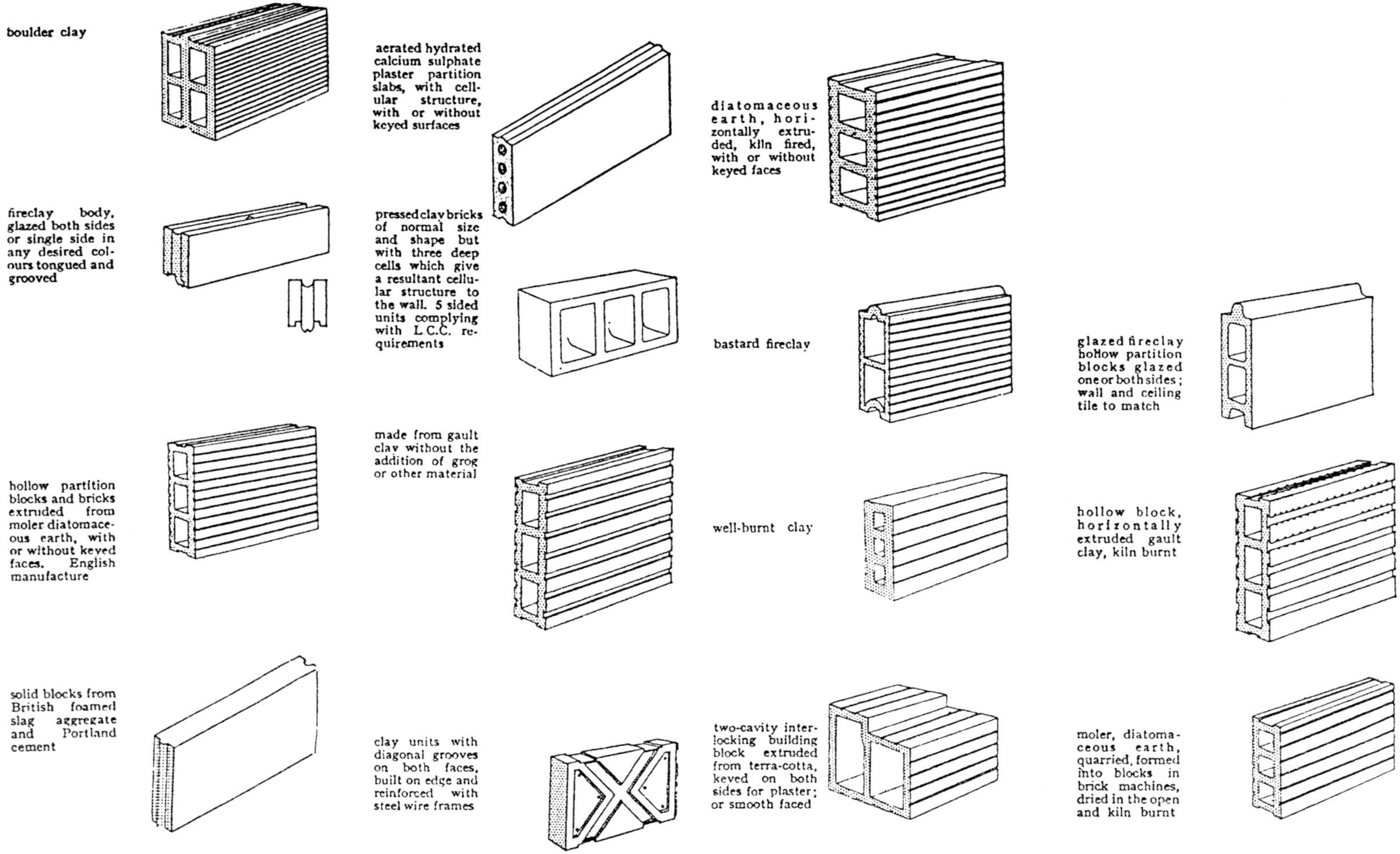

Figure 1.12 Mid-twentieth century hollow terracotta wares. Some of the many thousands of special purpose products available in 1938, some of which are to be found in historic structures and traceable to specific manufacturers.

liquid clay of specific property would be applied, perhaps in conjunction with mechanical texturing, producing a skin which could be virtually integral with the body of the brick. At its simplest minerals and pulverized fuel have frequently been added to surface sand to create colour effects.

A concomitant of industrial production is a wider pattern of transport and marketing. In consequence published information has been produced, much of it ephemeral and now rare. Nevertheless where the problem of identification occurs the trade literature of the period and the available instruction books become an increasingly important source of information to allow the identification of materials which can in turn assist decision-making in repair or replacement. As architectural salvage grows in importance so the ability to cross reference materials becomes more significant. Knowledge of the distribution network and market outlets of specific and highly finished products may lead to good resourcing, particularly where materials salvaged can be identified and catalogued at an early stage. Knowledge of the history of brickmaking on a local and sometimes wider scale is, therefore, a fundamental resource in conservation, affecting decisions in material derivation and selection.

2

Brickmakers, bricks and brickwork

Dried earth blocks preceded fired brick by millennia, but in dwellings built of mud brick, writing tablets, votive statues, crockery and storage utensils were commonly available. All these had been fired in kilns. As kilns grew larger for pottery and similar ware, the era of brick building approached.

The earliest fired materials used in buildings were not bricks but special items – tiles, finials and terracotta roofing slabs. The earliest manufacture was in ovens open-mouthed at the top, such as are still used traditionally for breadmaking in the Middle East. These arose out of the use of open fires contained within progressively higher walls to create and contain more heat. Pot-shaped, these structures rose and grew bigger acquiring fire holes at low level to improve draught, producing temperatures up to 1000°C which were sufficient to calcine limestone, fire clays and melt some glasses. By reduction of the mouth at the top of the oven the first pottery kilns came into being. These were simple pot-shaped updraught kilns later improved by opening air holes at the lower level.

The firing process consisted of burning sufficient fuel within the kiln to reach the required temperature after which it was allowed to cool and the fired objects were retrieved from the ashes.

2.1 Early kilns and clamps

This simple kiln had two derivatives. The first, a chamber superimposed in the core of the fire chamber where the requisite temperature was achieved by conduction of heat, and the second, the down-draught kiln in which a partial chamber was formed within the first by building a half height (or bag) wall within the outer perimeter behind which fire holes were

Figure 2.1 The tanour or tandour. The domestic version of the primitive oven derived from enclosure of a fire, which led to the evolution of the kiln.

Figure 2.2 Clamp-burned, wood-fired stock brick. The lowlands of England produced many bricks of this type, the semi-glaze resulting from the flue gases producing a distinctive grey sheen and sometimes reflecting the stacking (i.e. skintling) as in this case.

provided. In a further improvement the top was closed. To exhaust combustion gases central floor openings led by a tunnel to a separate chimney. Combustion within the fire holes carried the fire up against the sides of the kiln, the inner surface of which became an intense source of radiant heat while the flame path itself travelled down and out through the centre of the floor. Wares positioned in the floor centre were at the focus of radiant and conducted heat, which could be controlled by manipulation of the fuel and draught at the fire holes. There was an important difference in effect between the two systems. In the separate chamber there was no combustion and, therefore, the one-fifth oxygen content of the air remained and could react with minerals in the wares at high temperature. In the down-draught kiln the wares were heated in the flame path where the oxygen was partially consumed by the fire and was not, therefore, fully available for reaction with the clay minerals. In this 'reducing' atmosphere the same temperatures would produce markedly different results. In some circumstances combinations of atmospheres were achieved at different times or by different positions in firing. To produce oxidizing conditions excess air was allowed over the top of the fuel. For reduction the air inlet over the fire was closed once full temperature had been attained.

Small early kilns were more used for the production of tiles than for bricks. The production of bricks on the scale needed for substantial building was advanced by a different technique – the invention of the clamp. In this the unbaked brick itself is required to form its own oven, the bricks being stacked with channels between them allowing secondary airways between the individual bricks themselves. Fuel in sufficient volume in the brick matrix within these channels can, if the arrangement is judged correctly, provide sufficient heat to achieve temperatures in the region of 1000°C when the clamp is sealed. To achieve this with any degree of efficiency a clamp must be large because, as size increases, the volume increases cubically whereas the external surface increases only by the square, resulting in an improving ratio of volume to surface as size increases. The loss of heat occurs through the external surface and the critical factor becomes the amount of heat generated internally in relation to heat loss. Uneven firing and the unavoidable temperature gradient, falling off quite rapidly towards the outside face, means that clamp firing produces uneven heat patterns with over-firing towards the centre and locally on the brick faces in the flue ways while under-firing occurs towards the outer face. A further problem is the physical size of a clamp –

usually 2–3 m in height and of substantial length. The bricks being stacked one upon another produce considerable deadloads on the lower layers and these layers when at the maximum temperatures can soften, allowing the entire mass to settle and causing the production of a substantial proportion of misshapen brick. Except for trace holes containing fuel and fuel in layers between courses, bricks are set with minimum spacing. Space for flame and waste gases was created by firing shrinkage. Clamp builders have generally been content with losses as high as 25% or 30%. The under-burned and over-burned products require sorting before they can be marketed as low-grade material.

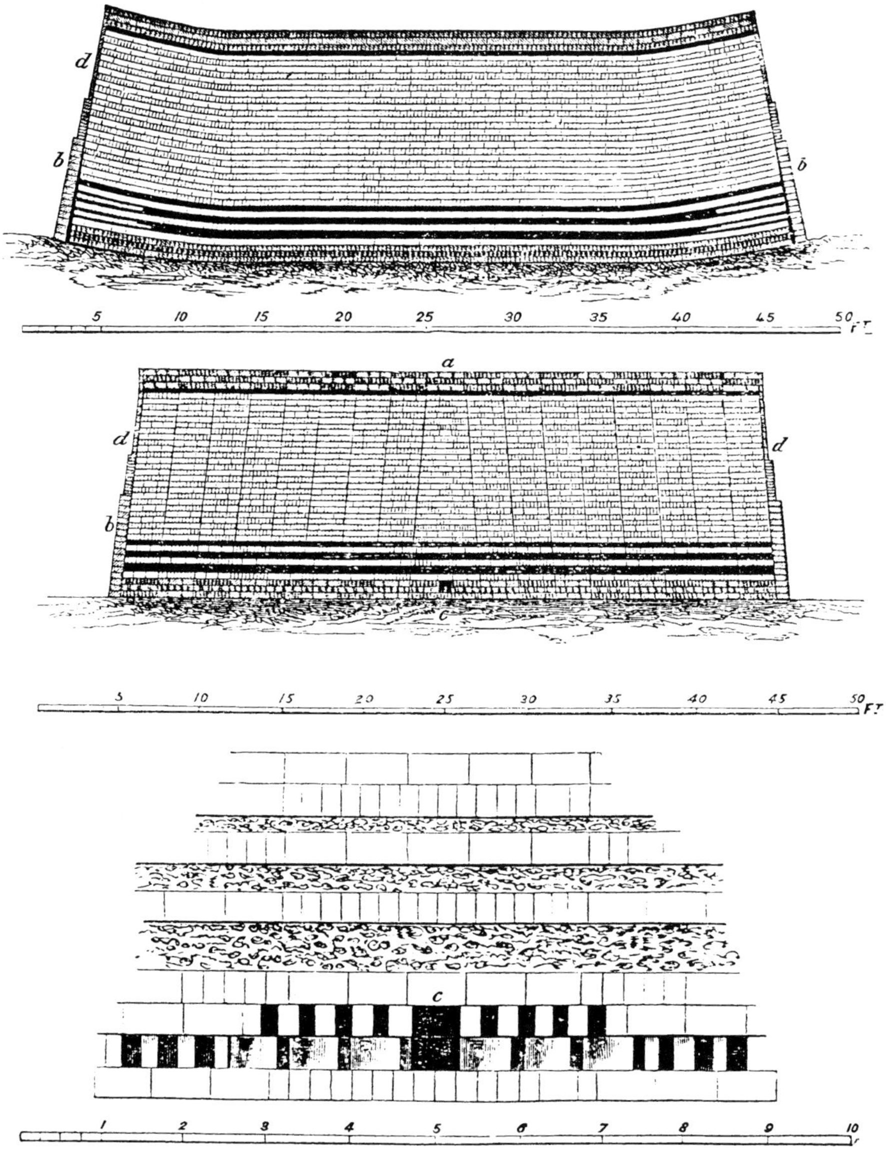

Figure 2.3 A brick clamp. Taken from Edward Dobson's *Rudimentary Treatise on the Manufacture of Bricks and Tiles*, 1850. These sections through a typical coke-fired clamp show the general method of laying and placing fuel.

Figure 2.4 A hand-made brick from the plain of York. The rugged texture of the cheaper eighteenth century hand-made bricks led to attempts at improving the finish by striking a sharp line in the mortar, as here, and to tuck pointing.

Historic buildings will commonly contain over-fired and misshapen bricks in foundation work and in the cores of external walls while under-fired bricks have traditionally been used internally and for cutting or shaping. Occasionally bricks are cut or in some way remodelled and fired a second time. Where a glaze is to be applied to a brick surface this is a recognized procedure and the second firing may be carried out in controlled conditions to achieve precise colours.

Since clamps were often local to the building, and sometimes related only to the one structure, it is on occasion possible to trace the proportions of bricks produced and to gain an insight into historic production methods. Clamps have no flue and their inefficiency in combustion causes air pollution.

2.2 Industrial kilns

Clamps were particularly a feature of southern England, but simultaneously updraft open top Scotch kilns and later closed rectangular kilns were in use, predominating in northern areas, and partly dictated by the nature of the clays.

Clamp burning having become a general method of producing bricks in large volume, a further reversal of technique gradually occurred as chamber ovens with chimney stacks became larger. Coal superseded wood as a fuel and the advantages in efficiency of a tall permanent chimney stack became apparent. One stack could serve a number of combustion chambers since the process was intermittent. As industrial production techniques gained ground massive multi-chambered kilns were associated with tall chimney stacks on the principle that each chamber required several days for loading and unloading, several days for cooling and a slow heating period. Of the total time cycle as little as one-fifth might be associated with full combustion activity, the heat from chambers in firing being passed forward to preheat and dry newly set chambers while cooled chambers to the rear were emptied. Improved firing produced much higher proportions of satisfactory product justifying investment in large kiln structures and mechanized preparation. With the improvements in the processes went patenting and a series of secondary inventions relating to the manufacture and handling of the products.

The further radical change in production has relatively little bearing on historical brickwork other than in the manufacture of replacement material. In earlier processes the point of firing was fixed and heat was introduced over a period of days reaching a maximum temperature sustained for a critical period measured in hours after which a cooling period of some days was necessary. Industrial firing was restricted to coal and coal-based products. In contemporary tunnel kilns zones of temperature are maintained at a constant level, often using gas or oil as fuel. Through these zones the bricks are carried mechanically on cars at speeds determined by the firing required. The control achievable and the ability to make effective use of waste heat coupled with further improvements in precision of manufacture have commended these processes to manufacturers. Bricks, carried on cars, can now be fired in precise temperatures and atmospheres, taking the uncertainty out of earlier processes. The cars are protected from the flame path – in some cases being submerged in the chamber.

2.3 Clays and brick earths

Clay, being a natural material, is not consistent and much of the skill in producing fired material lies in overcoming and exploiting variations. These marks of manufacture become characteristic of historic periods and the historic material. A knowledge of the processes involved in a particular field of work and of local techniques will have an important bearing on the conservation methodology applied in any particular case. Faults in brickwork may stem from inherent defects in the materials and processes well before the brick ever leaves the brickyard.

The purest clays make dense material. China clays (kaolins) purified after the decay of granites need the addition only of fluxes to produce porcelain which has a minimal application to building works. Other clays (wind- or water-transported kaolinites, illites, etc.) are variously mixed with silts and sands in their naturally occurring states. Calcite is a further common component in discrete form or as calcium oxide, typically in the Keuper marls. While the technicalities of the choice of clays are the prime concern of the manufacturer their characteristics should be known to the conservator because behaviour and weathering performance are related to manufacturing techniques. Earths suitable for making brick and other fired building materials are clay rich and the percentages of clay in the material are generally higher than is ideal for the formation of dried earth blocks. Clay may be less than 30% by weight of the raw brick, very occasionally rising to as much as 75% for terracottas. Clays which are used as a basis for brickmaking may be obtained from previously compacted clay-rich deposits, solidified in the earth's crust as shales or mudstones. The use of such semi-rocks as a basis for brickmaking is generally associated with the industrial era and was often a product or by-product of industries such as coal mining. Pre-industrial bricks have almost always been prepared from a suitable easily accessible brick earth and the material will usually have been obtained within a relatively short distance of the building site. In some areas it was common practice to excavate and fire the bricks on the site of the building itself. In nineteenth century Britain the excavation was commonly used for cellars and tankage.

All brick earths and shale have been laid down under water or formed as air blown deposits. While the latter are almost always uniformly mixed, water-deposited earths, sands and shales often vary considerably in content, even in very localized deposits, because the rate of water flow determines the deposition of particles. As large earth-burdened bodies of water slow down, moving into more static conditions, they deposit first their sands, then the silts and finally the clay: hence the variations. In consequence selection in excavation and choice of earths has been essential and local variations common. Many successful yards exploited deposits of varying content in accordance with the product required – brick, tile, pipe, pots, etc.

A further constituent of fired clays is ground waste from previous firings – sometimes called 'grog', less common in brickmaking but important in more sophisticated forms of ceramic manufacture, especially refractory materials.

2.4 Shaping and forming

A critical period in the manufacture of fired clay components is in the initial shaping and modelling. In almost all historical periods this has involved water but from the nineteenth century there have been processes in which dried powders of mixed clays and siliceous silts have been compacted to form bricks (sometimes giving them specific lightweight properties). Some processes have involved the cohesion of powdered clays containing natural oils or other hydrocarbons which burn off and fuel the firing process. Some brick earths naturally include hydrocarbons, as in the Oxford clays, or particles of fossilized plant life, as in the shales of the coal measures. The formation of the very large bulk of clay products prior to firing is, however, achieved with the plasticity induced by water. It is a crude general rule that the higher the clay content the greater the plasticity but the type of clay and the nature of the quartz particles are also important. Some brick earths are found in a condition of moderate water content and suitable mixtures of clays and silts so that they can be shaped without tempering. So fortunate a combination is rare and was much sought after because it saved the expense of one stage of the manufacture. This stage consists of improving the raw material by additions to adjust the balance of sand, silts and clays, adding, in some cases, carbonaceous material as fuel and providing the appropriate amount of water to give that particular brick earth sufficient plasticity to be worked. The extent of this preparatory process, therefore, involves digging clay or quarrying shale, perhaps in association with coal mining operations from which it may emerge as a waste material. In many instances both with shale and earths but particularly the latter it has been conventional to allow material to weather for a considerable period, from six months to three years. The effect upon shales is principally a softening. Weathering usually follows initial crushing so that water readily penetrates the clay structure. A typical shale will readily become slimy on the surface after wetting, as water is able to lubricate the clay molecules. The effect of weathering in earths is to improve workability, i.e. plasticity. The clay particles under pressure move into alignment under the electrostatic charges on the face of each clay platelet. With the plates better aligned there is less internal friction. At the end of this weathering or maturing stage the material will be fully plasticized. Substantial stones and lumpy material must be removed and any additional component must be added. Such components have traditionally included coal ash, sometimes wood ash, lime and rarely feldspars to produce fluxes, with sands, silts or earths needed to temper an over-rich clay. Sufficient water will be included at this point in the mix. A relatively short further period of maturation may be given and the clay then moves to the moulding point as anything from a plastic mix to a creamy slurry. Lime in the mix reduces shrinkage in firing due to expansile reaction as calcium silicates are formed, and this, to some extent, balances the shrinkage of clay molecules as they re-form during firing.

Historically the mixing was achieved manually by forking over, treading by animals, treading by human feet (allowing large particles such as pebbles to be detected and removed) or later by the use of a pug mill which originally was a vertical shaft in a conical cylinder, the shaft carrying a series of blades, which, being set at an angle, drove the plastic, raw material downward through the brick-sized orifice mixing as it went, compacting and forcing out air as the material was compressed. The vertical mill suited animal power but a horizontal arrangement was more common as engine power became readily available.

These principles were generally applicable throughout the world with variations in practice from place to place depending upon tradition and local conditions. The faults that could arise in this stage in the process could relate to the quality of materials, uneven mixing and a failure to separate out pebbles or other foreign matter. In some instances the mix would be wetted into a slurry and a sieving process would be used to screen out over-large components whose presence, as unshrinkable items in the mixture, would cause the clays around them in contraction to develop cracks radiating out from the obstruction. This fault would carry through into the finished product. A pebble exploding during the firing process would disrupt the brick or tile and might damage neighbouring pieces but would not reach the building site. A brick with shrinkage cracks, however, might reach the site with a source of future failure built into it. If near the surface the presence of such a fault would be indicated by an obtruding hump after firing.

Uneven mixing can follow the striations of the blades in a pug mill as excess of added material is drawn in but more normally it will follow the careless addition of components in uneven quantity or their unsatisfactory preparation: e.g. ground chalk inadequately screened or ashes containing foreign materials such as iron which had not been eliminated. Chalk in lumps too large to be absorbed in the fluxing of the burning process will form quicklime in nodules within the fired component and this will slake with results that shatter or damage the brick or tile. Ferrous inclusions can cause brown staining. Earths and clays may be reduced to a semi-fluid or even a fluid state as a slurry in order that at this stage they can be screened with certainty or in order to separate out larger sized particles, to be subsequently fed to settling tanks to provide a suitable material for shaping or casting. Unless the manufacturer is extremely fortunate in the supply of raw material this process is normally essential in preparing the finer earths used for firing terracotta.

Shaping the component before firing can involve a series of dissimilar processes all of which aim to model the amorphous material into a rigid object dry to the touch before it is introduced to the kiln.

Pottery techniques are used to shape larger and generally hollow components. The clay is delivered in a plastic or modellable state where hand or semi-mechanical processes are used to raise the shape. No water is lost during the forming process so the material remains plastic when formed and must be moved with care until further drying brings it to handleability. More complex shapes can be achieved by pressing sheets or slabs of plastic material against a former or mould. This is an operation performed both by hand and by machine to make complex facing materials such as embossed terracottas or merely simple slabs of clay laid on a slightly curved former to make 'hatched' or cambered plain tiles, the slight cross camber being designed to throw the water towards the centre of the underlying tile. Hand-moulded plain clay tiles may be given a longitudinal or transverse camber after the original flat moulding has dried to the leather-hard stage. This may be called cricking.

Solid bricks with ornamental faces have long been made by embossing, even where the ultimate design has been enhanced by carving the finished fired material. Adhesion of the plastic clay to the mould against which it is pressed may be countered by a dusting of dry sand on one or both surfaces and this sand provides colour and texture to the finished product. The alternative method of overcoming adhesion is to provide an excess of water on the surface of the clay – a process known as 'slop' moulding. Modelling in this way has been a widely practised traditional method of shaping bricks although the most common process involves the dusting of sand in the mould and on the clay lump (clot or clod) followed by the delivery of the lump itself at some speed so that its momentum carries it into all corners of the mould. The surplus clay is then carried away by a wooden or metal tool (known in England as a strike), by a flat lath or alternatively by a bow-wire. A projection in the face of the mould (the 'frog' – a term transposed to mean the depression in the brick itself) assists in directing the plastic material into the corners and the momentary movement of the clay across the face of the mould can produce curves or wrinkles in the surface often regarded as a valuable contribution to the finished walling. The frequently arising curved wrinkles are sometimes known as a 'brickmaker's smile'.

Other forms of plastic modelling involve the use of sustained pressure on the clay body. By the use of hand or mechanical leverage clay can be compressed into bricks or formed into tiles or blocks in moulds, the moulds being designed to release the component when the pressure is removed. Higher densities are obtained in the stiff plastic clay partly because the material can be supplied to the mould at a lower water content. A mechanically compressed component will, therefore, have been more solid and readily handled at this stage. This is a technique to be associated with the period 1870–1950. Crushed and reconstituted shales and some naturally occurring stiff clays are the most widely used in processes of this type. Mechanical pressure on semi-dry clay (as in Flettons) produces a fairly open texture which releases natural oils without disruptive force in the firing cycle.

A further application of mechanical forces comes in the extrusion process, which offered the advantage of densification in the clays. Extrusion may be continuous under the action of impellers similar to those used in a pug mill or discontinuous in the case of a charge of clay placed in a box or cylinder and propelled forward by a piston or plunger. The compressed clay is extruded through a nozzle, sometimes including metal dies, to emerge as a column of the profile required, this column then being sheared to length. The shearing action is usually carried out by the movement of a fine wire across the orifice. This process and the edges of the dies leave a distinctly striated surface indicative of the process involved. By the use of elaborate dies complex shapes could be formed. Many of these were used in conjunction with early reinforced concrete construction. Other processes have employed rollers to shape the emerging column and carry the clay forward. The discontinuous extruding machine, known as a 'Stupid', was rarely used for bricks but is still used by small manufacturers of pipes and of profiled clay roof tiles, particularly in less-developed economies. Higher quality bricks could be made by repressing extruded blanks in a process known as double-pressing. The repress gave polished surfaces and clean arrises; it was much favoured where smooth, bright, hard surfaces were required in work to very precise standards and tolerances.

Bricks formed by extrusion contain a pattern of holes through their centre to provide additional key and to reduce the clay weight and volume as well as improving drying and firing procedures. Blocks can be formed similarly to very elaborate profiles containing a high proportion of internal voids. They may be designed for lightweight walling and flooring where they may act as concrete formwork. By virtue of the mechanical pressures employed the extruded clays emerge at an advantageous density and are relatively handleable immediately on production.

At the extreme end of the plasticity spectrum brick earths and clays may be produced as slurries to make slabs, large terracottas, tiles and bricks. An advantage of the method is the improved mixing and ability to screen out unwanted material. The disadvantage is the higher porosity and lesser density coupled with greater shrinkage which comes from the employment of larger volumes of water and its drying out. This lesser density may, however, be an advantage particularly where the tile or brick has to be mechanically cut or handworked after firing or where it is to act as a body for the subsequent application of a slip or a glaze. Drying will initially be into the atmosphere or by absorption into the surface on which the brick is formed, sometimes in Plaster of Paris moulds. Shrinkage will normally then draw the brick or block away from the edge moulds allowing its release and as soon as it is sufficiently dry to be handled it can be positioned so as to allow the water to evaporate from all surfaces.

In a process derived from pottery a slip or sloppy suspension of clays (usually prepared for terracotta work) can be poured on to absorbent surfaces. Absorbent plaster moulds can be filled with the slurry to form a stiffened dewatered layer from which the surplus fluid in the centre is poured. The removal of water by absorption renders the deposit stable and handleable having taken up the shape of the mould. By this means embossed items such as elaborate chimney pots and hollow terracotta components may be produced. As moisture is extracted in the mould shrinkage occurs but this does not happen in the plane of the finished surface and, provided that the clay is sufficiently thick, shrinkage is taken up within the wall thickness. Stresses, if evenly distributed, assist in producing a coherent body of material. However, if the shrinkage produces stresses beyond the tolerance of the plastic cohesion of the material strains will occur producing tears or cracks which may not become apparent during drying or even during initial firing, although they may be revealed in a subsequent firing or in use. Inherent stresses contained within the material may be latent for long periods before causing a fracture. These faults are the result of differential contraction due to the moisture gradient from centre to skin of the drying body.

Drying terracotta blocks, bricks, tiles, etc. almost invariably involves moving them as semi-soft or relatively fragile objects. In modern mass production continuous flow techniques provide reliable methods of transport, using belting and trays on rollers. Historically, bricks were moved on carefully sprung barrows or carts often running on trackways surfaced with iron or timber slips. Tiles and bricks were dried under the shelter of light timber covers (hacks), with a double pitched top surface or in open sided buildings sometimes heated by flue ways under the floor. These hypocausts ran from a furnace at one end to a chimney stack at the other. In temperate climates air drying was slow in the winter and the provision of a heat source allowed continuous production throughout the year. Pipes, pots and tiles were usually dried in heated chambers and in many works the waste heat from kilns was utilized – sometimes by a bypass flue – between firing chamber and chimney stack. By careful calculation and experience of the moisture content of the clay and the length of the firing cycle it was possible to omit in some circumstances the separate drying operation and to move the moulded product directly into the kiln where residual heat provided the background temperature required to bring the clay to a sufficiently dry condition. This was more practical with thinner materials than bricks but where shaping could be achieved at low moisture content it avoided the separate handling costs and delays attaching to the reduction in moisture content of the brick earth from mouldability to dry biscuit, when it would be strong enough to support the other materials superimposed upon it in the kiln. As the kiln or clamp was uneconomic unless it was sufficiently large the lowest level of 'green' material had to be strong enough when stacked to carry the superimposed load of the upper layers. This load in turn dictated the height to which stacking was practicable with a further direct consequence on the economic size of kiln. Once the maximum stacking height had been achieved (about 2.5 m) the manufacturers' only recourse was to expand laterally. The kiln or clamp thereafter took on a flat slab-like shape in which the economics of the increased size were less advantageous than in a near cubic shape because of the ratio of surface area to volume. Lighter materials could not be stacked as high as more robust components such as brick, nor were the firing techniques required identical. Where mixed firings occurred the lighter materials would be superimposed on the heavier, but general practice dictated separate kiln types and separate firings for terracotta, tiles and pots.

During the drying stage poor handling could produce misshapen products. Uneven and over-rapid drying could cause cracking which, if it were microscopic, might not easily be detected though producing inherent weaknesses in the fired material. Excessive loading could deform the product even before firing began.

2.5 Behaviour in firing

Brick must perform its structural function adequately, stand up to weather, neither produce nor react adversely to salts, have no physical defects and contain no material (such as lumps of quicklime formed in firing) which will decompose during use. Fired clays are required to perform in many different ways – under intense heat as firebrick, under wear as floor tiling, under water saturation in foundations, as exposed elements subject to wide fluctuations of temperature as in roof tiling, terracotta and glazed tilework and as structural or decorative brickwork perhaps carved or modelled after firing. Some brick earths will be more suited than others for shaping and the performance required of the final product while the techniques of handling the clay, forming the component and firing will have a critical effect.

Once satisfactorily shaped and dried the historic brick passed through a process of slow warming while moisture was finally expelled and dimensional changes took place succeeded by the attainment of temperatures in excess of 900°C but generally less than 1150°C. This heat was maintained for something between eight and 15 hours except in the case of highly vitrified products, for which peak temperature was maintained for anything up to 30 hours, followed by a cooling period of up to 100 hours or more. For components of thinner material and other types of clays differing times have been applicable. These problems, originally those of the manufacturer, have become relevant to the conservator whose business it is to distinguish between the quality of product in use, its sources, origins and the nature of replacements. Methods of forming and handling are important indications of origin.

In the firing process liquid water and water of crystallization are driven off. Organic components are carbonized and then driven off as water and carbon dioxide in an exothermic reaction. As temperatures rise dehydrated aluminium silicates and silicon dioxide form and clays convert to metakaolin, which is amorphous, together with silicate compounds of sodium and potassium oxides and other impurities, particularly iron. In the presence of calcium oxide at upwards of 1000°C these materials become semi-molten. The presence

Figure 2.5 Over-firing I. A block of six bricks which had been subject to temperatures in the region of 1250°C and had fused into a single mass while retaining their general shape.

Figure 2.6 Over-firing II. A fragment of brick which reached unusually high temperatures in the kiln – perhaps 1400°C – and was sintered. In its plastic state gas bubbles formed within the matrix, remaining visible on cooling.

of alkaline fluxes affects the temperatures at which the semi-melt occurs, the metakaolin partially collapses and the mass contracts. In forming brick it is important that temperature is maintained at the semi-melt level for a period of some hours depending upon its composition and that subsequently the material is allowed to cool slowly allowing the amorphous (non-crystalline) vitrified materials to accommodate internal stresses by the process of slow readjustment common to all glasses. The liquid glasses having absorbed impurities discharge some of these as crystalline material. Alumino-silicates form fine needle-like crystals of mullite which, as a matrix, gives cohesion to the brick, together with the amorphous arrangement of the non-crystalline vitreous material.

As with glass, the physical chemistry of the performance of the components of brick earths in this state of semi-melt are both complex and uncertainly defined. The general characteristic is that a silicate matrix permits the movement of other polyvalent ions which take up an irregular arrangement giving an amorphous linkage of structures embodying regular crystalline material fixed in position on cooling. While it might be thought that a successful firing and cooling cycle would produce an inert structure this is true only in the physical sense. Metallic ions present, such as sodium and potassium not fully incorporated in the glassy melt, can dissolve very slowly in water. The variations in composition and the wide range of metals and semi-metals interacting in this silicaceous melt presents a mechanism for very slow decay of the structure of the brick.

2.6 Survival of historic methods

While traditional and quite primitive methods of brickmaking are still to be found in extensive use in the under-developed countries, it is only recently that such methods have ceased to be used in the industrially developed world. In low wage economies the capital intensive methods of mechanized brickmaking are not justified, while even in high wage circumstances small volume and specialized production, coupled with pride in the craftsmanship,

Figure 2.7 Ashburnham brickworks. The wood-fired kiln under its roof with timber stacked to the right and hack covers in the foreground. (Courtesy Kim Leslie.)

give rise to important hand-made brick and tile production.

2.6.1 Case study: Ashburnham brickworks

The traditional early processes of hand-making bricks, tiles and terracotta may be illustrated in the history of a yard in Sussex worked from the fourteenth to the twentieth centuries, possessing when it closed the last wood-fired brick kilns in the country. These were the brickyards on the Ashburnham Estate, four miles east of Battle, in East Sussex. This estate had also had extensive iron workings and lime kilns, the techniques and craft knowledge being passed from generation to generation. The methods of work in its brickfield and tilery remained simple, even to the use of a horse-powered pug mill right to the end.

There is no evidence that there was a continuity between the tile- and brickmaking of the Roman era and medieval brickmaking. In the south east of England pottery continued in production through the Dark Ages and the Saxon Kingdoms but the major changes of land ownership with the consequent introduction of continental crafts and craftsmen in the Norman era renders suspect any claim to full continuity as opposed to the introduction of continental skills. There would probably have been little to differentiate the works at Ashburnham from others across the Weald and throughout the country prior to the industrial era. The yard itself was moved in 1840 to the site where it continued until closure in 1968; and in 1847 reached a peak of production when a brick and tile machine (Ainslie's Patent) was purchased and brought into use. The products were bricks, tiles, drainpipes and flowerpots. The kilns were of the Scotch type being rectangular, open topped chambers built in brick capable of holding a stack of green bricks in the order of 3 m cube (10 ft × 10 ft

× 10 ft). Channels under these chambers led to firing holes into which fuel in the form of cordwood (branch wood) or the fiercer burning faggots (bundles of twig wood) was introduced and control of the kiln was effected by covering it with layers of fired brick and earth by which means the outgoing gases could be controlled.

Two different types of brick earth were used. The material from which the bricks themselves were made was dug from a bank on the site and was a highly silicaceous loam, the silica (sand) content exceeding 75%. In consequence the drying shrinkage at a little under 4% was unusually small, a figure of approaching 10% being more common with Wealden clays. The clay content of the loam was about 15% and by composition it would have made a very good building earth having a 2% calcium oxide content. With organic combustibles in the order of 3% it had some self-firing advantage. No additional material was required to temper the loam and the preparation was straightforward. After removal of the topsoil the loam or brick earth would be dug and traditionally was allowed to mature over the winter. It could, however, be used without the initial weathering process. Over the period of a day it was pounded, smoothed, cleaned of lumps and worked into a plastic mass – the pug. The consistency of this pug was critical and it was always prepared by hand.

For moulding, a supply of local sand, sieved to take out larger material, was the only further requirement. On the moulding board a former or wooden stock was fixed and over it rose a square moulding box (the mould). The interior was sanded and a suitable semi-shaped lump of pug was thrown into the box with sufficient force to fill the entire void, the top being immediately levelled off with a wooden tool. Stock and mould were then moved into an upright position, the stock being hinged. It was then returned leaving the mould containing the new green brick. Two wooden slips were then used to secure it in position in the mould and it was thus transferred to the hack barrow where the mould was removed. The hack barrow when full was taken to the drying area, being run along a track of iron plates. The bricks were set out on edge on base boards with air spaces between them, being covered

Figure 2.8 Ashburnham brickworks. The front of the kiln sealed after being brought to firing temperature on the final occasion of its use. (Courtesy Kim Leslie.)

with low loose timber roofs or hack covers made of very thin creosoted softwood. They were handled with wooden slips. High winds could wreak havoc in the hack yard blowing off these covers and damaging many weeks' production. The bricks were sufficiently dry after three to six weeks to be carried to the kilns and stacked on the slotted floor over the fire tunnels. The loading into the kiln would be done from one side and in rainy conditions a temporary cover would be laid over the kiln top. Twenty thousand bricks would go into a kiln, being stacked on edge, finger spaced, leaving open flue ways to carry the hot fire gases upwards. Finally a triple layer of burned brick with gaps between would be laid over the top of the stack and the opening through which the kiln had been loaded would similarly be closed up with burned brick laid

Figure 2.9 Ashburnham brickworks. A load of green bricks on the hack barrow being moved to the kiln for the final firing. (Courtesy Kim Leslie.)

in sand and sealed with clay – the wicket bricks. In principle the finished structure was not very different from a clamp, though most clamps would have covered a much wider area while being no higher. A clamp usually had fuel layered with bricks and, in the main trace, the flues on the base. Clamps were lighted at the windowed end and fire travelled slowly to the other. However, these Scotch kilns were lighted at firemouths on both sides, firing completely at the one time. For the initial four days fires would be burned at a very low level in a period known as water smoking. In this way the kiln temperature would be raised gently to about 200°C. Then in less than a day and a half the temperature would be raised to 1000°C by continuous firing, the temperature being judged by the colour of the fireways. A maximum temperature of 1100°C would be maintained for eight to ten hours – about half or one-third of the time often given in the firing cycle. This particularly amenable clay was perhaps one reason for the success and longevity of this works, illustrating the variability of local circumstance and technique. At this stage of maximum temperature the completion of firing was gauged by the settlement of the entire mass as measured by a line across the top of the kiln sighted on a brick stack in its centre. The contraction of the entire mass of heated bricks could be anticipated within precise limits and when achieved firing ceased and the kilns were allowed to cool for a period of five to six days when the flue ways were opened and air was allowed to circulate to cool the mass gently prior to unloading. The bricks produced had a red body and a grey glaze on the headers set into the flue ways. This colour, typical of wood-fired Wealden bricks, derives from the vitrification of the sand

Figure 2.10 Ashburnham brickworks. The top of the kiln glowing at 'orange heat' in the final firing. Note the 'on-edge' bricks used to assess the quite visible settlement which takes place at full temperature. In the background the drying yard with hack covers. (Courtesy Kim Leslie.)

dusted onto the surface and the silica within the brick itself fluxed by the potash in the wood ashes carried up the flues. The ash adds alkali and lowers the point of vitrification to about 900°C while the presence of calcium oxide in the clay would be a sufficient stabilizer to balance the vitrification producing a durable glaze of fused silicates. The same phenomenon does not occur unless an alkali is present as the viscosity of semi-melted silica is far too high to produce a glaze at the temperatures of brick firing. The glaze on such bricks is, therefore, an indication of wood firing in the kiln on a sand dusted or highly silicaceous brick.

The irregular nature of this semi-glaze is not simulated by means other than the original manufacturing process and bricks of this type of manufacture very often display marks as bars of the same glaze on other faces, sometimes at 30–45 degrees to the alignment of the brick in the kiln. It is known as skintle setting and improves stability in the kiln. These marks reflect the potash deposits in secondary flue ways formed between the angled bricks above and below that on which the effect appears.

Bricks manufactured elsewhere often display similar patterns due to similar setting, although the coloration is different. The Severn Gorge works were typical in this regard. From sixteenth century onwards bricklayers utilized the varied colours thus obtained in headers and stretchers in forming patterns in their face work. Deliberate diapering was achieved, or chequer patterns were made, but many patterns were simply the consequence of the occurrence of headers in bonding.

At Ashburnham the brick earth was not regarded as suitable for making tiles, pipes or pots, for which purposes an earth much richer in clays and much less silicaceous was obtained from another local source. The sand content of this clay was less than 25%, the clay itself having a high alumina content. The tile clay was dug in winter from a bed about 4 ft deep after the removal of topsoil. It was exposed in a brick lined depression throughout the winter and in the spring was prepared by thorough wetting over some days, being chopped and turned and then ultimately being fed into a horse-driven pug mill – a process well described as churning. The resultant lumps from which air had been excluded by the pressure of the mill were then moulded in much the same way as bricks in a sanded shaped mould. The clay would shrink by 12% in drying and firing. It was dried in an open sided shed but the through draught was controlled to ensure that drying was not too fast and several weeks would normally be allowed. The tiles were burned in the same kiln as bricks but always in the top one-third. Therefore, they were fired at temperatures a little over 1000°C. Tile-making was a summer operation. This divided firing was common elsewhere.

Land-drain pipes likewise were produced towards the end of the summer season, being extruded from the same clay by a hand operated lever machine. These products when green were always at risk from frost. Seasonal production, therefore, relates both to this aspect of the drying process and the overwintering of the raw material. The same clays were used for the making of flowerpots but in this instance the clay was reduced to a very much more liquid condition, run across a settling tank where larger particles would be deposited and was then extracted from a further settling tank as a finer clay-rich material – in practical terms a terracotta clay. Its drying shrinkage was 14% or 15% (in linear terms) in this condition and the pots were fired at the topmost level of the kiln. The Ashburnham works are not recorded as having produced terracotta in any other form than flowerpots.

In this remarkable survival of a traditional industry the skills and techniques which produced so much of the content of historic buildings have been available for contemporary study.

2.7 Coal-fired bricks and industrial processes

From these basic hand-making methods the industry has evolved by a series of progressive improvements, and most contemporary processes are recognizable in terms of the original and basic methods. Dry moulding emerged in the industrial era and the technique of adding fuel to the brick earth or finding a brick earth containing fuel amended the original techniques of firing by generating heat within the body of the clay itself.

Manufacturers with access to brick earths or shales containing a substantial proportion of oils or of fossilized organic matter (effectively particulate coals) were able to reduce the costs of production and increase their market share in consequence. Where such earths contained the right natural balance of clays and silts and had the additional advantage of being without impurities and the right plasticity when quarried, the manufacturer enjoyed even greater advantage. This circumstance accounts in large part for the success of the industry based on deposits at Fletton near Peterborough in England. Brickmaking has always been a very price-sensitive industry and in consequence the character of brick used in each locality has reflected the economics of brickmaking.

Coal measure shales with inherent coal particles were not self-combustible and the carbon could cause black cores, looked upon as a fault. Added coal, however, was useful as an energy source, but oil-bearing clay, which is more nearly self-combustible, was particularly advantageous, requiring minimal labour in preparation.

Competing manufacturers added crushed coke and clinker or coal ashes to their earths with the objective of even heat distribution in firing, usually producing mottled surface textures as a result. The inert part of such ashes might be clinkered but being the result of the burning of the shales in which the coal measures had been originally laid down the base materials were intrinsically similar. The imperfect combustion of domestic and hand-fired industrial furnaces produced a plentiful supply of ashes and in consequence of the availability of coal, coke and ashes at rail heads, and the advantages of transport for the finished product, a pattern of brickmaking grew up around the railway network.

Historically the brickmaking process seen through the eyes of contemporary writers in the middle of the last century summarizes the circumstances of the making of the large proportion of historic brickwork in Europe. The following account is taken from the *Cyclopaedia of Useful Arts*, c.1860. It provides a direct and comprehensive insight into the manufacturing processes from a contemporary standpoint.

The operations connected with ordinary brick-making are briefly these: digging the clay in autumn; leaving it to mellow by frost during winter, the masses being frequently turned and broken up, to expose them more completely to the action of the atmosphere: throwing the crumbled clay in spring into shallow pits, where it is watered and soaked; then tempering the clay by treading and kneading by the feet either of men or oxen, or by means of a horsemill: next conveying the kneaded clay to the bench of the moulder, who takes a lump and dashes it into a wooden or iron mould, striking off the superfluous clay with a strike or smooth piece of wood. The bricks are delivered from the mould, and ranged on a barrow or on the ground, until they are firm enough to bear handling, when they are trimmed with a knife. They are then built up in long dwarf walls, with sufficient space for the air to penetrate in every direction: these walls are thatched as a protection from the weather, and thus the bricks are left to dry until they are in a proper state to be consigned to the kiln.

The various argillaceous earths used in brick-making are generally mixed with some other substance, being for the most part unfit to be used alone. Some are almost pure clay or alumina, and are strong, and exceedingly plastic, but cannot be dried without splitting. Others, being light sandy clays or loams, are too loose to be made into bricks without the admixture of lime as a flux, to bind the materials together. Others again are natural compounds of alumina and silica; but these if free from lime, magnesia, or metallic oxides, are exceedingly valuable clays, being from their infusible nature adapted for making fire-bricks for lining furnaces, for making crucibles, glass-house pots, &c. Fire-clay is found throughout the coal-measures, and occurs in abundance, and of excellent quality, at Stourbridge, and also in the vicinity of Newcastle and Glasgow.

Bricks for ordinary uses are known as "place bricks," "grey and red stocks," "marl-facing bricks," and "cutting bricks." The place-bricks and stocks are the ordinary wall bricks. The marls are very surperior bricks, made in the neighbourhood of London, and used on the outside of buildings. The finest kind of marls and red bricks are called cutting bricks, and are used in arches over doors and windows, being rubbed to a centre and gauged to a height. The red bricks made of Hedgerly loam, from a village of that name, near Windsor, are used as fire-bricks about furnaces and ovens. Foreign bricks are *Dutch* and *Flemish* bricks and clinkers: they are similar in quality, and of a dirty brimstone colour. The first two are used for paving yards, stables, &c., and the clinkers which are most baked are used for ovens. Place-bricks are also used in paving dry, or laid in mortar, and they are put down flat or edgewise. If they are laid flat, 32 of them will pave a square yard; if edgewise, twice that number are required. Ventilating bricks are an invention of modern times. They are double the size of common bricks, although they contain only the same quantity of clay. They are hollowed out at the sides, so that when two are placed side by side, a circular opening is left between them, which, when tiers of similar bricks are laid on, forms of course a tube within the wall, and this may be applied to the purpose of either warming or ventilation.

The first process in brick-making is the tempering of the clay, which, as we have said, is the work of early spring, after it has lain exposed to the frost during winter. Great care is then taken as the clay is being turned over and tempered with water, to remove by hand every stone that can be discovered in the plastic mass; for the presence of even a small pebble in a brick causes it to crack in drying. Of course this hand-picking is impossible where much gravel occurs: in such cases the clay must be washed in a trough filled with water, until it becomes liquid enough to pass off through a grating into pits prepared for its reception; the gravel meanwhile being retained by the grating. In the districts where veins of skerry or impure limestone abound, it is found desirable to grind the clay between rollers, which crush the limestone, and thus obviate the evil which arises when even a small piece of this substance remains in a brick, and by the carbonic acid driven off from it in burning, forces a hole in the brick, and destroys its usefulness.

For the marl, or *malm* bricks, made near London, and used for the best outside work in houses, the clay is dug in autumn, ground to a pulp at once in a wash-mill, and mixed with chalk previously ground to the consistence of cream. This pulp is run off through gratings, and allowed to settle until it is firm enough for a man to walk upon it: it is then covered with finely-sifted ashes, and allowed to remain all the winter to mellow. In the spring the ashes are thoroughly mixed with the clay and pugged in a pug-mill. This is a conical wooden tub, having the larger end upwards, with an upright revolving shaft passing through it, armed with a number of knives, which cut and knead the clay, and force it through the mill, which is constantly filled at the top from the barrows of the work-people, while the clay continually issues from a hole in the bottom, where it is cut into convenient pieces and piled up for future use. The pug-mill is extensively used where

Fig. 235. PUG-MILL.

the demand for bricks is large, and where the brick-earth is favourable in quality; but in many country places, the indispensable labour of kneading the clay is still performed by the treading of men's naked feet, which become by constant practice sensitive to the slightest roughness in the mass, and able to detect the smallest stone or impurity.

When the clay has been reduced by one of these processes to the necessary state for brick-making, masses of it are successively brought to the moulder's bench. The mould is without top or bottom, and the workman's art consists in dashing a piece of clay with such force into it, as completely to fill it, and then cleverly striking off the superfluous quantity, and turning out the brick on a pallet, which is placed by a boy on a hack-barrow, which when loaded is wheeled away to the hack-ground, where the bricks are built into long low walls to dry. By another plan, the bricks are shifted at once from the moulder's bench to a drying floor, from thence to the hovel, or drying shed, and from the hovel to the kiln.

The moulder's bench is a rude kind of table, often provided with a trough for water, as well as a heap of

Fig. 236. MOULDER'S BENCH

sand, the mould being either dipped in water, or sanded, between the making of each brick, that the clay may not adhere. If water is used, the process is commonly called *slop-moulding*, if sand, *pallet-moulding*. In the neighbourhood of London, women commonly take part in the operations. Fig. 236 represents a woman thus engaged. The moulds were formerly of wood only; they are now sometimes made of brass, cast in four pieces, and riveted together, or of wood lined with brass, sometimes of wood with the edges of iron, sometimes with the two longest sides of iron. Brass moulds do not require wetting or sanding; but they are expensive, and the edges soon become worn. Wooden moulds therefore continue in some districts to be largely used. A good form of mould is a wooden mould lined with brass; the wood as well as the brass being in four pieces, and attached by rivets at the angles. This mould costs about twenty-five shillings, and was formerly still more expensive. The brass overlaps the wood at the edges, where it wears out rapidly, and

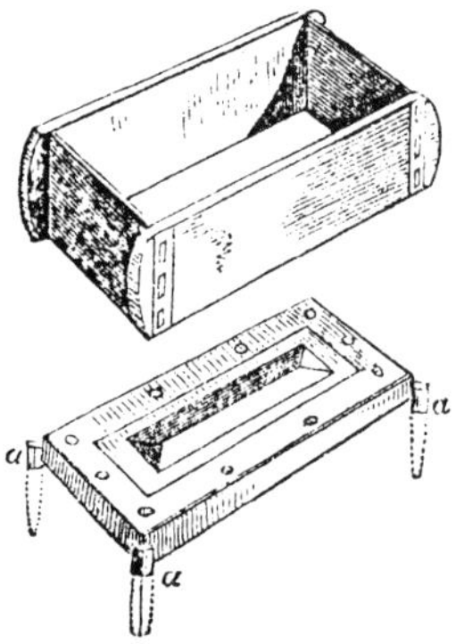

Fig. 237.

the cost of repair is nearly as much as the original price of the mould. Fig. 237 represents a mould made of sheet-iron in four pieces, riveted at the angles, and strengthened with wood at the sides. The bottom is separate, and is called a *stock-board*. This is fastened by pins at the corners to the moulder's bench. It is very common at the present time to make bricks with a hollow underneath, both for the sake of lightness, and to leave a bed for the mortar. This is managed by fastening a piece of wood called a *kick* to the upper side of the stock-board. The mould being placed on the stock-board, (which easily and accurately fits it,) and the clay pressed into the mould, a hollow space corresponding to this kick is of course formed on the under side of the brick. The pallets are pieces of board ⅜ths of an inch thick, of the same width as the mould, but a little longer. Six-and-twenty pallets form a set, and three sets are required for each moulder. Fig. 238 represents the bricks

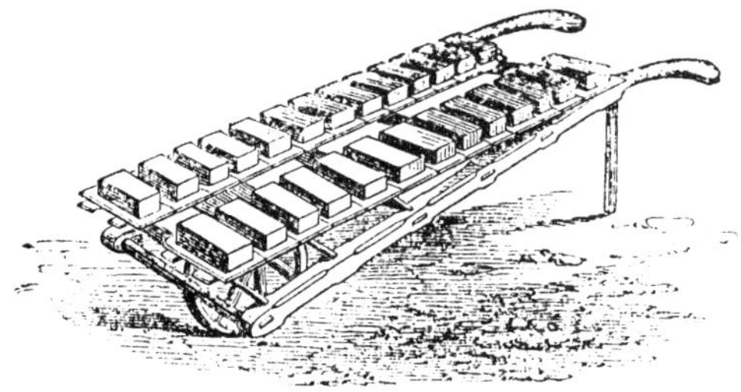

Fig. 238. HACK-BARROW.

on one set of pallets, ranged on the *hack-barrow*, which has a flat top of light frame-work, fit to receive two rows of bricks, thirteen in each row. Three of these barrows are required for use at each moulder's bench, one being constantly loading there, another unloading in the drying ground, and the third being wheeled to and fro. The low walls of bricks in the drying ground are called *hacks*. These are built two bricks wide, and eight bricks high, and the bricks are generally placed slanting, and not at right angles to the length of the wall. When the bottom row of one hack is formed, the workman begins a second hack, leaving the first to get firm before it has to bear the weight of a second row. Plenty of straw or hay is at hand to cover up the bricks at night, or in bad weather. For the finer descriptions of bricks, drying under cover is adopted, and in some instances flues are carried under the floors of the drying sheds, and currents of air are carefully excluded.

Where the demand for bricks is very large, brick-moulding is performed by machinery. A number of machines have been invented for this purpose, and some of them have answered the end very well. But it is doubtful whether the pressure employed is really an advantage. The density of the bricks is thereby increased, and they are smoother, heavier, and stronger than other bricks, which for some purposes is desirable, but they do not adhere so well to the mortar, they are difficult to dry well, and their weight adds to the expense of carriage, and prevents the workman from laying so many in a given time, as of the hand-made bricks. Machine-made bricks are also frequently disfigured by a ridge caused by the clay rising a little way up the sides of the piston, in the space which, without careful workmanship, is apt to occur between the piston and the mould.

Two brick-making machines much in favour in this country are Ainslie's and Hunt's. The latter has been extensively used in the execution of large contracts, and consists of two cylinders, each covered with an endless web, which are so placed that they form a sort of hopper on their two upper cylindrical surfaces, the ends being enclosed by two iron plates. The tempered clay is thrown into this hopper, and at the lower part it acquires the form and dimensions of a brick. Beneath is worked an endless chain, by the movement of the cylinders, and at various marked intervals are laid the pallet-boards under the hopper; the clay is brought down by a slight pressure, and enters a frame, which has a wire stretched across it, which projects through the mass, and cuts off the requisite thickness; this is immediately removed by the forward motion of the endless chain; and this operation is renewed as often as a new pallet-board is advanced under the hopper. Such a machine produces about 1,200 bricks per hour, and is worked by two men and three boys. By this plan less pressure is given than in most machines, consequently the bricks are less difficult to dry equally. Machine-work is cheaper than hand-labour in the moulding of bricks.

Machinery has been recently employed for making bricks and other articles of clay nearly in the state of a dry powder. The clay is subjected to heavy pressure in strong metal moulds, and is by this means reduced to one-third its original thickness. It retains just sufficient moisture to give it cohesion, and the bricks thus formed can be handled at once, and taken direct to the kiln. This method was devised by Mr. Prosser of Birmingham, and is highly useful for making ornamental bricks, floor-tiles, &c. By an experiment made on a nine-inch brick of this sort, it was found that the resistance to a crushing force is immense, ninety tons having been sustained without injury.

The final process in brick-making is that of burning the bricks in a kiln or in clamps, the former being the old and the best plan. The kiln may be a simple rectangular chamber, built of old bricks and rubble stone, with a narrow doorway at each end, and narrow fire-holes lined with fire-bricks in the side walls exactly opposite each other. The workmen introduce through the doorways a quantity of bricks, and stack them loosely but with considerable art in cross courses, within the walls, leaving openings that shall act as flues throughout the whole mass, and thus distribute the heat from top to bottom. When the kiln is filled, the top is covered in, and fires are lighted in the fire-holes. The fire is at first got up gently, that the moisture in the bricks may be gradually evaporated; but in two or three days, when the steam, ceases to rise, the heat is raised, the doorways are bricked up, and the temperature continued till the fire begins to appear at the top. It is then slackened, and the kiln allowed to cool. The heating and cooling are then repeated, and in about 48 hours the bricks are thoroughly burnt. An ordinary kiln will hold 20,000 bricks. The fuel consists in some places of fagots of furze, heath, brake, &c., in others of pit coals. When bricks are burnt in a clamp, they afford to a great extent their own fuel, for a clamp is an immense pile of carefully arranged bricks, in which *breeze*, (the technical name of ashes,) has been mixed with the clay in their manufacture. But layers of breeze are also added, and the whole is set fire to by means of fireplaces and flues filled with wood, coal, and breeze. The burning of a clamp continues from two to six weeks. The art of clamping well exhibits no mean degree of skill in the workman. They first build an *upright* or double battering wall along the centre, and then arrange a number of other walls in an inclined position on each side, corresponding in length and height with the central wall, and supported by it. The sides and top of the clamp are eased with burnt brick, and the lower courses of the central double wall are of the same material. There are numerous live-holes left in a large clamp, and these are fired in succession. The bricks near these live-holes are burnt too much, and generally spoiled by running together in masses called burrs, and the bricks at the outside of the clamp are not burnt enough, and are laid aside for reburning in the next clamp. Much judgment is required in apportioning the fuel to the size of the clamp, for the whole may be easily underburned or overburned, and so deteriorated or rendered comparatively useless. The burrs and clinkers, or shapeless masses of fused brick, may often be recognised in the rock-work of suburban gardens, while the pale underbaked bricks, sold at low price, are used in the inferior unsubstantial erections which disgrace the neighbourhood of the metropolis.

By the time this summary was published in 1850 there had appeared a comprehensive account of brickmaking technology by architect Edward Dobson, with a contribution by R. Prosser, a civil engineer, describing in detail the process of manufacture of bricks and tiles. (Republished as *Ceramic History*, 1971, ed. Celoria). Dobson practised widely as a civil engineer, living latterly in New Zealand. It was by publications of this type that a uniformity of knowledge spread through the industrialized world and access to similar brickmaking equipment was suddenly available across entire continents such as India with long histories of brickmaking of their own. The adaptations and merging of techniques produced a sudden alteration of brick manufacturing world-wide.

Another element of change was the evolving architectural profession which was taking a progressively more dominant role, even to the point of aesthetic evangelism, determining standards and taste by arguments in journals and responding to idealism and competitive nationalism. Architects in Britain complained (particularly in London) of the diminishing standards of brickmaking, particularly in relation to the widespread unregulated addition of coal ash and cinders, chalk lump and poorly tempered clay, as well as under-firing. There was social concern at the bad conditions of work, the employment of children in brick work (halted by legislation in 1866) and bad handling of the product which led in turn to higher specifications and demands for very specific bricks. Under these pressures manufacturing changed. Colour and character were scrutinized and, more importantly, innovation gave rise to improvements in quality. Definitions of brick types grew more complex. No longer was the 'place' brick the ordinary and the 'stock' the better class product. Greater selectivity gave rise to more categories. Pressing and double pressing were introduced. There were commonplace bricks (house commonplace for very ordinary), common stocks, stocks, rough stocks, grizzles (under-burned), shippers (deformed, and often exported as ship's ballast, presumably to be disposed of to colonies or settlements at minimum cost), seconds (a well-burned brick but off in colour, also used as a pavior) and cutters (well made but under-fired and soft enough to cut readily). Among the high quality bricks there were the special colours achieved by varied firings and additions to firing and bricks formulated with care prior to firing, including the expensive polished facing bricks. The polishing took place at the 'green' stage. New processes came in. Prosser, the engineer, had by 1840 proved that tiles could be manufactured by pressing dry clay and patented processes followed. The first British patent for a brickmaking machine was issued in 1741 (Pat. no. 575) and within a hundred years there were some seventy further patents involving devices based on plungers, revolving patterns, cutters and even hydraulic formers. Alternately working cylinders were introduced to provide continuous extrusion (patented by Fred. Ransome and John Warren) and leather-covered rollers were brought in to shape columns of clay rather than to force clay through dies with the consequent problems of clay deformation caused by exploding air bubbles.

2.8 Sources of clays in England

Colour in brickwork became a passion among some architects, particularly in France and England, by the mid-nineteenth century. The complex geology of northern Europe favoured variety. Natural materials were available for every conceivable colour of brick and with the industrial resources and equipment available brickmaking was no longer limited to superficial alluvium and estuarine clays. Confining the examination to England – in itself a complex subject – the superficial clays used in medieval times all had a post-glacial origin, having overlain Pliocene clays. Broadly they derived from the Holocene period and generally contained a broad distribution of ferruginous material as a result of their geological history of mixing during surface weathering. They were exploited mainly in eastern and southern England, producing mid-reds, which fired darker if well burned and orange if under-burned. They were almost always used directly and without additives. These superficial deposits from the geologically recent Holocene period provided the surface loams and estuarine clays used in the fifteenth and sixteenth centuries.

By the seventeenth century deeper deposits were being exploited. The Eocene strata around London and Reading were the most used, the principal brick works being in Essex, Kent, Middlesex and Berkshire. Many of these clays contained lime and burned to a soft pink and sometimes grey or yellow. By liming some clays or extracting deposits which had acquired calcite by washdown from the chalk overburden the firing could be modified to produce yellow and yellow-purples – the well known 'London stock'. These clays were known as 'malms' or 'marl', a term quite widely applied to other clays of a loamy nature. Sea coal, extensively used in London, provided a source of ash for amelioration and was also available as a fuel. Part-burned ashes from the coal-fired grates of the metropolis were supplied to brickmakers for addition to the clays and as fuel for their clamps. This practice was widespread in the eighteenth and nineteenth centuries.

South of the Thames, in the Weald, the Lower Cretaceous clays laid down beneath the chalk deposits had been exposed by glacial erosion often exhibiting blue streaking with ferruginous bands. In preparation the iron had become distributed to produce clays which burn to crimsons, terracotta, reds and orange.

Figure 2.11 London stock brick. A typical product of early coal-fired clamp burning, pointed subsequently with a careful sharp-edged near-flush joint.

The admixture of coke breeze after the railway building era produced dark mottling on the face of stock bricks which contained a wide colour spectrum in contrast to the reds of previous wood-firing. The mottled multi-coloured brick was the result of the reducing (deoxygenating) atmosphere between contact faces which changed the state of oxidation of the iron oxide. Reduced ironstone nodules melted to give iron spots or splashes.

Further north and west younger (Upper Cretaceous) strata have been exposed intermittently across central England from Yorkshire to Dorset. The clay is bluish, and calcitic laid down between 65 and 150 million yeas ago when life was abundant and shell-forming creatures, as well as Saurians, inhabited the earth. Bricks made from it are yellow due to their chalk content and found favour in the eighteenth century because their colour approached that of Bath and other sandstones.

Parallel to these clays ran another range of calcitic clays often confused with the Upper Cretaceous but deriving from the Jurassic strata laid down in the limestone beds before the Cretaceous period (up to 200 million years ago). These clays are found through Lincolnshire to Wiltshire. They contain the Lias beds which provide iron-rich soils, some of which contained organic material, reducing fuel requirements in firing. Burned for brick-making, they turn a bright strong mid-red, characteristic of the architecture of Leicestershire and Nottinghamshire. They were heavily exploited in the nineteenth century being transported widely by rail. Lias occurs also as shales – near, for instance, Bridport. The Oxfordshire clays, from the north of the county, are Middle Oolitic. Oolites are shell formations of primitive creatures rich in iron or heavily calcareous. In mass they create limestones and ironstone and in clays produce calcitic material which burns to a soft red, pale terracotta or yellow. Their importance in English brickmaking is that the greyish deposits of hard oolitic clay in north Bedfordshire and southern Lincolnshire yielded a material which could be fired immediately and without preparation into a strong pinkish brick. By the use of heavy pressing equipment shaping was possible without changing the water content. The material on excavation is at about the shrinkage limit of clay and is therefore usable without extensive further preparation. The deep beds at Fletton near Peterborough were quarried mechanically, the clay being crushed, screened and pressed immediately. The green Fletton brick is subject to only modest movement on drying. Immediately it is processed it is stable and capable of supporting moderate loads. It contains organic material which, at up to 10% by weight, reduces the fuel requirement significantly. The consequent commercial advantages, and a convenient geographical location have brought the Fletton to predominance in the British brickmaking industry, particularly in modern times when efficient transport of heavy goods by road has reduced the proportionate costs of carriage and eroded the advantages of local brickmaking. Its historic importance is localized.

Deeper workings associated with coal mining made older measures available. Permian clays, some 250 million years old, and the slightly younger Triassic clays were obtainable from pits in the English Midlands. The Permian beds yielded Etruria marls which produced one of the most remarkable brick and tile industries of all time. These iron-rich clays, fired red at high temperature and in a reducing atmosphere produced the dense, almost metallic, steel blue bricks and tiles used for water resistant paving, engineering and roofing. These bricks, principally known as Staffordshire Blues, are a product solely of the industrial era. The associated coal measures provided suitable fuels for the achievement of the high temperature firings required. Being deeper deposits these were drier and more compressed than the superficial earths. They were won as shales which required extensive working and pulverizing.

Bricks from the Triassic clays – the Keuper marls – are strong and light buff or salmon pink in colour. They are found across a wide belt from Bristol to Cheshire and eastwards into Nottinghamshire. Occasionally both Permian and Triassic measures are grouped under the general terms Permo-Triassic and New Red Sandstone periods (240–280 million years ago). Beyond that period, from 280 to 340 million years ago when the higher animals were evolving and plant life was rich, the era is known as the Carboniferous. It contains the coal measures laid down in sedimentation which produced extensive clay beds of quite widely varied character, all of which have matured into shale and rock under the temperatures and pressures that produced the coal seams. The quality of these clays varies from nearly pure kaolinite used as fireclay, to hard grey shales containing a high proportion of illites. Having been laid down before the period when the shell producing creatures flourished widely, they tend to be free of calcite. A high iron content is reflected in the strong mid-red colour of bricks made from most carboniferous shales. They require selection, grinding and extensive preparation particularly to distribute the iron pyrites (ferrous sulphide, FeS_2) produced in anaerobic conditions in the swamps in which the coal seams were formed. The disadvantages of preparation were offset by the presence of a fuel content deriving from fossil plant material in the clays. They were won as a by-product of the mining operation. Their occurrence in conjunction with coal provided both the fuel for firing and a substantial demand within the colliery itself to say nothing of the township that grew up around it. Many a coal owner became a brickyard proprietor and the landlord of great estates of housing built with his own bricks to house his work force. Historically their use is associated with the nineteenth and twentieth centuries.

The crushed shales from colliery workings and low grade coals available as colliery by-products gave rise to thriving industries whose better products were very much more accurate and generally denser and stronger than earlier brick. Coupled with the newly available cements of the nineteenth century they became the mainstay of much more precise techniques of civil engineering which enabled engineers to create brick structures for buildings, viaducts, aqueducts, harbours, forts and redoubts on a scale and of a strength equal to the best natural stones.

In all the long history of brickwork, methods of building had always been based on experience and presumption. Testing to destruction, where it occurred, would have been accidental and the results though serving as an awful warning would not have been applied. It was with the eighteenth century and advances in scientific analysis that the strength of structural brickwork was first subject to analytical scrutiny and as so often in technical advance the initial stimulus was given by military requirements. Civil engineers such as John Smeaton had already undertaken work on the development of mortars and this was followed by experimentation on brickwork.

2.9 Scientific brickwork

Faced with the threat of invasion from Europe by an enemy equipped with sea-borne fire power sufficient to destroy its established defences, the British government, through its Board of Ordnance, set out to plan and build a series of defensive towers (based on experience in Corsica) capable of withstanding the heaviest naval guns. In parallel with military strategies went analysis of the materials to be used and the most effective forms of construction. These analyses included testing sample structures specially built for the purpose, identifying the results, drawing conclusions, repeating the tests and then using the improved results in the final work. Thus, for the first recorded time, was introduced into brickwork the further ingredient of scientific trials of the product as opposed to trials of the manufacturing process.

The selected design was a circular tower, strengthened by having a substantial column at its centre, a ring barrel vault joining the column and the walls to carry a roof or gun platform. The base of the tower was the equivalent in reverse with an inverted ring vault as foundation, often in gravel. Circular test walls were built at the arsenal at Woolwich on the Thames by the Royal Corps of Engineers. Well-fired bricks of Thames clay set in a hot hydraulic lime mortar proved to answer well under artillery bombardment. The walls were built to radii of approximately 15 m externally and thicknesses of some 2 m. In the ultimate designs they were battered and the circular form was adapted to a slight ellipse. Further, the ellipses of the interior and exterior profiles were not concentric, producing a thicker wall on one side – the side facing the sea from which the heavy bombardment would be expected.

Success lay in the balance of characteristics. The shapes and dimensions chosen, the density and nature of the materials, their elasticity and their compressive strength were shown by experiment to react satisfactorily to the type of impact to which they might be subjected. Technology was still far from the point when the performance of materials could be used to produce a theoretical design which, when executed, would marry theory and practice, but the first steps had been taken and the comparative performance of types of brick, different structural profiles and that crucial constituent, the mortar, had been evaluated. Mortar was made with sand, pulverized sintered clays and residual ashes from coal burning mixed with eminently hydraulic quicklime slaked into the mixture to produce a powerful cement made workable by the introduction of tallow (animal fat). The heat of slaking allowed the small amount of fat to meld with the more gritty materials producing a cohesive mix and the heat gave it an initially rapid chemical activity. These mortars have proved powerful and durable, adhering tenaciously to the relatively porous bricks to produce coherent masonry masses capable of withstanding long years of exposure to the sea.

From these beginnings technical skills in the manufacture of brick and other fired products have grown consistently with parallel improvements in structural design, consistency, standards of testing and assurance of quality. Specialized shapes, special production methods and the use of specific materials have combined to increase performance first under the stimulus of railway engineering and later to meet the ever-widening applications in civil structures required for industrial application, to the point where British engineering brick was used in the calculated substructures of the now historic launch pads at Cape Canaveral. Materials other than earth or clay have been used to make bricks for building, but of these the only fired material of historic significance is slag, which is the melt of siliceous impurities arising as a by-product of smelting, particularly from iron furnaces. This material was run into moulds prior to running the liquid iron. At that stage it would be at a temperature of several hundred degrees centigrade above the normal temperature at which bricks are fired and if run into a sand mould would aggregate with the immediately available surfaces, initially melting the sand. In the absence of an alkali to provide a flux the full melt of silica might not be achieved. Such bricks contained gaseous voids, have a highly vitrified, dark appearance, were frequently fractured under thermal contraction and were prone to contain materials capable of being drawn into solution. They were, therefore, unlikely to be durable in other than mass construction conditions and are of minimal importance in conservation.

Of other special clays used in brickwork there remain two which have had little effect upon the visual qualities of historic buildings. Of these the first is the lightweight brick of which there have been a number of proprietary types used for sound absorption, thermal insulation and in structural situations where light weight was demanded. The best known of the early lightweight variants was an Italian material – first made in the eighteenth century – which produced a brick with a specific gravity as low as 0.6 (as light as a soft wood) used occasionally for vaulting. The raw material is diatomaceous earth, a fine cellular structured soil of silicaceous shell laid down by microscopic algae. An English version, known by its trade name of Moler brick, was advocated as a partitioning material and may be encountered in this form. The brick is weak and friable. The better diatomaceous earths are not available in Britain and Danish imports now meet current needs – primarily for furnace backing.

Another brick earth produced the light coloured brick known as fireclay. Also called refractory bricks, their essential merit was a very high melting point. Industrial furnace bricks were fired at temperatures in excess of 1320°C for extended periods. Fireclays vary as widely as red brick clays. Many low iron fireclays were very plastic and had a very high clay content. Many are rich in potash and fire at 1100°C. In practical terms their prime importance was for lining furnaces, fire ways and kilns but they have been used as facing material when a light coloured brick is required. They were dense and tended to be smooth-faced with high compressive strengths. For decorative use they were superseded by faced bricks which provided similar colours on a more conventional core material. Occasional seams of fireclay were worked specifically for the quality of the material but frequently it was derived in association with coal mining. In manufacture, apart from the high temperatures involved, a significant problem was control of the shrinkage of the clays, rich in alumina and generally of the kaolinite ranges. To overcome this problem wasters and material deliberately pre-fired would be crushed and ground to be mixed with neat clay in the moulding process. This 'grog' was used in furnace bricks primarily. The effect was to reduce shrinkage and provide an accurately dimensioned brick, but the penalty paid for the stability of the mixture during moulding and drying was the initial stressing of the matrix in shrinkage internally. The material was consequently fissipareous and relatively porous. Raw fireclays are low in iron, calcium and magnesium oxides. These clays fired to a buff colour and offer an alternative to Cambridge gault bricks or even London stocks.

Calcium silicate and concrete have also been used for brick manufacture. These materials are not the same, although frequently thought to be so. Calcium silicate is the product of autoclaving an intimate mixture of sand and calcium hydroxide under pressure. The resultant crystal structure is not that of the setting of cementitious material in which the principal linking constituent is tricalcium alumino silicate. However there is little apparent difference between these two forms of non-clay brick other than colour. The calcium silicate brick is inherently white but may be pigmented. The Portland cement brick is inevitably grey which only dark pigmentation will amend. Both these types tend to have a smooth face and precise shape as a result of pressing during manufacture. Bricks made with white cement are available and may be indistinguishable on sight from their calcium silicate equivalent. Both types are consistent, strong and durable. Their place in historic structures is minor.

3

Historic brick

3.1 Dating and characteristics

Just as the fundamental characteristics of pottery have provided archaeologists with a clear pattern of dating, bricks carry with them the distinctive stamp of their origin, their variation in colour and often the brickmaker's name. This variety is particularly evident in the temperate zones. Here the rate of decomposition of clay materials is significantly less than in the tropics where the predominant laterites, which represent the final stages of clay breakdown, produce bricks with a natural uniformity and an evenly brown colour. In the temperate zones, where, for reasons of climate and industrial development, brickmaking was more widely and energetically pursued, the variation in type, colour, texture and therefore the variation in quality of brickwork has been far greater. Bricklaying has contributed many distinctively local or specific techniques. Because of this inherent variety historical periods can often be associated specifically with particular types of material and workmanship.

Dating and analysis of sources in brickwork can be rewarding because the structure is permanent and can reveal reliably the way it originated. It can allow the historian to elucidate building periods accurately, provide sequencing for historic construction, and give the conservator information which may be vital for decision-making. The ability to make such analyses is, therefore, crucial to investigation. A particularly important method of achieving an historical understanding is to list with care the types of brick which are found in a locality, by size, colour range, clays and details such as 'kiss marks' (skintling), facings, flue marks, drying marks, methods of firing, cutting, stacking, handling and moulding. Such information gives a typology against which bricks can be identified and from which phases of construction can be dated.

Sequencing periods of building construction and dating the work will depend upon radically different techniques. On the one hand specific dates may be given by external analysis or documentation and on the other a combination of technical and stylistic features can define periods. Great precision may be achieved when these periods overlap but are not coincidental. The identification of changes in the work can often be related to other known events. The process of stylistic and technical analysis tends to rely on interlocking of probabilities. Thus a precise date may be achieved for a structure which is known to have been built (e.g. from documentary records) in a specific year or in which the timber has been precisely dated by dendrochronology. A date range, however, may be achieved by knowledge that a particular type of brickwork was produced during a specific dynasty, under legislation that controlled the workmanship or by a particular maker. It may be narrowed down by workmanship indicating the availability of bricklayers foreign to the area and known to have been present from a given date, while the style of construction may provide a convenient correlation with known buildings elsewhere. Brick may be recognized as an article of trade. Thus part or all of a building may be identified as of a particular period of commercial activity, dependent perhaps on the opening of a canal or railway, or as having been part of a trade limited by a particular event, such as a war or change of government or even an edict or new law and can be contrasted with work with different characteristics. From the days of Nebuchadnezzar onwards, fired brick has been impressed with the name of the ruler, owner or maker or with some specific attribution and the products of brickyards are distinguished by the type of clay, the size of brick, the methods of firing, the type of surfacing used and even by the shapes of special bricks. Colour can be a particularly deceptive source of attribution considered on its own since many kilns, and clamps in particular, produce bricks of different colour depending on methods and degree of firing. It is commonly recognized that where, as in most clays, ferruginous materials are present the least well fired bricks are light red in colour and the most thoroughly burned dark red or even blue. The ranges are commonly from a light brown or soft orange through the middle reds to purples in kilns where on cessation of firing heat is maintained by sealing the fire holes and therefore reducing the throughput of air causing the bricks to cool in an atmosphere of reduced oxygen. Such colorations are also dependent upon additives and the nature of the fuel used.

Where coke or other fuel has been added into the clay the localized effect of the higher temperature achieved at the point of burning will produce a different colour and form of vitrification, perhaps on account also of fluxes within the additive, with the result that the surface of the brick will carry a distinctive pattern. The micro-climate of firing in a reducing atmosphere which is associated with the coal or coke additive can produce colour changes in the form of mottling. Other patterns are induced by the deliberate addition of minerals such as manganese, perhaps simply by virtue of using a specific sand containing an ore as a surface coating even in trace form. While the superficial nature of the brick will provide important evidence of the origins of the material even down to the specific load, or batch derived from a particular firing, broad generalizations are often unhelpful to the point of being misleading. Conclusions can be reached only by building up a pattern of careful knowledge based on analytical experience. Any such pattern of localized knowledge is dependent upon examples which are firmly dated. It is easy to generate circular arguments based on a presumption that a particular brick type originated at a period popularly thought to carry a date and then to date others of the same type on the same basis – ultimately dating the manufacture on this evidence and apparently confirming the first presumption thereby. Because analysis of this sort is often directly comparative and depends upon linking examples together, a conclusion may have no greater validity than the evidence of one particular original until a sufficient basis of corroborative evidence comes forward. Carefully matched brickwork used in the past may well go undetected on the basis of colour, texture and brick type. A further danger is a presumption that the surface material is contiguous with the remainder of the structure. Brickwork has often been faced or faked. Cutting through a wall may well reveal periods of restructuring, thicknessing and facing out. Final uniformity may have been achieved by repointing. Perhaps only a tell-tale shift in bonding may remain to indicate that the structure is not as complete and whole as first appearances might suggest. Finished brickwork can betray its origins in the techniques used in the pointing and finishing of brick walls. Slight differences in workmanship may betray nothing more than two different brick layers, but the way in which a mortar joint has been struck may sometimes be distinctive of a particular period of workmanship and to be compared with the date attributable to the material itself. Sometimes older work will survive in protected positions or in positions which were subsequently inaccessible due to later rendering concealed by subsequent covering up. Where the original technique of finishing the mortar by later work has been superseded except for the occasional area which has been protected, careful searching may provide sufficient evidence for the restoration of the historical pointing as well as providing a clear indication of the date of that particular part of the construction.

It is now possible to determine the methods of firing and temperatures achieved in historic brickmaking by analysis of the chemical composition and physical characteristics. Hematite for instance generally forms between 700°C and 1380°C with peak indicators of temperature which can be correlated to the originating clays to give results within an accuracy of 10°C. This information in turn can show whether bricks originated in different firings. Porosity is a further factor indicative of temperature in a specific clay, the shrinkage which takes place above 900°C being caused by a fusing or sintering which affects the ultimate pore size of the fired product. The formation of CO_2 in the decomposition of calcium carbonate is an important factor here. The analysis of total porosity, median and average pore radius and densities can indicate the breakdown of sodium and magnesium salts with their consequent gaseous products and the resultant pore sizes which in turn demonstrate firing temperatures and clay type; irrefutable evidence of source and kilning which in turn locks into other evidence of date.

Methods of manufacture which are visible on the finished brick include the indentations formed in the green brick when stacked, the equivalent markings due to stacking in the kiln (kiss marks), marks made in striking and in removal from the mould. The consequences of extrusion are always evident and surface finish, by accident or design, gives a wide variety of options beginning with the differences between water and sand release, pressing and deliberate surface burnishing marking or modelling. To face a brick with a slip or powder coating which gives a finish different from the main body tends to be a modern technique with the exception of the age-old technique of glazing, often with a brilliant colour.

Each brickmaking area possesses its own tradition, for brickmaking and the use of brick is essentially related by raw material and practice to the locality, interlocking progressively with mechanized production. The archaeologist and historian builds up an increasingly precise knowledge of local characteristics. This knowledge is an essential input which is neglected at risk of serious errors in understanding the history and therefore the importance of the brickwork being conserved.

It is common to regard the size of bricks as being a clear indication of date and it is true that the commonest outward differentiation is by size – immediately post-medieval brickwork being conventionally regarded as thinner than 'modern' brickwork and Roman bricks being recognized as being thinner again. Responsible authors have issued salutary warnings over facile presumptions of date based on size alone, but in face of the increasing volume of knowledge and published examples no responsible architectural historian will set aside the assistance that this type of identification offers, although in use the immediacy of results must be tempered with circumspection.

In brickwork familiarity of both maker and user with an established size has tended to act as a control so that variations have been introduced gradually rather than by sudden change. Exceptions are manufacturers' and users' responses to the introduction of specific codes or taxes and the requirements of foreign markets. When shipment of tiles or bricks (perhaps as ballast) became commercially viable, convenient or regular, the availability of differing or specific sizes of product can be seen to have exerted a sudden influence.

In the British Isles, of which England is the preponderant brick-using country, the initial introduction of Roman brick sizes – succeeded by a hiatus in production – was picked up with a brick of not dissimilar proportions after a gap of six or seven hundred years and the 1½"–2" dimension in face thickness was increased only slowly in the early medieval period. It is perhaps worth noting that this thickness is about the maximum that can conveniently be formed by cutting slabs of rolled or beaten pug rather than by moulding. Under the influence of manufacturing and taxation pressures these sizes have been increased at various times but continual retraction from bricks which were outsize both as to length and thickness, and in recent times the unpopularity of the modular brick based on standard metric sizes, suggests that the use of bricks in the size commonly recognized (between 210 mm and 220 mm in length by 100 mm to 110 mm in breadth and approximately 65 mm in thickness) has an important visual relationship to building. That this order of size should be so profoundly popular suggests that its merit is both handleability and aesthetic satisfaction. Significantly it would appear that the 'natural' size of the fired brick is different from the dried earth brick and from the block of earth or concrete. The brick wall has developed a powerful aesthetic quality of its own and the natural scale of a wall given by coursing and brick sizing is apparently narrowed quite precisely by aesthetic considerations.

It is worth noting that some brick-producing countries habitually use a wider range of sizes than the United Kingdom; Holland, Belgium and Germany notably. Their standards, however, all range about the common norm.

3.2 The characteristics of historic bricks

3.2.1 Early bricks

Antiquity has given rise to a considerable variety of brick sizes. Among the large are found the Qin dynasty bricks of Northern China whose hollow blocks could be as much as 27½" (700 mm) by 15¼" (390 mm) by 6¾" (170 mm) (length × breadth × thickness). These were made by a people whose manufacturing skills allowed them to produce life-size terracotta figures – their production techniques approaching those achieved by modern technology although with far more primitive equipment. In western Europe the first consistent brickmaking was possibly Greek and certainly Roman. In classical antiquity brick sizes were influenced by the manufacturing methods of earth brickmaking in the Mediterranean and by the way in which the bricks were used. The Romans had achieved technical command of lime mortars and the pozzuolanic additives which caused them to set hydraulically. In consequence they were able to use their mortar in thick beds. They were, therefore, content to use large flat bricks – in effect slabs – and with the objective of tying their bricks back into thick walls they made them in substantial widths. Their methods of laying were related to the production of the particular brick size.

Figure 3.1 Roman brick forming a horizontal band tying in the lifts in stone-rubble walling. The burning of brick and tile provided a source of brick dust used as a pozzuolan as well as yielding the coursing material.

In addition to handleability production methods – particularly those relating to cutting or moulding, and stacking in the kiln – have much to do with the sizing of bricks. A requirement affecting the relationship between length, breadth and thickness is that the green brick shall be handleable without failure and stack readily, allowing passage of the heated air and providing adequate strength to carry the superimposed loads of a full kiln. Although this factor is not readily apparent to brick users and, curiously, often not to historians, it has done much in a practical sense to determine the size which brick manufacturers found convenient. The ratio in which the width is half the length and the thickness is a little less than two-thirds of the width provides a stacking arrangement which is coherent but loose fitting and allows sufficient flue ways between the bricks. The setting pattern and spacing has much to do with the final colour and surface texture of the brick.

3.2.2 English historic medieval bricks

An understanding of structural history will derive from the development of the techniques of manufacture and use in any specific area. Eastern England provides an exemplar sequence of evolution in medieval times.

Little brickwork is found in Britain prior to the late medieval period. Earlier a few large but thin bricks were used, sometimes for special purposes mixed with smaller bricks. Like Roman bricks these were also broad, thin and long, being slabs more than bricks in a conventional sense. Of these the earliest recognized survivors still in the positions for which they were made may be at St Mary's Church in Polestead, dated to AD 1153, where they are used principally to form the Romanesque arches in the arcades. These are thin and rather narrow at 13" (330 mm) × 5½" (140 mm) × 1½" (40 mm). Shortly after this date the abbey undercroft at Little Coggeshall, Essex (and the Abbots lodging there) included bricks 12" or 13" (300–330 mm) × 6" (150 mm) × 2" (50 mm) and at St Mary Somerton, by AD 1200, bricks 12" (300 mm) × 2" (50 mm) × an uncertain breadth were in use. At Waltham Abbey the tower and the great barn include bricks, some 14" and 15" (355–380 mm) × 7½" (190 mm) × 1¾"–2" (45 mm–50 mm), but this was by the late fourteenth century when new standards were being introduced. There seems to have been no strong pressure for standardization in the Romanesque period and immediately thereafter but, perhaps under the influence of some Flemish and Baltic traders, the British industry acquired a new status with the establishment of major brickworks producing material on a fairly continuous basis.

On the north bank of the Humber the monks of Meaux had manufactured tiles for roofing and flooring and probably also brick from the time of the foundation of their Abbey in 1151. The Abbey lay near Beverley, about seven miles from the river where the monks held shore lands. Their brickworks and tileries have been excavated and it is perhaps to these workings that the continuous evolution of the modern brick industry can be dated. The house was Cistercian and following the Dissolution its buildings served as brick quarries. Although nothing of it now remains it was the equal of many of the larger abbeys in the country. As a result of the brickmaking activities in the area Hull and Beverley were probably the first towns in the country to be predominantly brick-built, although London, Southampton, Ipswich and other East Anglian ports may have been using brick extensively in conjunction with timber framing at about this date and in part under similar influences and for the same reasons. Continental influence, the availability of fuel, the presence of good estuarine clays providing an excellent brick earth and the lack of an immediate source of stone were the obvious imperatives and there is some evidence that it was at about this time that clamp burning arrived as a means of achieving the larger volumes of production. Many early bricks were formed by cutting up slabs of pug which had been pummelled out on a bed of straw. The rectangular cutting produced distinctive edges and determined the thickness. It was neither convenient to roll nor to cut a very thick slab and the process of cutting produced bricks with quite wide fluctuations in size. Such bricks are known as 'place' bricks and virtually disappear by the mid-fourteenth century, although the term itself continued in use for a low grade of brick until recent times, and gave rise to the term commonplace.

Figure 3.2 Late medieval brickwork. Holy Trinity Church, Kingston-upon-Hull.

Flat roof tiles of this period tended to be larger than later plain tiles. Both rib and square holed peg-tiles were made, typically up to 13½" (340 mm) × 7" (175 mm) × ¾" (18 mm).

An important factor in the emergence of local brickmaking on the Humber Estuary at Kingston was that both Beverley and Kingston-upon-Hull were planned thirteenth century foundations – bastide towns – laid out to a grid pattern and built by organized labour in an area deficient in building stone. By the fourteenth century the bricks appearing in this area are of a reasonably consistent size, varying between 9" and 11" (230–280 mm) × 4½"–5¼" (115 mm × 135 mm) × 2" (50 mm). A brickyard further south, on the northern coast of East Anglia at Waveney, was, in the last half of the thirteenth century, producing bricks of virtually similar face and breadth dimensions but consistently thinner – 1⅜"–1¾" (35–45 mm) in thickness. These survive at the Abbey of St Olaves and the church at Ashby. It is possible in view of the sizes that the initial bricks were imported from the Humber rather than produced locally or, more probably, that the brickmaker came from Kingston-upon-Hull. The production at Hull provided the materials for the walls of the city which have now disappeared but the north gate at Beverley and some other parts of the fortifications of 1409 still stand. The yards in Hull, however, were making large numbers of bricks for the building of the city and its walls a century earlier and it is recorded that in 1303–4, 54 350 bricks were made, production rising in the following year to 92 000.

The Church of St Mary which stands just inside the North Gate at Beverley is stone built but it too was to have been built in brick, at least in part, for a complete burning was rejected in 1427 (10 000 bricks). Brick was also being supplied for the building of the Minster at Beverley, particularly for vaulting the earliest stages.

The crucial survivor of this early brick florescence is, however, Holy Trinity Church in Kingston-upon-Hull, the greatest of the churches in the medieval city. It has stone dressings and columns and the interior is largely rendered. Medieval Hull was smaller than Beverley but this was a very substantial church, begun relatively soon after 1300 and continuing for nearly 250 years. The brick sizes prove consistent for each historical period but change sequentially, enlarging noticeably as the fourteenth century went on. One brickyard was a municipal enterprise which reached a maximum production of 150 000 bricks per annum, at which stage a standard brick of 9¾" (250 mm) × 4¾" (120 mm) × 2⅜" (60 mm) average was predominant.

This outline of the development of manufacture could be repeated geographically. Many areas in Europe were ahead of Britain – the Hanseatic states in particular.

3.2.3 Europe and its brickwork

The extensive evidence of continental influence includes the import of tiles from Flanders

Figure 3.3 The fortress-like east end of Albi cathedral, a massive medieval construction. Brick-built, unusual in material and origin, the form exploits the nature of the material. (courtesy W. L. Weller.)

to ports on the south and eastern coasts of England and the presence of continental craftsmen and the property ownerships of English patrons in France from the Pas de Calais to the far south west. Bricklaying shows relatively little foreign influence in the fourteenth century if judged by the persistent use of English bond.

In many senses England was a backwater in Europe, and this was distinctly true of brick building. In contrast with the long continuity of brickmaking in the Mediterranean from Constantinople to Seville the established traditions of the northern shores of the Continent were burgeoning in the late medieval period with a dramatic emergence of new structures – market halls, city walls and ecclesiastical buildings, among which, and far to the south, rose what must be the greatest of these medieval prototypes. The immense fortress-like Cathedral of Albi was built by the imperious Bishop Bernard de Castanet in the last quarter of the thirteenth century after the destruction of the Cathars. Fortress-like with good reason, its exterior is a series of massive bastions with high walls, minimal windows and a nave 100 ft high by 60 ft wide and some four times as long. Building went on for more than 100 years, the great undecorated brick exterior being finished for the most part in stretcher bond and showing no sign that its builders ever intended it to be faced or enriched with stone.

Brick church building, however, is much more typical of the northern coastlands of Europe from Calais to the Baltic where it was in consistent use long before its general use in England. Brick may have been used in part with stone either as a core or with stone dressings or as the sole structural material from the beginning of the Gothic period. It is found in the greater churches such as St Bavon in Ghent (1274) and in churches in the smaller towns and villages where brick walls and vaulting were typical of the Gothic period. One of the greatest of these brick churches was the Marienkirche – St Mary's – in Lubeck, begun in the thirteenth century (1276–1310) and remodelled as it was enlarged. It was a potent influence in the area at a time when the civic pride of the prospering Hanseatic cities was expressed in rivalries including their massive brick churches. In every locality a growing consistency of brick manufacture forms the essence of a regional style most clearly expressed in the Low Countries, northern Germany and the Baltic. This was the austere brick Gothic of the Klosterkirche in Berlin (1271), St Nicholas at Stralsund (1311), St Ausgar, Bremen and the Marienkirche at Weimar. Thin detailing was relieved by the external and internal use of glazed tile, of which a distinctive survivor is the blue tiling in the spandrels of arcading in the Rathaus (town hall) of Lubeck.

Figure 3.4 A tiled doorway of the late sixteenth century in Santo Domingo. An Islamic technique absorbed into Christian Spain crossed the Atlantic with the Spanish conquerors, displaying still the disciplined forms of its origins.

3.2.4 Continental influences and the effects of legislation

Despite increasing native manufactures the history of English brickwork in the sixteenth century remained susceptible to foreign influences in the more sophisticated areas of brick building. Consistently the smaller Flemish bricks imported in the thirteenth century for, among others, the Tower of London (1278) at a size of 8½" (215 mm) × 4" (100 mm) × 2" (50 mm) appear in various places and guises throughout the fifteenth century. Some of these suggest that they were copies made by English brickmakers. The use of moulding and carving techniques to form special bricks grew in this period until the early sixteenth century when under Italian influence moulding and shaping reached the stage of high elaboration in a period distinguished by its important terracotta, used by Royalty at Nonsuch and the aristocracy at houses such as Sutton Place.

Although terracotta is normally thought of separately there is no absolute division between a structural brick with a decorative moulded face made in a fine clay and a pressed, modelled, decorative element applied with no structural intent or function. The achievement of the precision required for such work follows a period in which standards had been set – the earliest in Hull – for sizes of bricks, made specific by the provision of sample moulds and patterns and then by written stipulation, specification or statute. Ordinances of 1570–1 attaching to the by-laws of the Charter of the Tilers and Brickmakers Company are believed to have set the dimensions of 9" (230 mm) × 4¼" (110 mm) × 2¼" (60 mm). These standards applied to London and its environs, although they had become widespread by the late sixteenth century. Under Charles I (1625) the specification was confirmed, with a slightly greater width, and in 1725 the width was increased to 4½" (115 mm) and the thickness to 2¼" (57 mm) for 'place' bricks and 2⅝" (66 mm) for stocks. By a further act George II reduced the face dimensions fractionally and in 1769 an enactment under George III brought them down to 8¼" (210 mm) × 4" (100 mm) × 2½" (65 mm). If these earlier standards were varied for no very evident reason there followed an immediate contrast in 1784 when bricks were taxed in order to provide funds to pay for the wars in the American Colonies. To mitigate the effect of the tax, brick manufacturers then enlarged their bricks. As each brick was taxed individually, bricks up to 11" (280 mm) × 5" (125 mm) × 3¼" (80 mm) are recorded. Governments of the day responded with increased levels of taxation in 1794, 1797 and 1805, the tax rising from 4/0d per thousand to 5/10d per thousand eventually, although for practical reasons no greater brick sizes emerged. The legislators of George III put an end to the process by defining the maximum volume (150 cu in) of the brick upon which a single tax would be levied. Ultimately the tax was regarded as damaging to the burgeoning industry of Victorian England, being repealed in 1850 when the standard dimensions of 9" (230 mm) × 4½" (115 mm) × 3" (76 mm) were determined. Much historic brickwork is based on this nineteenth century standardization but bricks manufactured by industrial processes in the north of England have tended to be of this full thickness or more while in the south lesser thicknesses have been used. This practice was reflected in a British Standard which focused on 8¾" × 4⅙" × 2½", subsequently being given metric definition at 215 mm × 102.5 mm × 65 mm. These bricks appear in some few buildings now regarded as historic and the metric modular bricks intended to facilitate industrialized building processes may do so also. These larger modular bricks have lengths of 290 mm or 190 mm × 90 mm × 90 mm or 65 mm. They have not been widely used.

This broad history of the evolution of brick sizes in England illustrates the social, geographical and political influences which may be traced by the historian to the advantage of the conservator. The increasing breadth of local history and research into industrial archaeology makes increasing sense of the expanded microcosm into which the local history of brickwork fits. Local peculiarities of manufacture and size differentiation may define building periods, sometimes with remarkably close tolerances. Other influences are national and international.

Figure 3.5 The post-medieval emergence of brickwork in England: the gatehouse at Layer Marney, Essex.

3.2.5 Brickwork in the Americas

The trans-shipment of building techniques from Europe to North America illustrates the social strands by which skills were transmitted with varied but broadly contemporaneous results. The Spaniards were the first to carry a building tradition into the Americas, their city at Santo Domingo in Hispaniola being a fully developed urban structure early in the sixteenth century. In Santo Domingo it is still possible to see the continuation of Roman brick traditions, Roman tile and mortar-making and of vaulting in brick in the Gothic tradition. Brick dust was used as a pozzuolan, which, with the many fragments of brick and tile to be found in the earliest mortars (1504 onwards) may have been the residual material from trans-shipment rather than the product of local brickfields. In fort, monastery and palace the Spanish Colonial building industry marched in step with Europe. Sixteenth century bricks were thin slabs up to 330 mm × 330 mm × 45 mm thick but by the seventeenth century the brick in the order of 260 mm × 120 mm × 70 mm had been adopted. By this time European building techniques were spreading extensively into the Americas.

In the mid-seventeenth century between 1629 and 1675, from England alone three distinct regional cultures appear to have been transported. The East Anglian Puritans arrived in Massachussetts between 1629 and 1640. Royalists, gentry and servants populated the Chesapeake Bay area between 1640 and 1675 and towards the end of this period the Quakers from the English Midlands established themselves in Delaware.

The East Anglian arrivals in Massachussetts carried with them a powerful tradition of timber framing which had developed through the evolution of the medieval 'hall house' to the sixteenth century 'baffle-entry' house in which a great brick chimney stack stood at the core of a timber framed building flanked on the one side by the entrance and on the other by a stair. Brick noggin or infill in the timber framing was by then common in East Anglia, though the houses were regularly sheathed over with plastering, or pargetting as it was locally known.

With this transference of culture went language and place names. The skills of brickmaking were carried over and incorporated into the timber framed buildings of the period. The powerful post-medieval brick building traditions of the city of Cambridge were thus repeated in Cambridge, Massachussetts and brick became the medium for educational and religious structures. Harvard Yard owes much philosophically, architecturally and in brickmaking skills to its predecessors in Cambridge, England. The tradition flourished, absorbing further immigrant waves on a pattern in the English mould.

Further south, around Chesapeake Bay, the Royalists brought a totally different social tradition – the building of great houses. Sir William Barclay built the Governor's house, Green Spring, about the year 1647, a great brick building immensely solid and symmetrical, of two storeys, demolished after 150 years. Stratford Hall, which was the birthplace of Robert E. Lee, a similar building on an H plan, followed in 1725, among a plethora of major houses begun in the late seventeenth and early eighteenth centuries. Here the tradition was not educational but squirearchical and the techniques of construction in the eighteenth century in English and English Garden Wall bond indicate clearly the origin of these great Virginian mansions. This was not the way of the Quaker societies from the Midlands and the south of England who focused on Pennsylvania – one of them, William Penn, of Warminghurst in Sussex, giving his name to the state. They carried with them the tradition of solid brick town housing and so it was that the early buildings in Philadelphia were modelled on the reticent formality of urban building in the communities of the Friends.

In every region brick building exemplifies such continuities of usage and workmanship, and these sequences can be of great help in understanding the built tradition. Other essential resources of the conservator include detailing which called for specially shaped bricks and methods of handling standard components by bonding. These reflect on local expertise which inevitably reflects tradition.

3.3 Historic bonding

Bonding describes the way in which bricks are laid in a masonry structure. Such detailing is not only a source of strength – or weakness – in a wall; it also has a significant effect on weathering. Brick bonding is a direct consequence of the deliberate decision of designer and workman but it also relates to brick size. Much ingenuity can be expended in achieving junctions and openings in complex structures, and although bonding is normally thought of in terms of the superficial appearance of the wall its significance extends right through a wall. Bonding should always be thought of in three dimensions.

The fundamental form of bond is that in which the brick is superimposed (not neces-

Figure 3.6 Colonial Williamsburg, Virginia, USA. The Wythe house, the eighteenth century mansion of a Professor of Law. Unmistakably English in style, the Wythe house is typical of the visible transplantation of style and technique from continent to continent. (Courtesy John Cannon.)

sarily centrally) on the joint of the brick beneath, the process being carried on consistently. With the standard size brick this is known as stretcher bond and it is commonly the form in which Roman brick was laid. In this form of bond the lap, i.e. the projection of one brick across the other, is normally uniform. Where the vertical joints between the bricks are left open to permit ventilation, the bonding is normally known as honeycomb walling.

Bricks of conventional current size and shape may be laid with any one of their three faces outwards and it will readily be appreciated that if combinations include the use of the flat (and normally base) side of the brick as the outer face a wide variety of bonds is possible.

Stretcher bond has become predominant as the result of the use of cavity walling in which a single (half brick) skin is presented to the weather. The single leaf or 'half brick' wall is approximately 4½" in thickness and is tied back to the inner wall with mechanical fixings – rods, wires or straps. A wall which is one brick thick will accommodate a standard brick laid lengthways across the wall and is, therefore, approximately 9" in thickness. Thus the thinnest wall possible is half brick on edge and this itself is used sometimes to provide a cavity wall. A half brick leaf permits little variation in pattern unless snap headers are used and it offers only modest structural strength since acceptable slenderness ratios are quickly exceeded.

Historically there are two broad subdivisions of bond in walling. The first, known as English or American bonds, involve variation course by course so that one or several courses of headers are succeeded by one or several courses of stretchers all being laid broken joint. In the alternative range of bonds – Flemish – the wall pattern achieves a consistency by the repetition of the same sequence of laying with each successive course.

Historically English bond has been regarded as a very strong form of construction. In basic form it consists of layers of header bricks succeeded by layers of stretcher brick, which produces a distinctly striated character in the wall, particularly where the firing has produced a variation in colour between header and stretcher faces. English Garden Wall bond involves multiplying the stretcher courses by three or sometimes by five. A number of variants are known, including one in which the stretcher course is laid on edge to provide a partially hollow wall. This is known as Dearnes bond.

The alternative is the Flemish bond group in which headers and stretchers are consistently alternated. Thus Flemish bond shows headers and stretchers alternately with equal laps. Variations on this bond include the Flemish Garden Wall in which three or even more stretchers appear between each header. Flemish bond is not particularly characteristic of the Low Countries nor is Flemish Garden Wall bond (also known as Sussex bond) particularly characteristic of the English south coast.

Flemish stretcher bond alternates four courses of stretchers with the header of Flemish bond, while Monk bond (also known as Yorkshire bond) brings two stretchers between each header, each successive course being consistent, with equal lap, bringing headers into regular arrangement.

The variant of Flemish bond which uses brick on edge is known as Rat Trap bond and this can be extended to Garden Wall Rat Trap bond by increasing the number of stretchers. Being largely hollow, such walls contain less brick than their solid equivalents although they offer improved water resistance and economy of brick. They are, of course, less strong but are adequate for two-storey domestic construction and in theory can be used for three storeys subject to adequate tying.

The strongest single brick wall is achieved by header bonding in which the bricks are laid regularly and uniformly in each course in a transverse position across the wall with equal lap. Variants which introduce stretchers into this pattern are also known.

In Herringbone bonding stretchers are laid at an angle of 45° to the horizontal. Apart from being used in paving this bond was most commonly used as a filling to panels of timber framing. The reason for this particular application is that by varying the angle of inclination the bricklayer could adjust the width of the panel to suit the varied timber spacings with a minimum of brick cutting and a maximum of decorative effect. This technique is known as brick noggin.

Many situations involve the use of special bricks. Some, such as half length bricks (batts) and closers (third or quarter size), will often be cut on site by striking a brick sharply with a cutting edge. Where the fractured face lies within the general mass of brickwork this is not necessarily bad practice but where a cut brick has for some reason an edge exposed to the weather the rapid deterioration of the brick from the cut face inwards is clear evidence of the inadequacy of the technique. Much depends upon the quality of the brick itself. A brick made or moulded to the size required is a better object, though variants in firing techniques in consequence of the size change frequently cause difficulties in matching surface texture and colour.

English work is neither peculiar nor particularly remarkable in its use of bonding. Some of the most intricate forms of pattern-making by bonding are found as early as the Seljuk period (twelfth to fourteenth centuries) in present day Iran, and some neighbouring territories. Their precursor was the complexity to be seen in the mausoleum of Ismail the Samanid at Bokhara (*c.*AD 910), in which complex facework patterns bond back into a brick core adapted to each circumstance. The interlacing diaper vaults of the Seljuk mosques and their walling demonstrate an aesthetic relish in bonding patterns which far exceeds any comparable European craftsmanship.

3.4 Worked bricks

Some types of brick can be worked satisfactorily after firing. These bricks will be fired more gently than most and will be worked by rubbing or non-percussive cutting to provide

Figure 3.7 Faced brickwork simulating stone: Venice, probably eighteenth century. Juxtaposed against Istrian stone quoins, brickwork formed to facilitate rendering is revealed by the failure of the stucco below but still surviving upper right.

the required surface. They will be less robust than general brickwork and are best used in protected situations, but where the working has been gentle, even and uniform they can weather very satisfactorily and prove durable. The essential difference between these and normal cut faces is that whereas the cut face sustains the effects of sudden rupture introducing invisible fractures, the worked face will not have sustained such intense shock and does not, therefore, contain the fissuring which increases weather penetration into a weakened surface. In all cases, however, they have lost their fireskin, although in soft brick a fireskin is minimal. Such worked faces include those of soft bricks which have been rubbed down to form voussoirs in brick arches or to provide the gaged (gauged) brickwork of which eighteenth century architects in Europe were so fond. They were used also for carved brick panels which were the alternative to terracotta ornament used to form boasts in the sixteenth century and in the London revivalist styles of the nineteenth and the early twentieth centuries.

Carved brickwork has been used extensively in Islamic architecture. Under the Abbasid dynasty surface carving was used extensively. Much of this work was lost in the Mongol invasions of the thirteenth century when cities such as Baghdad were largely destroyed. The technique initially used on stucco was translated into brick. Patterns were carved in the

Figure 3.8 A detail of the facing of the tomb of Ismail the Samanid. The rich surface texture is a product of the laying of cut brick bonded into the wall construction.

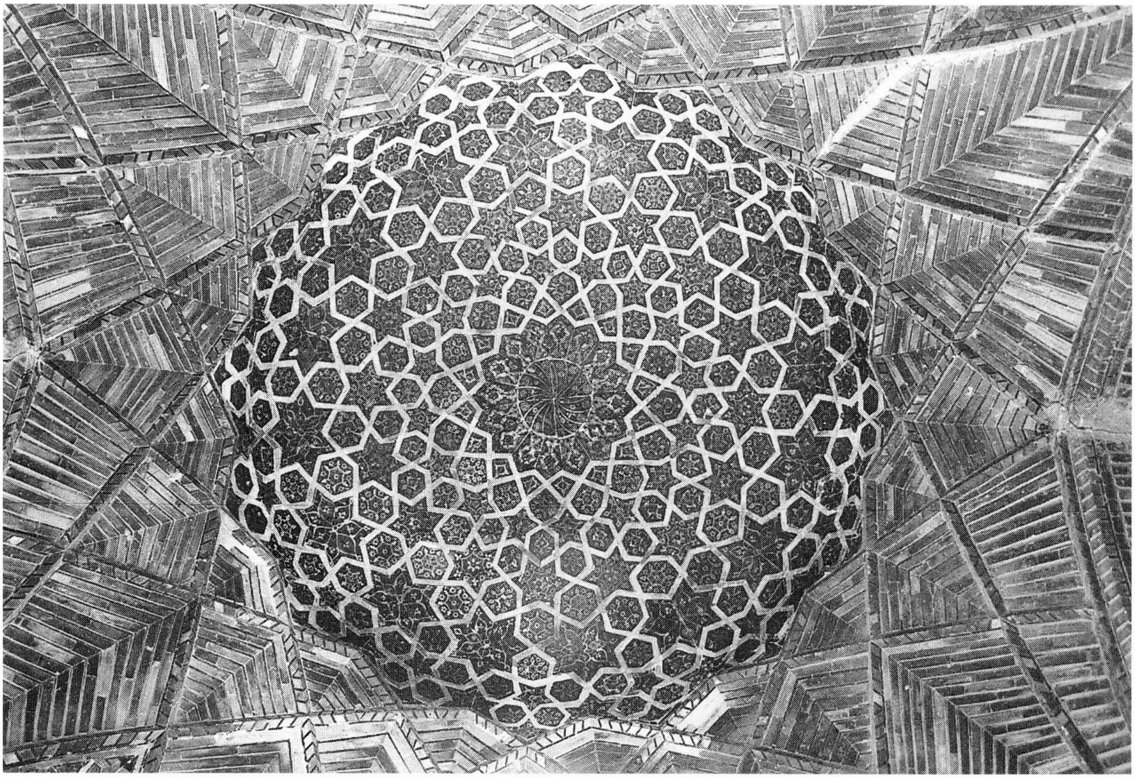

Figure 3.9 'Brick magic', Bokhara. An intricate dome structure of brick liernes and panels, every brick cut to its individual shape. Within the central medallion brick slips divide individually cut and carved brick inserts bedded in mortar on a brick dome.

most minute detail into the surfaces of soft yellow bricks forming filigree patterns, spiral columns, intricate spandrels and arabesques. The bas relief surfaces might be partly or completely glazed. Cut and carved bricks were used in mosaic panels and formed the elaborate three-dimensional components of much 'muqarnas' (stalactite) vaulting, but such has been the destruction in the capital cities of Persia and Mesopotamia that its calibre must often be judged by the secondary and often inferior work which is to be found in outlying regions. Across Iran and into Central Asia, into Egypt and the western Islamic world these techniques spread developing locally and regionally into pattern-making in which brickwork was exploited as a base medium for carving and modelling of the greatest subtlety and intricacy. In these regions brick was exploited for its capabilities as a base material, both cut to shape and carved in the complexity of the bonding needed to achieve vaulting in the three-dimensional geometry in which Muslim architects had no equal. This tradition has continued unshaken to the present day even though it may be commented that much of the later work is imitative and lacking in the inventive dynamic of earlier periods.

3.5 Identification of historic bricks

There is no universal scientific test which will do for brick what dendrochronology, at its best, will do for timber – identify the year of its production (felling), but a material that is the subject of so many variations in origin, manufacture and artefactual techniques must present a great deal of evidence about its origins if the available material is properly marshalled and analysed. The general nature of the clues has been set out and information of this nature has been used with increasing thoroughness to provide identifications for historic brickwork, but rarely is this evidence used in isolation. Usually stylistic evidence, other building materials and techniques and documentary evidence is available to help define the date of a building or its brickwork. This perhaps is the reason that brickwork typology has not advanced as quickly as it might have done. A survey of brick from within any one brick-using area can encapsulate its history in the microcosm of the one material – particularly if tile and pottery are regarded as legitimate ancillaries.

Case study: a model brick typology

The purchase by Essex County Council of an important medieval site, Cressing Temple, has provided the County's historic building division with the opportunity to analyse the bricks found on the site from Roman times onwards and to compare them with others from the same area. Extracts from the exemplary results of this analysis, published by Essex County Council, are reprinted here as a case study.

A brick and tile typology for Cressing Temple

by Pat Ryan and David Andrews
(extract from *Cressing Temple: A Templar and Hospitaller Manor in Essex* (ed. D.D. Andrews). Essex CC, Chelmsford, 1993.

Introduction

The typology presented here has been set within the context of the development of bricks in Essex, being based upon observations made of dated buildings elsewhere in the county. The history of brickwork has attracted considerable interest over the years, and attempts to establish new typologies could be regarded as re-inventing the wheel. What is new to this study is the proposition that by considering not just dimensions but also the characteristics of a brick, it is possible to date it. Of course, there will always be exceptions, but in general terms an accuracy of 50–100 years is attainable.

This typology has been broken down into chronological divisions, within which there are numbered types. Where possible, common names, such as Garden Wall Tudors or London Stocks, are used for the bricks as being more readily comprehensible and memorable than a classification distinguished by numbers or letters alone. The scope of this paper has been extended to include flooring tile and brick, and roof tile, related categories of material which were produced in much the same way as brick.

Characteristics of brick

In dividing the bricks into types, the following criteria have been used: fabric, dimensions, regularity of shape and appearance, sharpness of arrises, smoothness of faces and bottom surface, and the presence or absence of such features as sunken margins, pressure marks and frogs.

The qualities of the clay or brick earth influence the characteristics of the finished product. The bricks described here, with the exception of the Coggeshall bricks, some if not many of the 19th–20th century bricks, and the flooring bricks, seem to have been made of unrefined brick earth which gives a sandy fabric containing occasional large pebble and angular flint inclusions. This impression is in agrrement with the more scientifically reached conclusion of the Firmans (1967) that early bricks were made from superficial deposits of clay and brick earth without any significant addition of other materials.

Colour is very variable, depending not only on the constituents of the raw material but also on how a particular kiln or clamp was fired, and on the conditions obtaining in the different parts of a kiln. The final size of a brick also depends on how much shrinkage has taken place in the drying and firing, and this is also influenced by conditions in the kiln.

Size has also been determined by various regulations which attempted to standardize the manufacture of bricks. Amongst the more significant of these were the following (cf. Lloyd, 1925; 12). The Tylers' and Brickmakers' Company charter of 1571 stipulated a size of 9 × 4¼ × 2¼ inches (230 × 110 × 60 mm). A statute of 1725 fixed a size of 9 × 4¼ × 2½ inches (230 × 110 × 65 mm) for place bricks, and 9 × 4¼ × 2⅝ inches (230 × 110 × 70 mm) for stock bricks. During the 18th century, smaller bricks were allowed within a 15 mile radius of London, a statute of 1729 fixing a size of 8¾ × 4⅛ × 2½ inches (225 × 105 × 65 mm), and one of 1769 a size of 8½ × 4 × 2½ inches (215 × 100 × 65 mm). The latter dimensions were extended to the rest of the country in 1776. However, the Brick Tax of 1784 led to the manufacture of larger and thicker bricks, though the doubling of the duty on bricks in excess of 150 cubic inches in 1803 checked this tendency. The tax was repealed in 1850.

Other characteristics are the result of manufactory processes. Whilst slop moulding (i.e. wetting the mould before making the brick) has been typical of the brick-making tradition in some parts of England, in Essex it has been the custom to sand the mould. The appearance of the bottom surface of a brick is an indication of where the freshly moulded brick was put to dry. In former parlance, bricks were known as 'place' and 'stock' bricks (Neve, 1726: 42–44; Gunther, 1928: 229). Stock bricks were made of stiffer clay, and were dried much more carefully than place bricks, which were laid on the ground after they were moulded and, being of a softer clay, acquired rough or pitted bases which may have straw or grass impressions. In contrast, stock bricks were laid between boards which were initially stacked three bricks high, hence their relatively smooth sandy bases. They were then set on their narrow edges to dry in open-sided shelters known as hacks. Rough bases are most common on medieval and 15th–16th century bricks. It seems that as the practice of drying bricks on the ground lapsed, the term stock became used for utilitarian bricks, whereas previously stock bricks had been superior to, and more expensive than, place bricks. (More recently, the name London Stock has been applied to a particular type of brick made for the London market, a usage so prevalent it has been retained as a common name in this typology.)

Sunken margins and pressure marks are also effects created in the process of moulding and drying bricks. Sunken margins are slightly stepped down top edges. They were probably formed by the brickmaker using the edge of his strike to press the clay firmly down to completely fill the corners of the mould. Pressure marks are impressions occurring in a stretcher face of the brick, and are caused by stacking bricks on top of each other to dry when they are still rather soft. Another distinctive feature are frogs, the familiar concavities present in the base of many modern bricks. They may be made in a variety of ways.

Kiln-burnt bricks often have light and dark stripes or 'kiss' marks caused by the arrangement of the bricks in the kiln, the exposed parts being burnt to a light colour and the parts where the bricks rest against each other to a dark colour (Mitchell, 1930: 136).

It has only proved possible to satisfactorily identify types where large areas of brickwork are built uniformly with one type of brick. It is important in identifying a brick type to have a large statistical population which illustrates the full range of variability that might occur within one type. When examining small numbers of bricks, it may be possible only to distinguish them in broad terms, such as Tudors or 19th century soft reds.

Brick development as illustrated by the types present at Cressing Temple

Apart from some fragments of Roman ones, the oldest bricks known at Cressing Temple are excavated examples of the Coggeshall type datable to the later 12th and 13th centuries (cf. Gardner, 1955). They are unusually large, about one foot long, and are noticeably sandy textured, sand temper having been added to the clay. They are also, exceptionally for an early brick, accurately moulded with square arrises. Such bricks are increasingly being identified in the fabric of churches where previously they have been mistaken for reused Roman material. Drury and Rodwell have explored the possibility that the Cistercians found a market for their bricks within a ten mile radius of their abbey at Coggeshall (Drury, 1981: 126). It is possible however that there were other manufacturing centres of these bricks. In view of their size, it is natural to compare these bricks with the Great Bricks which are known from 14th century and earlier contexts at Waltham Abbey (Huggins, 1972: 111–114).

The other characteristic type of medieval brick, the predominantly yellowish bricks imported from the Low Countries (Drury, 1977a: 84; Andrews *et al.*, 1990: 89), have yet to be identified at Cressing. They are known to occur locally, having been excavated at Rivenhall (Rodwell, 1986: 95–96) and being found in the fabric of St Nicholas church, Chipping Hill, Witham.

Bricks definitely attributable to the 15th century are also so far wanting at Cressing, In most cases, it is not easy to distinguish them from 16th century Tudors. They can be smaller and better fired, with more vitrified examples. They are generally 9 inches (230 mm) or slightly less in length, 4¼ inches (110 mm) wide, and 2–2½ inches (50–65 mm) thick. Early 15th century bricks can be less than 2 inches (50 mm) thick (e.g. Writtle, Pleshey), whilst late 15th century bricks (Rochford church tower, Fryerning church) may be up to 9½ inches (240 mm) long.

Sixteenth century Tudor bricks tend to be large and rough in appearance, often distorted with very creased faces and sides that are by no means parallel. They are at least 9 inches long, usully 9¼–9½ inches though some are up to 10 inches. The width ranges from approximately 4¼–4¾ inches (110–120 mm), usually about 4½ inches, though exceptionally it might be as much as 5 inches or more. As might be expected, the thickness is less variable, though still in the range 2–2½ inches (50–65 mm).

The most plentiful Tudor bricks at Cressing are the Garden Wall Tudors, common to the garden wall and the Wheat Barn nogging (Figure IA). These are a large rather crude Tudor, usually not very high fired and in the Wheat Barn nogging very soft and under-fired. They are place bricks, having rough bases. It may be assumed (and this is no more than speculation) that similar bricks were used for the Great House. There are at least no other distinct types of Tudor brick present in quantity on the site, and those that are recognizably reused are mostly of this type. It is possible that they were made in the clamps that apparently existed in the field across the road to the south of Cressing Temple (see note 1).

Other recognizable types of Tudor are few. The brickwork from the pond over the road includes bricks that are smaller than the Garden Wall Tudors, as well as Garden Wall Tudors. The other distinct group, the Granary Tudors, so-called because used in the nogging of that building, are not local to the site, having been used in a series of post-war repairs.

As the 17th century progressed, bricks became better made, with sharper arrises and a less distorted and irregular appearance, but retaining creased faces. In size, they became slightly smaller, though the limited number of reliably dated buildings for this period means that this tendency is difficult to assess accurately. They are generally not quite as thick as before, not being in excess of 60 mm. They are more likely to be stock bricks with smooth bases. They may also have diagonal pressure marks on their stretcher faces (Figure IC). Such pressure marks seem most characteristic of the 17th and 18th centuries, though they also occur on 19th century bricks. These improvements may reflect the increasing use of kilns rather than clamps, and possibly the use of stiffer clay. At Danbury and Stanway (Drury, 1975; Fawn, 1984, 1985), the only manufacturing sites in the county of this period which have been investigated, kilns were being used. Sir Roger Pratt, writing in the third quarter of the 17th century, observed that kiln bricks were the best (Gunther, 1928: 48). The only bricks of this type recognized at Cressing are those from a wall found under the bullock shed in 1991. To judge from their relatively large size, they date from early in the 17th century, or even the end of the 16th century (Figure IB).

By the end of the 17th century and in the early 18th century, bricks were becoming smaller and neater, more precisely made with smoother less creased surfaces (Figure ID). They typically measure 8–9 × 4 × 2–2¼ inches (205–230 × 100 × 50–55 mm). A large proportion of these bricks were fired to a higher temperature, with the result that they are darker in colour. Bricks in the plinth and nogging of the west wall of the Wheat Barn are of this type. Such bricks are not, however, common at Cressing, otherwise only having been identifed in two of the buttresses on the east side of the garden.

These more accurately made bricks made it possible to obtain the precisely laid brickwork with neat, narrow joints that is a distinctive feature of the better quality houses of the period. The more roughly made Tudor bricks could only be laid with rather wide joints in less precise courses, and were often rendered. Render was often conceived as a means of disguising brickwork as stone (cf. Andrews *et al.* 1986). What is difficult to assess is whether the bricks were roughly made because they were intended to be plastered, or were plastered because of their rather crude appearance.

By the mid-18th century, bricks tended to be thicker and much more accurately made with sharp arrises, parallel sides and smoother surfaces. Mid- to late 18th century bricks measure 8½–8¾ × 4–4¼ × 2½ inches (220 × 105 × 65 mm). These bricks tend to be consistent in colour (though this may be because they were sorted with greater care by the bricklayer).

This sets the pattern for 19th century bricks, which retain the same general characteristics, but vary slightly as to dimensions. Typically they measure 9 × 4¼–4½ × 2½–2⅝ inches (230 × 110–115 × 63 mm), although in the earlier part of the century there is a tendency for bricks to be a little smaller (e.g. 8¾ × 4 × 2½ inches, 225 × 105 × 65 mm). Later 19th century bricks are more uniform in size and appearance, with sharper, squarer arrises and smoother faces, though there are exceptions. This development is the result of the introduction of mechanical manufacturing processes from the first half of the century (Cox, 1979: 36). Such bricks are often referred to by builders today as Soft Reds, and this is a useful general term to describe them.

Two features can be of assistance in the identification of 19th century bricks. One is horizontal pressure marks (Figure 1E), which seem to largely supersede diagonal ones from a little before 1800. They probably continue into the early 20th century (as do diagonal ones, though they are very much less common). The other is the frog. Frogs are known, it seems, from the 17th century (Harley, 1974: 80), but in Essex do not become at all common until the 19th century. Early frogs tend to be crude and rudimentary; later ones are larger and more precisely made. After *c.*1875, the maker's name may be impressed in the frog (Figure 1F).

Of the new types of bricks that appear at this period, two occur at Cressing, the Suffolk Whites and London Stocks (Figures 1H and I). 'White' or whitish bricks made from Gault or lime rich clays, or clays to which lime has been added, have been called Suffolk Whites because that is the principal area in which, locally, they were manufactured. They were made from the 18th century, but in Essex are most common from the end of that century until the middle of the 19th.

London Stocks, as already mentioned, are utilitarian bricks made for the London market. They are rough in appearance, and yellow to purple in colour with black patches. They were made from a clay which contained naturally occurring or added powdered chalk. Fuel, either breeze from gas works or decomposed town waste with a high proportion of partially burnt cinder, ash and coal dust, originally known as Spanish soil, was also mixed with the clay (Hartley, 1939: 167–71). After moulding and drying, the bricks were burnt in clamps skilfully constructed from the unburnt bricks and fuel in the form of breeze, town waste and a small quantity of coal and wood faggots. They were originally manufactured in the environs of London and larger towns, and at coastal sites where cheap transport by Thames barge was available for both fuel and the finished product. The advent of the railways also encouraged the use of this type of brick throughout the county. At Cressing, they seem to have become available about the middle of the 19th century. An early dated example of their use is to be found a few miles away at Bocking where there is a house in Woolpack Lane built of London Stocks with a plaque bearing the date 1846.

Floor tiles and paving bricks

Also needful of consideration are flooring bricks and tiles. Three main categories of material can be identified in Essex: floor tiles generally five to six inches, large floor tiles nine or twelve inches square, though sometimes larger, and rectangular flooring bricks or pavers similar in size to a normal brick but invariably thinner.

The small floor tiles include the well-known slip-decorated examples found at churches and monastic sites. However, just as common and probably more so, are plain lead-glazed tiles of this size, and also those with a plain cream slip coating covering all or half of the tile (cf. Drury, 1977b; Drury, 1976; Drury and Pratt, 1975: 112). Typically, these are 14th–15th century in date.

Nine-inch square tiles were found by Gardner (1955: 24) in the chapter house at Coggeshall where they seem to be late 12th century. It is not until the 15th century that glazed and slipped or partially slipped tiles of this size are found. Those with nail holes in their bases are considered to be of Flemish origin (Drury, 1977b: 113). Those without, such as the partially slipped and glazed nine inch square floor tiles found in Great Sampford church in 1992, may be made locally.

Unglazed nine- or twelve- (or exceptionally fifteen-) inch square floor tiles are often known today as pammets, and this is a useful term for distinguishing them. It would seem logical that they derive from the large floor tiles discussed in the previous paragraph. However, the discovery in 1992 in a cellar of the former Tudor house adjacent to Hatfield Peverel church of a pammet which is essentially a double brick measuring 9½ inches (240 mm) square by 2 inches (50 mm) thick suggests another origin for them. Certainly early examples are usually in a brick-type fabric which tended to be too fragile for heavy wear unless the pammets were very thick. Later ones are often in a pinkish to cream fabric, with a marbled or streaky appearance when broken. Flooring bricks or pavers (Figure IJ) cannot as yet be taken back with certainty before the 18th century, though before then and indeed later ordinary red bricks were sometimes used for flooring but can never have proved very durable. Flooring bricks improve in quality with the passage of time like ordinary bricks, early ones being rougher in appearance and 19th century ones often

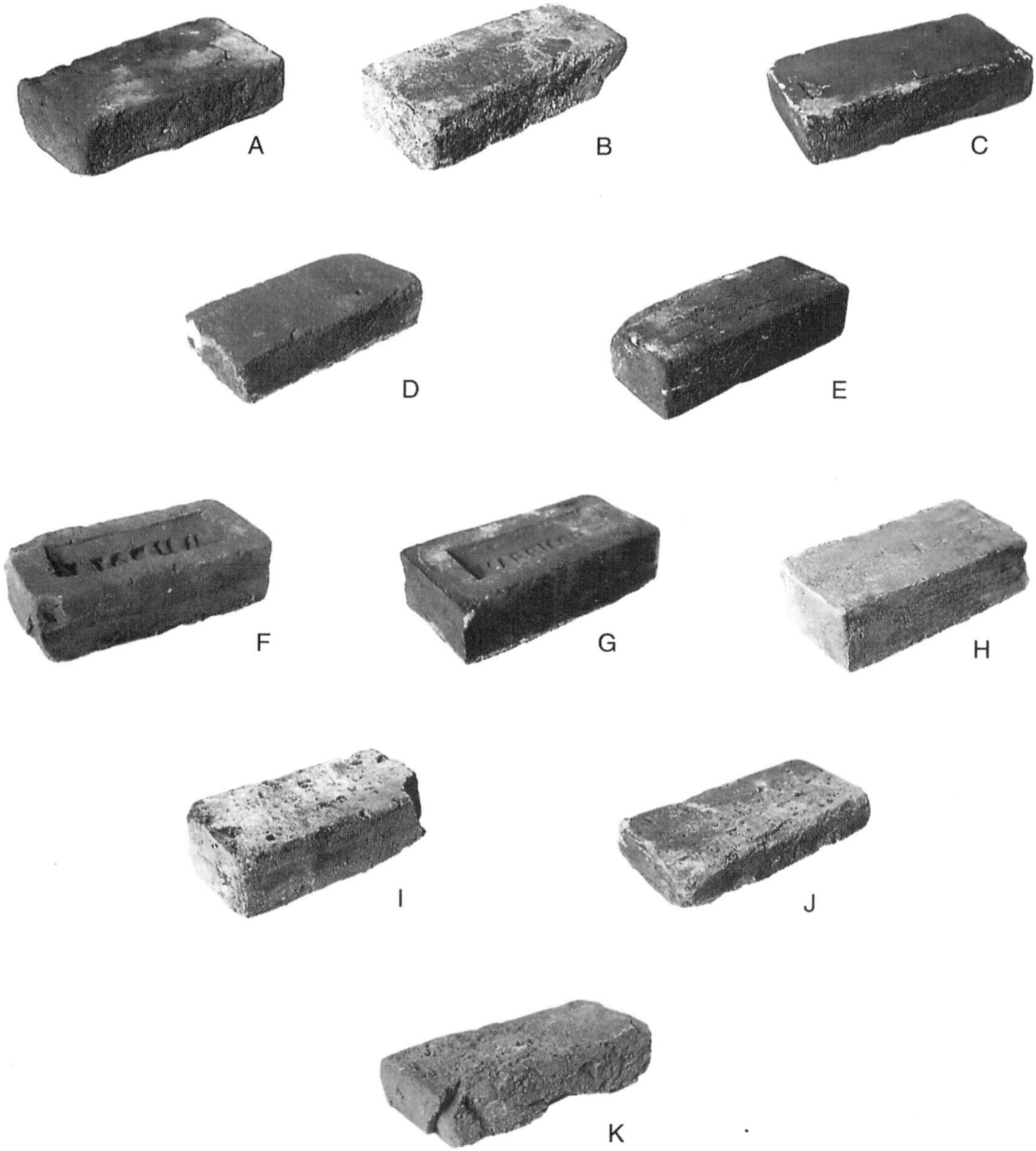

Figure 3.10 Examples of Essex bricks: A, Cressing Garden Wall Tudor (248 × 115 × 60 mm). B, Early 17th century brick from the bullock shed excavations (250 × 110 × 60 mm). C, 17th century brick from Bardfield Saling church (230 × 110 × 52 mm). D, Late 17th century brick from a drain excavated in the walled garden (220 × 100 × 50 mm). E, Early 19th century Soft Red from the dovecote at Berechurch Hall (220 × 100 × 65 mm). F, Late 19th century Soft Red made by W.H. Bott of Hatfield Peverel (235 × 110 × 70 mm). G, 20th century Soft Red from the C ressing walled garden made by the Marriage brickworks (230 × 110 × 63 mm). H, Suffolk White from Duke Street, Chelmsford (226 × 110 × 65 mm). I, London Stock from Beeleigh Mill, Maldon (230 × 110 × 65 mm). J, 19th century flooring brick (225 × 115 × 53 mm). K, Dutch clinker found at Cressing (154 × 65 × 35 mm)

being very precisely made. The fabric is usually pinkish to cream, streaky in fracture, and hard and durable. They were usually laid flat, but in areas of intensive use like threshing floors might be laid on edge.

Several small fragments of pammets or similar were found in the 1978–80 excavations at Cressing, including a glazed example from the chapel floor. Flooring bricks occur reused in some walls and can be a useful dating indicator. (The flooring bricks in the north room of the Granary are second-hand and were laid in 1991.)

A few Dutch 'clinkaerts' or clinkers have been found in the excavations (Figure IK). These are small misshapen paving bricks which were designed to be laid on edge. In 1850, Dobson stated in his treatise on brick and tile manufacture that Dutch clinks, clinkers or paving bricks were made at Moor near Gouda, from slime from the river bed and sand (Celoria, 1971). However he also referred to Dutch clinks made in England. Sir Roger Pratt advocated their use for stable floors in the 1660s (Gunther, 1928: 128). They have been found during fieldwalking over the site of Old Thorndon Hall which was demolished in the 1760s.

Roof tile

An archaeological sequence excavated at Harwich indicates that at that town roof tile or pegtile was not used before *c.*1300 (Andrews *et al.*, 1990: 89). In general, there can be little doubt that in Essex its use only became at all common in the 14th century. However, large ribbed tiles have been found in the 13th century contexts at a number of ecclesiastical and manorial sites throughout the county, including Waltham Abbey, Great Braxted church, Chelmsford Dominican priory, Pleshey and Great Easton. At the Grange Barn, Coggeshall, erected *c.*1180–1225, exceptionally large nibbed tiles measuring 13½ × 6¾–7 × ⅝ inches (342 × 170–180 × 15 mm) were found in the 1984–85 restoration. Areas of large nibbed tiles (12½ × 8 × ⅝ inches; 320 × 200 × 16 mm) were also found during the re-roofing of the Cressing Barley Barn, and there seems no reason to doubt that they were original and were made locally (see note 2). A patch of similarly large tiles exists on the roof of Barnston church, and is thought to have come from Tilty Abbey, which like Coggeshall was a Cistercian foundation.

It seems likely that the production of nibbed tiles declined in the second half of the 13th century and was replaced by pegtiles. Large flat pegtiles 13 × 7¾ × ⅝–¾ inches (330 × 198–203 × 15–18 mm) have been found at Cressing. The similarity in size of these pegtiles and the nibbed tiles suggests they are early pegtiles produced in the latter part of the 13th century. The standard late medieval pegtile has remained basically unchanged in dimensions into modern times. Tiles from the medieval tile factory at Danbury (Drury and Pratt, 1975: 111) measured 10⅝ × 6–6⅞ × ½ inches (270 × 150–175 × 12–15 mm). Some of these tiles were glazed, and a few glazed fragments of pegtile have been found at Cressing. A tile used to raise the height of the chancel wall of Bradwell-iuxta-Coggeshall church in the 14th century measures 10¼ × 6¼ × ½ inches (262 × 160 × 14 mm). A statute of 1477 fixed dimensions of 10½ × 6¼ × ⅝ inches (267 × 160 × 13 mm) for roof tiles (Salzman, 1952: 230). An intact pegtile from the 16th–17th century mansion at Copped Hall, Epping, measured 10½ × 6 × ½ inches (265 × 155 × 11 mm).

The Cressing Temple Brick Typology

Medieval Bricks

1. Coggeshall bricks. Reddish brown with reduced core. 318 × 165 × 40–50 mm. 12½ × 6½ × 1½–2 inches. Regular appearance, smooth faces, base and upper surface; granular fabric. Found in the excavation on the west side of the Barley Barn, and in the drain leading off the tank in north room of the Granary. Suggested date 1180–1220.

Tudor Bricks (Place Bricks)

1. Garden Wall Tudors (Figure IA). Orange to purple, but predominantly light in colour. Very rarely with blue grey glaze. 240–247 × 115–120 × 60–65 mm. 9½–9¾ × 4½–4¾ × 2¼–2⅝ inches. Irregular appearance; irregular, slightly rounded arrises; creased faces, pitted bases. Sandy fabric with pebble and angular flint inclusions. A large Tudor, exceptionally long and thick.
   Found in J. Hope's excavations (types A1 and A2); the main build of the garden wall; the Wheat Barn nogging (which seems to be made of very under-fired examples of the same bricks); and reused in the cartlodge plinth.
2. Granary Tudors. A distinctively orangey brick, paler and softer than those in the garden wall. 230–240 × 115–120 × 55–60 mm. 9–9½ × 4½–4¾ × 2⅛–2⅜ inches. Fairly regular appearance; fairly regular, rounded arrises; creased faces, pitted bases. Sandy fabric with conspicuous iron ore inclusions giving a somewhat speckly appearance, together with some pebbles and angular flints.
   Found in Granary nogging; rounded south-east corner of the garden wall; ?buttress by north garden door; nogging panels at west end of south wall of Wheat Barn. Suggested date 16th to early 17th century.
   The presence of this brick in repairs dating from the 1950s and 1960s suggests it may not be local to the site, but was acquired by Mr Cullen from elsewhere. It may also be significant that it has not been identified amongst excavated samples of bricks from Cressing.
3. Tudor brick from pond across the road. A small Tudor. Orange, sandy fabric with pebble and angular flint inclusions. 230 × 110 × 50–60 mm. 9 × 4¼ × 2–2⅜ inches. Fairly regular appearance; fairly regular, fairly rounded arrises; creased faces, pitted base with groove running length of brick.
   ?15th–16th century.

Late 16th to early 17th century Bricks (Stock Bricks)

1 Wall under bullock shed (Figure IB). Orange, sandy fabric with pebble and angular flint inclusions. 240–250 × 110 × 60 mm. 9½–9⅞ × 4⅜ × 2⅜ inches. Fairly regular appearance; fairly regular, slightly rounded arrises; fairly smooth faces, fairly smooth base, oblique pressure marks. Similar bricks with perhaps more pebble inclusions occur in the northernmost buttress of the west wall of the garden.
   *c.*1600–1650, but thickness suggests it is earlier rather than later, datable to the beginning of the 17th century.

17th Century Bricks

1. A small orange brick with a sandy fabric. 220 × 105 × 50 mm. 8¾ × 4 × 2 inches. Fairly regular appearance; fairly regular, slightly rounded arrises; creased faces, fairly smooth base.
   Third buttress from south end of the east wall of the garden. ?same as type B from J. Hope's excavations.

2. Farmhouse chimneystack bricks (as seen in roof void). 8¾–9 × 4–4½ × 2–2¼ inches. 225–230 × 110–115 × 50–55 mm. Fairly regular, fairly smooth faces. A few examples are vitrified. Some have diagonal pressure marks.

Late 17th to Early 18th Century Bricks

1. Bricks from drain along south side of the garden, and drain along the south side of the Granary (thought to be one and the same drain). Dark red, well fired, some vitrification. 8¼–9 × 4–4¼ × 1¾–2 inches. 210–230 × 100–110 × 45–50 mm. Fairly smooth bases, slightly creased faces. Regular shape, somewhat rounded arrises.
2. Purplish, dense, well-fired, sandy fabric, pebble and angular flint inclusions. 220 × 105–110 × 55–57 mm. 8¾ × 4⅛–4¼ × 2⅛–2¼ inches. Regular appearance; sharp, fairly regular arrises; smooth and slightly creased faces; kiss marks. J. Hope's excavations type D. A few examples also occur in the east plinth wall of the cartlodge, the upper courses of the west plinth of the Wheat Barn, whilst the nogging in the west wall of the Wheat Barn is entirely in this type of brick.

18th–19th Century Bricks

1. Suffolk Whites. A distinctively pale brick. Rather variable in colour: may be white, cream, grey or pinkish. Sandy fabric, interior of brick sometimes marbled. 220 × 100 × 60 mm. 8⅝ × 4 × 2½ inches. Regular appearance; regular sharp arrises; smooth and slightly creased faces, smooth base.
   Found in oblique south-west garden wall; east plinth of cartlodge; and south plinth of granary.
   *c.*1750–1850.

19th and 20th Century Bricks

1. London Stocks
   Yellow to purple with black patches. 230–235 × 110 × 65–70 mm. 9–9¼ × 4¼ × 2⅝–2¾ inches.
   Regular appearance; sharp, regular arrises; creased and cracked faces, smooth bases.
   Found in oblique south-west garden wall; patch in top courses of north garden wall near Wheat Barn corner; north wall of building next to Granary; west and north walls of wood shed next to Wheat Barn.
   *c.*1850-early 20th century.
2. Orange to dark red, dependent on firing. Sandy fabric with pebble and angular flint inclusions. Rather distinctive sandy faces.
   215–235 × 115 × 65–70 mm. 8½–9¼ × 4½ × 2½–2¾ inches.
   Fairly regular appearance; sharp arrises; smooth and slightly creased faces, horizontal pressure marks; some sunken margins, small frogs 45 mm wide and 15 mm deep. Found in central section of east garden wall; also in two buttresses against west wall.
   ?second half of the 19th century.
3. Marriage bricks (Figure IG).
   Dark red brown to purple, well fired.
   230 × 110 × 63 mm. 9 × 4¼ × 2½ inches.
   Regular appearance; sharp, regular arrises; smooth faces, smooth base; sometimes slightly distorted in shape. Frog with 'Marriage' impressed in it. Horizontal pressure marks. Found in paving of east terrace and boundary wall of terrace. Hugh Marriage is recorded in trade directories as making bricks at Hatfield Peverel between 1899 and 1922.
4. Rather soft reds, orange, red and purple in colour. 230 × ? × 65 mm. 9 × ? × 2⅝ inches.
   Regular appearance; regular arrises; some horizontal and oblique pressure marks.
   Found in rounded south-east corner of the garden wall. This was built in the 1960s, but the bricks are presumed to be reused. Probably late 19th–early 20th century.
5. Dark orange to purple.
   230 × 112 × 65 mm. 9⅛ × 4¼ × 2⅝ inches.
   Regular appearance; sharp, regular arrises; smooth and slightly creased faces, smooth base; horizontal pressure marks, shallow frog (160 × 50 mm; 6¼ × 2 inches), possible impression of brickmaker's name but if so very worn. Found in cartlodge plinth.
6. Dark soft reds.
   220 × 110 × 65 mm. 8½ × 4¼ × 2½ inches.
   Very regular appearance, sharp arrises. Horizontal pressure marks. No frogs.
   Plinth on west side of Wheat Barn midstrey. Bonded with 19th century gritty mortar.
7. An unusually thick brick.
   230–235 × 110 × 70 mm. 9–9¼ × 4¼ × 2¾ inches.
   Very square with very sharp arrises; smooth faces. Horizontal pressure marks and long shallow round-ended frogs.
   Plinth in the Barley Barn.

20th Century Bricks

1. Soft reds
2. 215 × 105 × 63 mm. 8½ × 4⅛ × 2½ inches.
   Regular appearance, very smooth surfaces which have a slight tendency to spall. Occasional traces of pressure marks. Present in an extensive refacing on the south side of the Wheat Barn plinth, and dated by an inscription: BUILT 1967.

Flooring Bricks or Pavers

1. White flooring bricks
   Dense fabric. Regular appearance; sharp, fairly regular arrises; smooth faces; smooth base.
   225 × 110–115 × 47–50 mm. 8⅞ × 4¼ × 2 inches.
   Present in cartlodge east plinth; south wall of the Granary which includes at least six such bricks; ?south-west oblique garden wall. 18th–19th century.
2. Dutch clinkers. (Figure IK).
   Cream, yellow and grey. dense, fine textured fabric. 150 × 75 × 40 mm. 6 × 3 × 1½ inches.
   Very misshapen; very irregular, rounded arrises; greenish glaze on some bricks. Signs of wear on one stretcher face. J. Hope's excavation, type E.

NOTES

1. The site of a clamp in the field to the south of the road (Great Warren Field) has been identified by Mr R. Martin. An area of scorched subsoil measuring about 17 × 12 m is evident in this field after ploughing. In the absence of any brick associated with this feature, it may be doubted whether it represents the site of a clamp, but it is difficult to suggest other explanations for it.
2. A tile kiln in a field to the north of the site (Ten Acre Barn) was excavated in 1946 by the farm manager Mr P. Bayliss and is indicated on the 1953 OS map. According to the OS Field Reports cited by the Essex

Sites and Monuments Record (TL 81/4), the kiln was located at TL8041/1903, and was 'constructed of brick and clay, measuring 33ft long by 18ft wide, with exterior flues which were filled with charcoal and wood ash. The floor of the kiln was approximately 3ft to 3ft 6 inches below present ground level'. It was dated by M.R. Hull to the 16th century. Only one kiln was excavated but it seems there were probably two more. A cursory inspection of this field in January 1992 revealed a concentration of rather thick crude-looking tile comparable to the early tiles found on the Wheat and Barley Barns.

Table of Dated Brick from Essex

NOTE: The majority of Essex brick is orange, red or purple in colour according to the temperature reached during burning. In the descriptions colour is only mentioned when it is other than these. The descriptions are arranged in the following order – colour, where applicable, general shape and appearance, creasing or otherwise of stretcher and header faces, kiss marks, pressure marks, characteristics of base, any further observations, and type.

Date	Building	Dimensions	Description of Brick
late C12th	Coggeshall Abbey guest house	13 × 6¼ × 1½" (330 × 160 × 40)	Regular, smooth faces, smooth base and upper surface, added sand temper, reduced core. Coggeshall
late C13th/early C14th	Maldon Friary excavation	8¼ × 4 × 1¾" (210 × 100 × 45)	Cream/pink, very irregular, creased rough base. Medieval
early C14th	Maldon All Saints aisle	10 × 4½–5 × 2½" (255 × 115–130 × 65)	Cream/pink/purple, very irregular, very creased, rough base, grass marks on base and sides. Medieval
early C14th	Dengie church	8½–9 × 4 × 1¾–2" (215–230 × 100 × 45–50)	Creamish to brown, green glaze, irregular, creased. Medieval
early C14th	Purleigh church tower	7¾–8½ × 4 × 2" (195–215 × 100 × 50)	Cream/buff, green glaze, irregular, used decoratively. Medieval
*c.*1340	Witham church tower and north aisle	9½–10 × ? × 1¾–2" (240–255 × ? × 45–50) ? × 4½–5 × 1½" (? × 115–130 × 40)	Yellow/pink/red, very irregular, very creased, possibly slop moulded. Medieval
*c.*1340	Lawford church chancel	7½ × 3¼–3½ × 1¾" (190 × 85–90 × 45)	Cream/pink/brown, green glaze, irregular, creased, used decoratively. Medieval
1369 (? reused)	Waltham Abbey gatehouse	14½–15 × ? × 2–3" (370–380 × ? × 50–75)	Fairly regular, creased, reduced cores. Medieval Great Brick
early C15th	Writtle, King John's Hunting Lodge	9 × 4 × 1¾–2" (230 × 100 × 45–50)	Irregular, creased, rough base, blue/grey glaze, some straw impressions. C15th
early C15th	Maldon, Moot Hall (D'Arcy Tower)	9–9¼ × 4¼ × 2–2¼" (230–235 × 110 × 50–60)	Irregular, creased, rough base, some sunken margins. C15th
C15th	Pleshey castle bridge	9 × 4–4¼ × 1½–2" (230 × 100–110 × 40–50)	Irregular, creased. C15th
*c.*1439	Faulkbourne Hall tower	8¼–9 × 4–4¼ × 2–2¼" (210–230 × 100–115 × 50–55)	Fairly regular, creased. C15th
1460s	Roydon, Nether Hall gatehouse	8–9 × 4¼ × 2½" (220–230 × 110 × 65)	Fairly regular, fairly creased, rough bases, sunken margins. C15th
late C15th	Dedham church tower	9 × 4–4¼ × 2" (230 × 100–110 × 50)	Irregular, creased. C15th
late C15th	Rochford church tower	9½ × 4½ × 2–2½" (240 × 115 × 50–65)	Irregular, creased, diaperwork. Tudor

Date	Building	Dimensions	Description of Brick
late C15th	Fryerning church tower	9+ × 4–4¼ × 2" (230 × 100–110 × 50)	Irregular, creased, diaperwork. Tudor
1489–90	Hadleigh deanery tower	9½ × 4¼–4½ × 2¼–2½" (240 × 110–115 × 55–60)	Irregular, creased, diaperwork. Tudor
1496	Billericay church	9–9½ × 4¼ × 2¼–2½" (230–240 × 110 × 55–65)	Irregular, creased. Tudor
1498	Gestingthorpe church tower	9½ × 4½–4¾ × 2¼–2¾" (240 × 115–120 × 55–60)	Irregular, creased, diaperwork. Tudor
early C16th	Thaxted, Horham Hall	9–10 × 4¾ × 2½–2¾" (250–255 × 120 × 65–70)	Irregular, creased. Tudor
early C16th	Layer Marney church	9–9¼ × 4¼–4½ × 2–2¼" (230–235 × 110 × 50–60)	Irregular, creased, blue-grey glaze diaperwork. Tudor
early C16th	Woodham Walter Hall	9–9¼ × 4¼–4½ × 2¼" (230–235 × 110–115 × 60)	Irregular, creased, sunken margins, rough base, diaperwork. Tudor (place)
1543	Chingford, Queen Elizabeth's Hunting Lodge, plinth	8¾ × 4–4¼ × 2¼" (220–230 × 110 × 60)	Irregular, creased, rough base. Tudor (place)
*c.*1540	Tolleshunt Major church	9½ × 4½ × 2¼–2½" (240 × 115 × 60–65)	Irregular, creased, diaperwork. Tudor (place)
*c.*1540	Leighs Priory inner gatehouse	9½ × 4½ × 2" (240 × 115 × 55)	Irregular, creased, blue-grey glaze, diaperwork. Tudor
*c.*1540	Ingatestone Hall	9½–10 × 4½–5 × 2½" (240–260 × 115–125 × 65)	Irregular, creased. Tudor
mid 16th	High Roding, New Hall barn and moat	9¼ × 4¼–4½ × 2" (235 × 110–115 × 50)	Irregular, creased, rough base. Tudor (place)
1556	Ingatestone church south chapel	9½–10 × 4½–4¾ × 2¼–2½" (240–255 × 115–120 × 60–65)	Irregular, creased. Tudor
1563	Woodham Walter church	9½–10 × 4½–5 × 2–2½" (240–260 × 115–125 × 50–65)	Irregular, creased, rough base. Tudor (place)
1586	Springfield church tower	9–9¼ × 4½–4¾ × 2" (230–235 × 115–120 × 50)	Fairly regular, creased. Tudor
1589	Chingford, Queen Elizabeth's Hunting Lodge, chimney	9 × 4¼ × 2¼" (230 × 100 × 60)	Irregular, creased, rough base. Tudor (place)
1594	Rivenhall church, Wiseman monument	9¼ × 4¼ × 2¼" (235 × 110 × 55)	Fairly regular, creased, fairly smooth, diagonal pressure mark. Tudor (stock)
*c.*1600	Audley End, College of St Mark	8¼–8¾ × 4 × 2¼–2½" (210–220 × 100 × 55–65)	Irregular, creased, some yellowish to purple. C17th
1605–1609	Markshall, Marks Hall (demolished)	9½ × 4½–5 × 2" (240 × 115–125 × 50)	Irregular, creased, rough base, some sunken margins. Tudor (place)
*c.*1625	Ingatestone church, north chapel, lower part of walls	9 × 4 × 2¼" (230 × 100 × 55)	Fairly regular, creased, kiss marks. C17th (kiln burnt)

Date	Building	Dimensions	Description of Brick
1658–*c.*1700	Stanway, Olivers brick kiln	8¾–9 × 3¾–4¼ × 1¾–2" (220–230 × 95–110 × 50)	Regular, fairly smooth base, diagonal pressure marks. Late C17th/early C18th (stock)
*c.*1665	Gt Maplestead, Hopwells, hall chimney	9 × 4½ × 2" (230 × 115 × 50)	Fairly regular, creased diagonal pressure marks. Late C17th/early C18th (stock)
1674	Stebbing, Quaker Meeting House	8–8¾ × 4–4¼ × 2" (205–220 × 100–110 × 50)	Fairly regular, slightly creased, diagonal pressure marks. Late C17th/early C18th (stock)
1680	London, St Anne Lutheran Church	8½ × 4 × 2¼" (220 × 100 × 55)	Regular, slightly creased. Late C17th/early 18th
1682	Black Notley church, buttress	8¾ × 4¼ × 2–2¼" (220 × 110 × 50–55)	Regular, creased, kiss marks. Late C17th/early C18th (kiln burnt)
late C17th	Maldon, Plume Library	8–8½ × 2–2½ × 4–4¼" (200–215 × 100–110 × 50–65)	Regular, creased and fairly smooth kiss marks, diagonal pressure marks, chequerwork. Late C17th/early C18th (stock, kiln burnt)
*c.*1693	Faulkbourne Hall (Bullock build)	8½ × 4 × 2½" (215 × 100 × 65)	Regular, smooth and creased, diagonal pressure marks, chequerwork. Late C17th/early C18th (stock)
1699	Great Saling, Saling Hall	8½ × 4 × 2¼" (215 × 100 × 60)	Regular, slightly creased, chequerwork. Late C17th/early C18th
*c.*1700	Terling U R church	8¼–8½ × 4–4¼ × 2¼" (210–215 × 100–110 × 60)	Regular, slightly creased and smooth, kiss marks, diagonal pressure marks, black headers used decoratively. Late C17th/early C18th (stock, kiln burnt)
1708	Little Baddow U R Church	8–8¼ × 3¾–4 × 2¼" (200–210 × 95–100 × 55)	Regular, smooth and slightly creased, diagonal pressure marks, chequerwork. Late C17th/early C18th (stock)
*c.*1718	Colchester, Hollytrees, east wall north & south walls	8 × 3¾ × 2¼–2⅜" (205 × 95 × 60) 8¼ × 4 × 2½" (210 × 100 × 65)	Regular, smooth and slightly creased, kiss marks, diagonal pressure marks. Rubbed red brick dressings, dark headers, tuck pointing. Regular, smooth or slightly creased. Late C17th/early C18th
*c.*1720	Belchamp Walter Hall, garden wall and stables	8¾–9 × 4¼–4½ × *c.*2¼" (225–230 × 110–115 × 60)	Regular, slightly creased, diagonal pressure marks, chequerwork. Late C17th/early C18th (stock)
1728	Maldon All Saints, north wall of nave	8¾–9½ × 4–4¼ × 2½" (220–235 × 100–110 × 65)	Fairly regular, fairly smooth, diagonal pressure marks, chequerwork. C18th (stock)
*c.*1730	Ingatestone, 23 High Street	9 × 4¼ × 2½" (230 × 110 × 65)	Fairly regular, slightly creased and smooth, diagonal pressure marks, chequerwork. C18th (stock)
1732	Terling church tower	8¾–9 × 4–4¼ × 2¼" (220–230 × 100–110 × 50)	Regular, slightly creased and smooth, diagonal pressure marks, chequerwork. Late C17th/early C18th (stock)
1734	Ingrave church	9 × 4¼ × 2½" (230 × 110 × 65)	Regular, slightly creased, diagonal pressure marks. C18th/early C19th (stock)

Date	Building	Dimensions	Description of Brick
1757	Woodham Walter Lodge	8½ × 4 × 2¼–2½" (215 × 100 × 55–60)	Regular, slightly creased. C18th/early C19th
1758	Chelmsford, Mildmay almshouses	8¾ × 4¼ × 2½" (220 × 110 × 65)	Regular, slightly creased and fairly smooth, kiss marks, diagonal pressure marks. C18th/early C19th (stock, kiln burnt)
*c.*1760	Markshall church	8¾ × 4 × 2½" (220 × 100 × 65)	White, regular, smooth and very slightly creased, smooth base. Suffolk White
1779	Hutton church, T. Hills vault	8¾ × 4 × 2½" (220 × 100 × 65)	Regular, smooth. C18th/early C19th
1789	Chelmsford, Shire Hall	9 × 4¼ × 2½" (230 × 110 × 65)	White, very regular, fairly smooth, horizontal pressure marks. Purchased from Playford near Ipswich (Briggs, 1991; 98). Suffolk White
*c.*1805	Maldon Friary	9 × 4 × 2½" (230 × 100 × 65)	Regular, smooth and slightly creased, smooth base. C19th
early C19th	Maldon, Beeleigh Mill	8¾ × 4 × 2½" (220 × 100 × 65)	Regular, smooth and slightly creased, smooth base. Late C18th/early C19th
early C19th	Maldon, Beeleigh Mill	8¾ × 4¼ × 2½" (220 × 110 × 65)	Yellow and purple with voids and black patches, regular, smooth and slightly creased, horizontal pressure marks. London Stock
1827	Woodham Mortimer, Elm Farm barn	8¾ × 4¼ × 2½" (220 × 100 × 65)	Very regular, smooth, smooth base. C19th
1832	Tillingham, Baptist manse	8¾ × 4 × 2½" (220 × 100 × 65)	Regular, smooth and slightly creased, smooth base, diagonal pressure mark. C18th/early C19th
1843	Chelmsford railway station	9 × 4¼ × 2½" (230 × 110 × 68)	Yellow and purple with voids and black patches, regular, creased. London Stock
*c.*1874–1890	Hatfield Peverel brick-works	9¼ × 4¼ × 2¾" (235 × 110 × 70)	Very regular, smooth, horizontal pressure marks, frog impressed W H BOTT. Late C19th/20th
1878–1895	Boreham brickworks	9¼ × 4¼ × 2¾" (235 × 110 × 70)	Very regular, smooth, horizontal pressure marks, frog impressed W CLOVER BOREHAM. Late C19th/20th
*c.*1894	Chelmsford, Old Court Nonconformist chapel	9 × 4¼ × 2¾" (230 × 110 × 70)	Very regular, smooth, frog impressed BEACH. Late C19th/20th
*c.*1900	South Woodham Ferrers brickworks	8¾ × 4 × 2½" (220 × 100 × 65)	Yellow and purple with voids and black patches, irregular, shallow frog. London Stock

The archaeological method, as will be noted from this detailed account, is to assemble the individual characteristics of all available brick-work on site and set it up in the context of wider local and documentary evidence. For conservation the use and reuse of brick tile and paviours whose source and age are identifiable can do much to determine the age and sequence of construction. This in turn may be fundamental to decisions prior to intervention.

4

Refined brick earths – terracottas and tiles

There is no thoroughly satisfactory demarcation between the materials used for the body of tiles and shaped, refined architectural elements made out of clay or brick earth. However the term terracotta (literally Italian, burned earth) is a substantive definition under which must be grouped a very large range of products in use in the building industry created by the firing of brick earths from which the larger particle sizes have been separated or ground down and which generally are richer in clays than the materials used for producing coarse brick. Such materials are normally formed to more precise sizes and in thinner sections than bricks though they may reach maximum dimensions far in excess of any brick known. They may be made hollow to achieve the required thin sections. Perhaps the greatest of such products, and certainly the most heroic, are the terracotta armies of Northern China, whose extent is yet unknown but clearly already far exceeds any comparable creation. It hardly matters whether strictly they may be described as architecture: they are certainly works on an architectural scale creating an architectonic environment. They share

Figure 4.1 The collapse in 1903 of the Campanile of San Marco in Venice. The lime mortar can be seen to have crumbled to powder in the movements and the symmetry of the failure is a clear indication of uniform basal overload.

Figure 4.2 A nineteenth century terracotta detail set against the scale of the facing brick. The natural component size in industrial era terracotta is much larger than the brick and is comparable with stone.

in common with all tiles and terracotta a firing process comparable to ordinary brick though the methods may vary and their substantial size is achieved only by forming each piece as a shell – a pot – to permit drying without the damage caused by shrinkage on a wet core.

4.1 Background and status

At its most robust terracotta may effectively be no more than moulded brick, and much of the earliest English terracotta (of the sixteenth century when it made its first great appearance in Britain) is brick which is structurally integral with a building but carries on its surface intricate designs or is shaped to provide architectural detail.

The body of terracotta may have any colour in the range of brick from white through to a dark purple, although the extremes of the range are uncommon, particularly at the darker end since the high firing involved endangers the stability of the component itself. By far the most common colour is a red or brown. The ferruginous clays which produce these colours reflect the degree of firing in the darkening of tone and with that change in colour goes an increase in hardness. Being derived from a natural material and subject to variations in firing, historical terracottas vary in colour even when produced in the same batch.

Many glazed architectural components should be classed as terracotta and must be understood and treated in the same way as glazed earthenware made for other purposes. Glazes may be introduced in the primary firing or added on a repeat firing. Some special materials which should also be classed as terracotta have been created by the incorporation of powdered or reground fired clays, including porcelains, which can be carried to a high temperature in the kiln to produce an extremely highly finished and durable architectural material. The most famous of these products in English terms is Coade Stone, produced in the eighteenth century for ornamental purposes, initially at Poole then in London. Its formula was a closely guarded trade secret but it is now thought to have contained a high input of reground kaolin-based fired ware and feldspars, kilned at unusually high temperatures. Its fluxing additions were non-plastic which significantly reduced the shrinkage and risk of cracking in the drying stage. Other potteries, noticeably Italian, have produced equivalently durable and refined materials for architectural purposes, including the maiolica ranges, and most famously the highly finished tin-glazed and richly coloured sculptures of the Della Robbia family. Such sixteenth century works have become art objects in their own right, prominent among a large class of artefacts whose conservation lies in the hands of museum conservators rather than builders.

There is an important but imprecise division between elements whose conservation is as art objects and those to be conserved simply as building elements. Slabs of moulded terracotta applied to the face of a large commercial building, as was common in Europe and America in the late nineteenth and early twentieth centuries, may be replaced individually in the name of conservation, the originals quite literally and legitimately being thrown away if they are not conveniently repairable. It would by contrast be quite unacceptable to destroy a wall panel of Iznik tilework of the sixteenth century or a contemporary Della Robbia plaque. Such objects must be treated as significant art works or artefacts in their own right.

There are two important perspectives which the conservator will have in mind in facing this problem. The first relates to time. A damaged item, which might quite legitimately be thrown away at one point in time might, by view of increasing rarity, a change in the art market or simply in consequence of increasing age, become an item for conservation as a museum object. Many Victorian church artefacts, terracottas among them, have moved into this category having been dispensed with in the early part of this century in a way which would be both unacceptable and probably outside present law in Britain, much of Europe and America.

The other perspective is one of size and relates to the perception of the conservator. A conservator of objects might find it unacceptable that a carefully finished historic element such as a fine glazed tile should be mounted on a building in a position that was vulnerable and likely to diminish its effective life. His view, focused on the object alone might be that it should be removed to a safe environment for reasons of conservation and this view might be strongly supportable in relation to its known history of manufacture, rarity, or archaeological significance. Taking the stand of a building conservator, however, the view might be that the historic significance of the building would be seriously diminished by the removal of a genuine artefact and its replacement by an imitation or by failure to provide a replacement. Thus two equally legitimate views on conservation could produce opposing conclusions in respect of a fine artefact on an important building, and a careful judgement would have to be entered. The level of consideration would have to be comparable to that applied by the librarian in charge of fine books of increasing rarity and value. There comes a point at which the librarian's decision will be to remove them from circulation and place them in a protected or restricted category and this in itself may be a judgement depending upon the effluxion of time. Many comparable judgements have had to be made over items which become collectable and hence articles of trade or museum reference or simply become more valuable until ultimately they have to be guarded more assiduously against thieves than from the effects of rising damp.

4.2 Terracotta faience and tile

In the context of the burned earths which lie in the broad category of terracotta one specific characteristic takes on an importance which is much less significant in bricks – the burned surface, known as a fireskin in some contexts and a glaze in others.

During firing, the surface of the piece is subject to physical and chemical circumstances different from the remaining body. It may indeed already be different before it enters the kiln by virtue of treatment at the green stage.

It may already have received a surface coating, perhaps of silicates or more widely ranging inorganic compounds such as lead or tin glazes aimed at producing colour or protective glaze or it may simply have been polished, tooled and worked producing a higher surface density than in the core. In the process of finishing a terracotta block prior to firing the hardened green material is often subject to burnishing, perhaps after superficial wetting, with the result that the surface material has been made more dense and this is reflected in the performance characteristics which finally result. Polishing and burnishing tends to orientate the plate-like clay crystals in the surface layer. Similarly the deposition of slip on a Plaster of Paris mould tends to align the particles. Plaster of Paris causes flocculation in clay slurry or paste creating a clean line of separation so that shrinkage can take place. Higher temperatures produce greater vitrification in the fireskin and the surface tensions 'frozen' on cooling produce surface stresses on a micro-scale which are characteristic of glass. A fireskin may be as much as 2.5 mm deep. During firing, the surface is the zone which receives the heating and in order to bring the core to full temperature the surface must exceed that temperature. This is unavoidable however slow the firing, because fundamental thermodynamics of heat transfer demand a thermal differential. The skin must therefore be brought to a higher temperature than the core. Further the gaseous conditions in the surrounding atmosphere will not be those within the core. The core material will attempt to follow the external regime but there will always be a lag so the surface will achieve a semi-melt first. As liquidity of any component of the piece is approached surface tension will exist in the melt and this in turn will exert a physical constraint on the surface of the piece which is different from that of the body. This is followed by a differential thermal contraction on cooling as a result of a reverse thermal gradient. The resultant compressive forces externally dominate the opposite tensions inside the body. For these reasons, and perhaps for others in addition, the surface of a piece of fired terracotta will differ from its substrate. For instance where combustion firing is employed the vitrifying effect of the reducing atmosphere in the wake of the flame arises because ferrous silicates melt at lower temperatures than ferric silicates. The durable protection and visible finish of the fireskin is a critical visual element. No element of terracotta should be conserved without careful consideration being given to the nature, the function, durability and response of the fireskin to any treatment proposed.

The term terracotta may be used conveniently in two senses: to describe all manufactured earthenwares derived from selected clays, purified or refined brick earths and more narrowly to describe such wares which are specifically shaped to be applied to a specific, and usually decorative or ornamental function. Thus a fired clay sculpture, a flower pot and a roof tile may be made of identical clay, fired identically in the same kiln at the same time. To describe the tile as terracotta would be very unusual, if correct; to describe the pot as terracotta would be understandable and normal and to describe the sculpture as terracotta would be inevitable. Perhaps it is only pedantry which demands greater linguistic precision.

There are a number of associated terms, however, which are more precise. Faience, derived from the Italian town of Faenza, refers to white bodied, glazed wares of porcelainous type made from kaolin-rich local clays and widely exported from that town in the Middle Ages. They are in some senses only an equivalent of modern sanitary ware which in itself raises the problem of the status of such material in the classification. The term Maiolica (or Majolica) was applied to the earthenware products of fifteenth and sixteenth century potteries at Faenza, Gubbio and Deruca. The term faience derives from the French use of the Italian description, 'porcellana di Faenza' to describe the tin-glazed earthenware made in France in imitation of Faenza maiolica and is now taken to be the general description of ornamental earthenware which is porous and coated with an opaque white tin glaze. The white ground was subsequently over-painted in translucent glazes and fired again. From Italy production spread, often with the movement of craftsmen to North European centres such as Rouen and Nevers, to various English and North European potteries as well as southern centres such as Lyons and Marseilles.

Encaustic tiles describe wares in which a pattern is carried in vari-coloured clay slips. The term encaustic derives originally from an ancient technique of painting in which the pigment is deposited in casein, gum or hot wax. On heating, the slips combine with the substrate. Encaustic, which literally means 'burning in', also refers to the production of fired enamels and it may be in this sense that multi-coloured tiles derived their description. The term then became applied to the manufacture of tiles in which a pigmented body was applied, usually into an indentation in the tile, and although technically this is usually a slip or waterborne pigmentation the term encaustic tile now embraces the range of medieval decorated floor tiles important in North European Gothic and Gothic Revival architecture. The technique of applying pigment in a hot wax to a tile subject to wear was likely to be less durable than firing a slip. The tiles themselves were sometimes made of a sandwich of different clays. If wax was used it burned out leaving pigment to meld with the vitrification of the body, whereas the slip provided its own vitrifying material.

Of the many trade names invented by manufacturers to describe their products the terms Coade Stone and Cararra Ware are the best known in Britain. Coade Stone was a late eighteenth century material manufactured at Lyme Regis and subsequently in London until the mid-nineteenth century (see above and below). Cararra Ware was a nineteenth century unglazed densified parian porcelain.

Porcelain is sometimes confused with glazed terracotta although it is distinctly different. True porcelain is a white vitrified ceramic material fired at about 1400°C and consisting of China clay and a fusable form of feldspar (petuntse), first made in China before AD 900. The glaze derived from the same feldspar produces a single integral material. There is no porous underbody. Soft paste porcelain and English bone china are also of impervious body obtained using fluxes – sand/lime, soda, flint or bone in various combinations at a firing temperature above 1100°C. Porcelain is little used in building work.

The Chinese, who had achieved such extraordinary efficiency in the production of terracotta by the third century BC, had moved to the production of pure porcelain by the early medieval period. It was a further 900 years before the factories at Meissen were able to achieve the first equivalent in Western Europe. Fragments of early porcelain, however, have turned up in potteries associated with the Round City of Baghdad in the eighth or ninth centuries in association with coarse slip-glazed earthenware made in attempt to simulate the Chinese work. While the Mesopotamian potters were unsuccessful in their attempts to achieve high quality porcelain they did produce the earliest lustre wares known and their techniques spread through Syria to North Africa where the lustre tiles in the Mihrab of the Great Mosque in Kairouan (ninth century) survive as one of the earliest manifestations of the quasi-metallic surfaces obtainable in reducing atmospheres.

Figure 4.3 Life size terracotta figures at Xian in China.

A convenient differentiation between terracotta and faience is that terracotta is modelled and shaped but unglazed and faience is the same material but coloured and glazed. Even here there is a difficulty: much historical ornament on buildings consists of part glazed terracotta.

Terracotta enters the history of architecture as an unglazed element of walling in medieval Italy. This is not to deny its presence in antiquity in China, India and the Greco-Roman World and in all these cultures the manufacture of terracotta roof tiling has been extremely important. No great civilization has failed to produce pottery, for the most obvious and practical of reasons, and it is a short step from making a cooking pot to making a roof covering. There seems no need to pursue postulated derivative connections between bark used as a primitive roof covering and the invention of alternating concave and convex tiles. The evolution of the concave sloping gutter joined by a convex shedding tile is direct in either medium. What is certain is that timber and wood products will have been extensively used for roofing for a very long period but tiles survive if made of terracotta and disappear irretrievably if made of bark or split timber.

While the wonderfully sophisticated stone masonry of the Athenian temples of the sixth century BC survives to be measured and drawn, the tiled timber roofs which covered them are known only by fragments of stone, marble and terracotta. From Roman times the European tradition has been continuous and the sites of Roman tileries yield good evidence of the techniques and products passed over to the native communities in the rest of Europe.

If the Roman art of tile-making almost died out in Britain during the Saxon period the same was certainly not true in the Mediterranean where the vigour of the Eastern Roman Empire maintained a continuous programme of building matched erratically in Italy and Southern France until their resurgence under the Normans.

4.3 Floor tiles

Apart from the production of roofing tiles, the great achievements of medieval terracotta-making are to be found in floor tiling in Europe and in the wall tiling of the Islamic world which evolved from the earlier uses of mosaic. Coloured and patterned floor tiles emerged as a significant art form in Northern Europe, where the distinctive use of tilework is probably simply to be attributed to the climate and nature of construction. A fundamental structural problem of the countries of Islam is physical erosion caused by rising salts in walls. A relatively impervious internal covering will protect walls against evaporation simply by reducing the rise of moisture internally. In northern climates rising damp may be a nuisance but fewer salts are deposited and the rate of decay in walls is very much less. In floors, however, the position is the converse. A slightly damp floor maintains its coherence in a hot climate, whereas in a northern climate it can become sufficiently wet to be both unpleasant and to wear quickly. It is, therefore, perhaps understandable that the contrast between northern Europe and the shores of the Mediterranean was in the preferential use of tiles on the floors in the north and the walls in the south. Valid or not this conjecture fits a pattern in which the north European development of floor tiling produced a rich tradition of patterned tiles where engraved and embossed terracotta surfaces were inlaid with coloured slips to provide a flat finished surface, sometimes salt-glazed, but always permanent and durable.

Those tiles in which the coloured slip was applied superficially or in which a pattern was applied under a glaze suffered from wear because the patterned section was raised. Clearly they would be less satisfactory than those in which the contrasting colours were finished on a single level surface. The self colour of the slips ranged through buffs and straw yellows to greens and greys, near blacks and occasional greys and blues imposed on variously brown, red-brown and dark brown base materials. Complex ranges of designs made use of the repetitive nature of impressed pattern-making and provided fertile ground for invention with figurative devices interspersed into foliate and geometric themes. Many tiles were designed with a multiplied geometry in mind so that one element in each design carried through into the next to form overall patterns, circles, crosses and diagonal lines and the floors themselves were designed with the placing of specific tiles to form a layout

Figure 4.4 A theme widespread in the Muslim world where the lowest zone of the wall is faced with glazed terracotta to resist rising damp. Above this dado other materials – stucco, plaster and timber – may be employed.

using borders, guards and specifically delineated areas. Many floor layers, however, were content with nothing more elaborate than an even chequerboard to provide a broad pattern. Others made use of calligraphic tiles to implant a message – usually some theological admonition, perhaps a salutary reminder of the fundamentals of life. Others made use of animal or figurative symbols which carried a specific meaning, occurring perhaps on the arms of a donor or in some other way being directly related to circumstance. These tiles were relatively large and easy to lay by comparison with their predecessors which had been smaller and laid as a mosaic. Monastic, and particularly the Cistercian, fraternity was particularly fond of this type.

The expertise of floor layers ran to exploiting natural variations in firing to make floor patterns. On a larger scale the patterned tile disappears under the stronger impact of the superimposed design. This same sequence of evolution occurs in glazed Islamic tilework where the techniques of mosaic and laying were superseded by the introduction of larger, painted tiles.

Much encaustic work, where the pattern was literally burned in by the second firing, was given a salt or other glaze which, being relatively soft, would fail on a tile continually wetted from below. Tiles laid on a mortar bed without a damp course suffered particularly. The visual softness of the unglazed surface is now valued and contrasts with the harder and sharper qualities of the more precise, efficient and highly glazed reproductions made in the nineteenth century.

For an extensive description of the conservation of tile and terracotta in floors, see Fawcett (1998).

4.4 Architectural terracotta

The application of decorative architectural terracotta was a European phenomenon in the later Middle Ages. Later Gothic architecture was rich in carved stone sculpture and it was inevitable that the modelling and firing of clay would be used as a short cut in the provision of sculptural and architectural detail. Italy was the fountainhead and as the Renaissance drove out the Gothic design, redolent of the barbarous north, the opportunity rose for sculptors to create works which achieved the flamboyance and humanistic forms denied to Gothic designers.

Important examples survive in Santa Maria della Grazie in Milan and in the great Carthusian church at Pavia, the Certosa, now only moderately valued but immensely influential in its day. Fifteenth century Italy produced a range of craftsmen whose skills were carried by itinerant craftsmen into northern Europe. In Great Britain their work is to be found in a south country florescence at the beginning of the sixteenth century which, for political reasons, disappeared as suddenly as it had arrived within forty years. Great palaces such as Nonsuch, now destroyed almost

Figure 4.5 Sutton Place, Surrey. Early sixteenth century: a bay opening into the original central court, and a detail. The use of terracotta apes the use of stone in mullions, cinquefoil and trefoil window heads, window dressings and parapets.

without trace, Hampton Court, and houses such as the palatial courtyard house (now U-shaped) Sutton Place in Surrey, the eccentric towers of Layer Marney in Essex, Westhorpe Hall in Suffolk (also destroyed) and East Barsham in Norfolk were echoed on a smaller scale by houses still with an aristocratic connection such as Laughton Place in Sussex. In the grander examples terracotta was made in the Italian manner and in the same way as sculpture in which the core would be carved away to leave a hollow piece with an approximately constant thickness. Much English terracotta of this period, however, is simply solid moulded brick carrying designs or architectural detail derived from the pattern books or repertoire of the leading Italian craftsmen. The architectural detailing is still thoroughly late medieval in concept. Terracotta spandrel panels on a scale – and to a design – which might equally well have been executed in stone or timber and the mullions and cusped window heads decorating virtually all this early sixteenth century English work are in no significant sense different from their counterparts in other material. The names of a number of the Italian artisans and artists who worked in England are known, e.g. Gerolamo da Treviso and Giovanni da Maiano, both of whom worked for Henry VIII, as did the more famous Pietro Torrigiano, principally a metal worker, who was employed on the building of the chapel of Henry VII at Westminster Abbey.

After this upsurge there was a lacuna, out of which there emerged in the eighteenth century the use of terracotta as a substitute for sculpture in stone. The closely guarded secret

Figure 4.6 A turret at Layer Marney (early sixteenth century). The small regular bricks are capped with elaborate terracotta crestings.

of the Coade Stone family all but concealed the origins of the material as a highly fired terracotta. In many instances and for many years examples have been taken for genuine stone. This and the very high firing temperatures involved produced a buff stone-coloured material of unsurpassed hardness and durability. Surviving examples are essentially to be classed as works of art. Sir John Summerson wrote (Georgian London 1945, q. Fidler): 'In many a weatherworn facade the Coade sculptures stand out crisp and firm and I have seen burned churches where the Coade monuments are the only ones still recognizable.'

In the nineteenth century the production of pottery and ceramics benefited from control given by industrial techniques, with the result that terracotta products were made to higher standards of precision and executed with greater certainty and economy. Markets widened in Europe and America, terracotta being especially popular for its resistance to the increasingly polluted urban atmospheres which damaged stone. The material was widely seen as an acceptable substitute for stone and so began the era in which it was greatly advocated and widely used as a facing material.

Whereas, with the exception of pure ornament, earlier terracotta had been integral with the building structure, nineteenth and twentieth century terracotta was increasingly employed as a skin material having no structural function to the point of being hung on or applied to the structural frame. Industrial European and American terracotta – that is moulded, modelled, fired clay – took on the decorative role and if by differentiation between structural and non-structural components only the shaped structural bricks are considered at this point, we would omit an immense range of purpose-made elements applied decoratively as cornices, pilasters, friezes, swags, boasts and dressings of every kind. Terracotta was becoming also a material for the creation of art objects in its own right as a result of which there arose new ranges of problems to exercise the minds of present conservators.

The techniques of manufacture also changed. While the sixteenth century moulded brick was the product simply of standardized techniques using a carved timber former against which the brick could be pressed, the later wares might be mechanically pressed or formed against absorptive plaster moulds. The use of hard plaster and even cast metallic moulds offered greater opportunities for precision which manufacturers were to take advantage of, producing materials which were hard, durable and often over-precise. The advantages offered by well made terracottas were mechanical and they appealed to the designer familiar with engineering techniques. They were fired at temperatures as high as 1100°C to 1300°C in a wide range of colours with accurate junction details and their precision was such that architects and engineers adopted the material with enthusiasm for use in conjunction with steel framing. It provided the opportunity to achieve a thin skin façade with the broad appearance of masonry including a wide range of typical stone details. They were light to handle and of a material robust

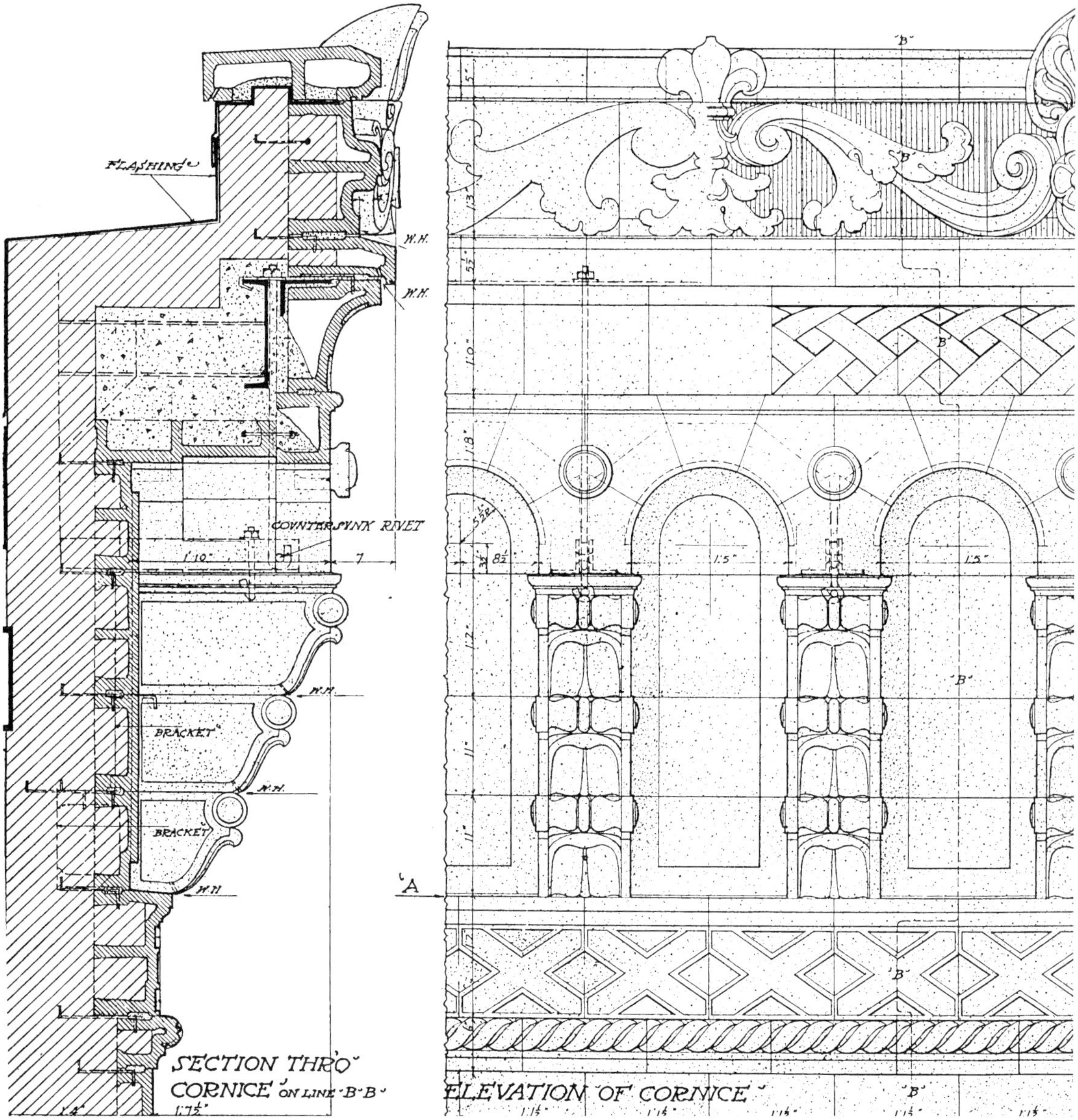

Figure 4.7 Typical fixing detail used in late nineteenth and early twentieth century terracotta work. Movements in the structure opens micro-cracks through which water enters to cause failure due to frost action and rusting.

enough to be used on building sites. It was widely used for forming fireproof floors, terracotta partitions and fireproof casings to steelwork in the twentieth century. Extruded terracottas were devised to fit many patent designs for flooring and internal walling in addition to external applications. As a facing material it provided a finished surface of low cost in view of its complexity, with the advantage of mechanical jointing to a high level of accuracy.

Problems were to arise with its fixing. Such facings were fixed back to metal bracketing, their hollow sections being filled with concrete and the fixings relying on copper, bronze or protected steel. Some of the highest quality terracottas were based on fireclays often derived from seams associated with colliery workings which, although durable, could not be jointed flexibly to take account of the varying behaviour of all the components. Further problems arose when builders treated them as permanent shuttering, filling the voids with concrete.

Differential expansions, structural movements and flexings, water entry through failed pointing, rusting of fixings and reinforcement and incompetent handling have led to many failures: but well-built masonry structures with well-made terracotta facings have proved durable, sometimes extremely durable, and use of the material has been vindicated in official reports in Europe and America. The department of scientific and industrial research of the British government in 1928 published a report of an extensive and lengthy investigation into the use of terracotta on buildings which began with these words: 'An outstanding feature of our investigations has been the relative scarcity of examples of faulty terracotta in service as compared with faulty stonework; the two years' work has indeed made it clear how well the majority of terracottas withstand the atmosphere of our towns.' While these words remain sustainable, a further sixty years' weathering have demonstrated very clearly that the resistance of terracotta to decay is not indefinite and is dependent upon the quality of the material and the quality of detailing both of which can give rise to serious failures.

Figure 4.8 The 'lavatory' brick – a white salt-glazed finish largely self-cleansing, durable and widely used for its hygienic and reflective properties. Selected well-fired bricks were refired after the application of the glaze which is carried from the weather face to a pointing depth on the return faces.

Even with the correct choice of materials for manufacture a series of faults could arise even before the terracotta block left the works. The manufacturing process involves extensive clay preparation in which the source material, derived from crushed shales and native clays known as marls or bore clay, may be combined with china clay (kaolin deposits from weathered granite) and grog (ground vitrified or semi-vitrified material derived from wasters or deliberately made in an earlier firing). The natural clays would be weathered and remixed before being allowed to sour, i.e. stand, sometimes for months to achieve maximum workability. Some were the product of slurries which had been allowed to pass through settlement lagoons and in all cases the plastic material entered a stage of thorough mixing when larger components and foreign materials were sieved out. This process allowed blending to take place. It also provided for the oxidation of iron compounds to ferric oxides and allowed the introduction of stains and fluxes. Some manufacturers would retain the blended material as a putty for months or even years prior to moulding. Faulty preparation might allow the survival of unwanted materials such as nodules of calcite, flint or iron pyrites, all of which would have a significant deleterious effect on the finished product.

The manufacturing process usually involves the preparation of moulds of Plaster of Paris. These are the inverse of the required design and are made over-size to allow for shrinkage

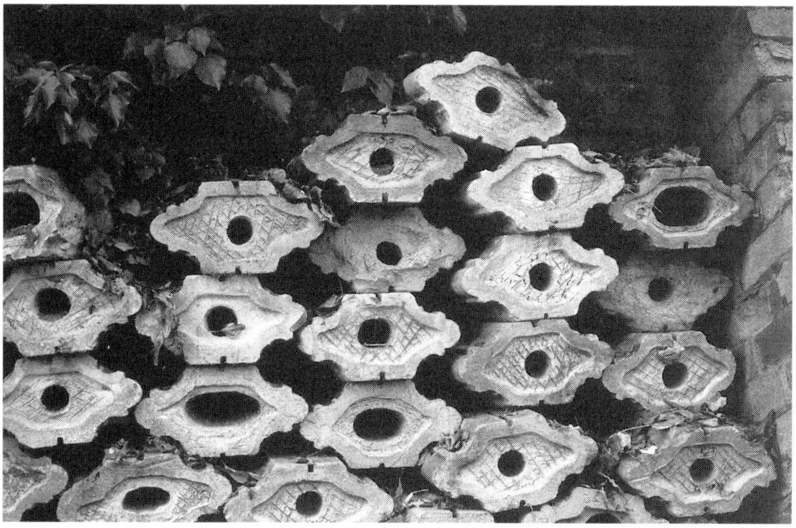

Figure 4.9 Terracotta replacement mullions and similar components held in store at a historic mansion. The hollow construction of large terracotta units is clearly visible.

in drying and firing. Depending upon the type of clays involved, the negative mould will be some 8–9% larger than the intended finished product. The shrinkage allowance might also differ with direction due to gravitational and mould effects. Limitation of the length of moulds was usually 0.5 m and the thickness of material was not allowed to exceed 50 mm except in the case of moulded bricks, where tolerances were greater. The design of both mould and finished product were dependent on a number of technicalities. Moulds might be made in multiple pieces and the shape of the final unit might have required compartmentalizing by the construction of thin webs of clay within the core. Loss of moisture into the plaster moulds over a period of two or three days caused the clay body to shrink away from the mould and it could then be removed with care to a drying chamber. Complex shapes might be divided into smaller units to give a stable face for placing during drying and firing. Poor handling at this stage could cause deformation, while poor mixing could cause uneven distribution of clays resulting in colour striation and warping in drying and firing. Poor compaction within the mould could produce a weak body in the terracotta.

On removal from the mould hand fettling would be undertaken. This involved cleaning the detailing and remedying any failures such as air holes or pockets which had not been filled fully by the clay. Joint marks in complex moulds would be smoothed off and sharp arrises rounded to stand up to firing. Where surface polishing was indulged in, the effect was to produce a higher surface density improving the fireskin. If necessary additional shapes or formations could be made on the terracotta at this stage. Dowel holes might be cut, drainage holes, grooves and tongues being formed. In this way a standard mould could be made to serve a number of variants. The degree of working would inevitably vary from place to place and, therefore, the compaction of fine particles would in turn vary. The process of water absorption via the plaster can leave particles within the surface zone in a very moderately compacted condition or even in a wavey arrangement reflecting their deposition in pouring the wet mix. Only skilled and careful burnishing can overcome this problem to leave a strong and dense fireskin on partial vitrification in the kiln. Burnishing also removes any deposition of salts which can sometimes occur on the surface as a result of the drying process and this is carefully wiped off before burnishing – a process sometimes known as leathering from the nature of the tool used.

Some types of terracotta (faience) are glazed prior to kilning. The application of a creamy liquid by dipping or by brush to the body itself disturbs the surface arrangement of the clay particles but the resultant reduction in density in this case does no harm because the

Figure 4.10 Early twentieth century pressed bricks used to form decorative panels in walling. These terracotta units are of the same clay and firing as the adjoining face brickwork. One brick tile is failing under frost action.

glaze itself forms a fireskin deeply bedded on to the substrate. The glaze will consist of minerals designed to pigment the finish and fluxes together with silica (water-ground calcined flint) to achieve the vitrification.

At this dried stage of fettling the piece may be adapted to a particular use, perhaps by the introduction of further webs or grooving of the existing webs so that they may be knocked out after firing. When the piece is moved into the kiln where it will remain for up to fourteen days the initial heating will be very slow, initially driving off residual water still present within the pores, followed by the expulsion of water of crystallization at about 500°C. As the internal structure of the crystals and the calcite break down any organic matter burns out, as does sulphur in pyrites. At these temperatures the metallic ions begin to oxidize – particularly the ferrous materials which are converted to stable ferric compounds colouring the body of the material. Finally, the semi-vitrification process is effected at temperatures in between 900°C and 1400°C dependent upon the composition of the material. As with the firing of bricks, the final shrinkage occurs at this point and the body, now having become soft (pyroplastic) will deform under pressure such as superimposed loading or excessive temperatures. Historically, firing took place in a pottery-type kiln – a down-draught operation, which gave the most even temperature throughout since the flue gases at the point of initial combustion are outside the kiln or behind a bag wall. Modern electrical and gas-fired kilns achieve greater precision, being in effect ovens.

This description of the processes involved in making historic terracotta summarizes previous manufacture and in many countries these methods are still in use. Variations in technique in advanced contemporary practice have been derived from the refinements of ceramic technology, and the modern works is a far remove from the historic factory.

The finished product may have suffered from a number of defects. Those which caused it to be rejected at the works include deformation during firing, perhaps due to pressure in the kiln or to internal stressing. Cracking may have occurred if the clay was unevenly wet on pressing or unevenly mixed or in consequence of over-hasty firing, or variation in fuel. Colour can be spoiled by uneven distribution of minerals or the presence of pyrites which can produce an excessive localized concentration of ferric compounds. Other iron concentrations in clays may be bands of ironstone (hematite). Nodules in fireclays are often mixed iron and calcium carbonates.

These iron spots will be visually damaging rather than damaging in performance. Arrises which have been left too sharp may develop hairline cracking through tensions in drying which become apparent in the cooling fireskin and glazes may fail in a number of ways, the principal being crazing or blistering. These faults can all be identified in the product which may be rejected at works or on site, but in historic buildings defective items can sometimes be identified as having failed due to long latent imperfections in manufacture, in which case the system of fixing is not to be blamed. Under-burning, leaving a product which is too soft, or failure to provide adequate compaction in the mould, leaving a product which will delaminate, are both faults which are not readily identified prior to fixing and many terracotta failures are attributable to these manufacturing defects. Careful examination will determine whether a unit has failed due to inherent faults of this type.

Other failures arise simply due to use. Chimney pots which may be adequate for several hundred years unused may delaminate after only fifty years in use due to sulphur contamination or chimney fires. Acid atmospheric conditions may have a similar effect, and in the coal-smoke pollution of large urban areas have failed for this reason coupled with the damaging effects upon mortars.

4.5 Fixing terracotta

Until the nineteenth century the fixing of terracotta on buildings was very straightforward.

The simplest terracotta – moulded brick – was simply part of the structure and was built in. Elements formed specifically as working components, such as mullions, columns, spandrel pieces and the like, were fixed in position in much the same way, being bedded in the same mortar as was used for the remainder of the structure. Perhaps in the case of superimposed elements – acroteria, balustrades, urns and sculptural ornament, they might be given the benefit of a metallic fixing, in which case a copper dowel might have been used. Wrought iron no doubt served for this same purpose, perhaps bedded in lead. This less-durable method of construction has generally failed due to rust expansion and we are probably unaware of the number of such fixings that once existed. Bronze would have been the only suitable alternative to copper and expansion tolerance would have been essential.

The nineteenth century brought with it new methods of construction. Cast iron frames came into widespread use, sometimes in combination with wrought iron, and even where frames were not used cast iron columns very frequently carried conventional floor systems and external walls. Industrialization also brought acidic air pollution which visibly disfigures and damages natural stones. As a result terracotta came into focus as the durable alternative. To sheathe a building is very different from building-in terracotta components and where the building is fully or partially framed the immediate answer is to devise a system of hooks and metal fastenings to tie the terracotta components back to the main structure even though they may, as a skin, carry their self-weight down the building on to transverse structural members or even to the ground. Hollow terracotta was often filled with concrete and this conveniently locked it into the structure. Unfortunately expansion coefficients differ, although not greatly. Moisture penetration would be minimized by sealing the joints between the terracotta with a durable mortar all too often with a cement content. The problems subsequently to emerge would relate to joint failure, surface breakdown, differential thermal movement and cracking, overloading under pressure, moisture penetration resulting in trapped pockets of water, salts formed during setting of cement and corrosion of the fixings. This catalogue of prime failures could be extended considerably by variations on or combinations of these prime faults and the various techniques of conservation must take account of situations which often relate to several simultaneous weaknesses or causes of failure.

During the nineteenth century the use of glazed brick expanded greatly and the material was bedded in the harder cement mortars which had become available. There was also a major resurgence in the production of glazed wares for building purposes and a number of tiling makers, particularly Doulton of Lambeth in London, came to dominate a vibrant architectural scene in which artistic creativity rose to a zenith by the turn of the century to dwindle again almost to nothing within fifty years. Europe and the northern cities of North America ran a parallel course.

Figure 4.11 High density vitrified terracotta components used to carry fenestration on a mid-twentieth century building.

Figure 4.12 Perhaps the most powerful use of architectural terracotta of all time, The Natural History Museum in London: serried ranks of columns echoing a Gothic cathedral porch (architect, Alfred Waterhouse).

4.6 Tile hanging

This extensive use of fired clay cladding materials was preceded by an interesting, if somewhat localized, use of terracotta for the same purpose in the eighteenth century, often not recognized as such. Tile hanging had long been common on buildings in the south of England as well as in many other parts of the world, usually in isolated areas for particular reasons. Just as on the Welsh, Cornish and Brittany coasts vertical slates on the outer face of exposed walls had been used to combat extreme weather conditions so in the less severe circumstances of rural southern England timber framed buildings panelled out with wattle and daub or with brick noggin could be made more weatherproof by the vertical application of roof tiling.

Timber laths nailed to the structure were sufficient. Warmth, weatherproofing and social prestige were gained and all at modest expense. Early examples of weather-tile simulating brickwork derived from the first quarter of the eighteenth century and the use of such tiling to simulate brickwork came even from the hands of architects as well known as Henry Holland, James Wyatt and Sir John Soane. It was, therefore, not a technique confined to the lower orders architecturally or socially, and whoever it was who had first hit upon the idea of forming a tile with an exposed surface resembling a brick bequeathed to his successors a system by which older buildings could in an instant be made more modish and the established tradition of timber framing could be transformed without structural alteration. Double hung sash windows and a pedimented door set in a façade of those tiles, which by the nineteenth century had become known as 'mathematical', lifted many a high street house from its medieval half-timbered appearance to the height of fashion at minimal cost. (The term 'mathematical' for these particular tiles can be traced to the late eighteenth century. They were made in header and stretcher sizes and a number of specials were made to form corners or gauged voussoirs. Sometimes they were exploited to clad a curved bay in an otherwise straight line brick façade. Sometimes they were glazed, particularly glazed black.)

All forms of brick bonding were imitated but a header bond was particularly favoured. The distinctive feature that makes them deceptively like brickwork in appearance is the pointing. Although the tiles were necessarily fixed back to the principal structural wall they were bedded in lime mortar and pointed up as though they were bricks. This gave them a structural validity in their own right at least to the point where they became a self supporting skin. Mathematical tiles were made in many colours from light yellow to glazed black although the majority were brown and red, occasionally with grey glazed flaring on the surface. Black glazing was achieved by use of a galena paste with fluxes achieving a lead glazed surface with manganese as a colouring agent.

Tile hanging is susceptible to damage due to its fixing. Mathematical tiles were generally built on to a structure rather than hung although both methods were known. Sometimes quoins were built in brick and rendered where detailing called for it. However, embedded battens or laths were likely to decay through woodworm, rot and the failure of nails, so the prudent builder, with longevity in mind, tended to build his tiles against a wall with occasional fixings rather than to rely on hanging. Often the tiles were stopped against a timber end piece or reveal and occasionally timber was used to form quoins in imitation of stone against which the tilework could be built. Despite its many potential defects the mathematical tile has proved durable and permanent.

It gained a short lived resurgence in the middle of this century when extruded tiles made to a slightly different design were extensively used for facing prefabricated buildings, although they were invariably hung rather than bedded on. Both extruded clay and pressed concrete were employed for this purpose on buildings, some of which now qualify for conservation as historic examples of effective systems of building. Some such 'system buildings' in CLASP construction are Listed under Statute.

The techniques of manufacture which have applied to terracotta have traditionally applied to tile-making, which being manufactured on a wider scale than architectural terracotta has been more associated with brickyards and brick works. Because tiles are thinner than bricks they have always required greater care and skill in handling. Firing has, therefore, usually been in intermittent downdraught kilns. Roman tile and pottery very often shows a blue core indicative of a relatively short period of firing during the 'burn-out' period when the available oxygen has been unevenly distributed through the material and at the core is insufficient to oxidize the ferruginous, sulphur and carbon content.

The advantage of the Roman system of tiling is the reduced overlap required by comparison with plain tiling and this advantage is increased where pantiles are employed. Less weight and therefore reduced structural timber have been the prime advantage. However taking account of lap and pitch, plain tiling makes less demands on the terracotta. Its historic popularity may derive more from simplicity of manufacture than anything else. Whereas bricks were dried historically in the open under removable shelters (hacks) while terracotta was dried in controlled internal conditions, tiles fell into either category, sometimes being dealt with like bricks but increasingly being taken into drying sheds where conditions could be more closely controlled. Sometimes these sheds had a small stoke hole at one end, perhaps under a lean-to roof and flue ways would then be carried under the floor in terracotta pipes or brick lined flues to a chimney stack at the opposite end. Such drying sheds would have low eaves and open or partly open sides. Drying sheds achieved a faster throughput and inevitably were attached to permanent sites based on reliable sources of high quality clay.

Floor tiling has been made in a wide range of sizes. The early imports to Britain were from the Low Countries and, therefore, tend to be identifiable along the southern and eastern coasts. In Suffolk they are known as pammets and became an article of traditional local manufacture. There is as yet no reliable system of differentiation as to source.

4.7 Ornamental roof tiles

If the subject of roof tiling appears to be relatively straightforward and mundane, some exceptions must be made: particularly in the arts of China and Japan (see Plates 4.2 and 4.3). There, from early times, mythology and symbolism introduced imaginative and often colourful wares on roofs and nowhere has there been greater zest and imagination in the handling of roof details than in the many temples and palaces of the Far East. Many centuries of Chinese art have left a legacy of glazed sculpture associated with crestings, ridges, terminals, eaves pieces and the simple tiles themselves. In terms of conservation such highly sophisticated glazed wares are worthy of treatment as museum objects and therefore the wisdom of retaining them in situ, open to the effects of wear and weathering must be evaluated. The techniques of conservation that apply are those of the ceramics and pottery conservators which, with present materials and methods, do not stand up to extremes of exposure. In consequence, the natural compromise is to renew the affected piece retaining the original as an exemplar. The ethical arguments for so doing are strong because the original builders themselves recognized that in placing such objects in exposed conditions they would inevitably be replaced; so conservation of the object itself goes hand in hand with replication which in itself is essential to the conservation of the building.

4.8 Glazed wall tiling

The history of glazed wall tiling, which is again virtually inseparable from the history of glazed brick, begins with a curious episode in Mesopotamia. In the ancient site of Uruk in southern Iraq (now Warka) a temple was built around the Ziggurat of Inanna in the third millennium BC. Walls built of small square-sectioned bricks characteristic of late Uruk architecture were coated with a mud plaster into which baked clay cones with coloured heads – red, white and black – were set. These cones were used to form zigzags, triangles, diagonals and other bold repetitive patterns, many of which recur in succeeding periods, forming mosaics predominantly over 4000 years old, at present one of the most inexplicable decorative elements in the whole history of ceramics.

Figure 4.13 Islamic tilework in full maturity – fifteenth century cladding of a tomb in the Shah-i-Zindeh, Samarkand, Uzbekistan. While some multi-coloured tiles are employed the pattern is built up essentially of individually shaped and single coloured units.

Figure 4.14 Tomb of Yahya bin Kassim, Mosul: a detail of the fourteenth century decorative panelling. Carved brick is locked into a geometric web of glazed terracotta ribs.

With our very incomplete knowledge of antiquity it has to be recognized that other examples of mosaic and tilework used as wall decoration may intervene in the period before the creation of rich glazed brick wall decorations of Babylon in the first millennium BC. Those moulded and glazed bricks were successors to the stone bas relief carvings of earlier civilizations but they gave the sculptors and their patrons the one important advantage of intrinsic colouring. The great gates of the city and the wall panels have been reconstructed, placing the salvaged fragments in context and it is clear that by the time of Nebuchadnezzar II (604–562 BC) the art of glazed wall tiling had reached virtual perfection. In the sixth century BC Darius I resuscitated the city of Susa in present day Iran, which lies a little to the east of the Tigris–Euphrates basin. His work is rich in figurative decoration. The friezes of arches illustrate in graphic detail military costume of the time and equally graphically the expertise of the potters who created it.

The great upsurge in the use of faience as a wall decoration, however, lies with the Islamic period in the later medieval phase. Baghdad and Raqqa were probably the original prime centres from the eighth to the tenth centuries and some early surviving tilework of this period, notably wares exported to Qairouan in North Africa, illustrate that glazed wall tiling was already a highly developed art – but it was to the late medieval period that on present evidence the greatest achievements must be attributed. Several techniques are involved, having been predominant in different areas or at different times and all have continued more or less without significant break to the present day.

The art of tile mosaic found in the Timurid empire centred in southern central Asia in the fourteenth century and was carried east into Moghul India and west to Morocco where

Figure 4.15 Cut tile mosaic: Fez. Replacement tesserae cut by hand adze by contemporary craftsmen.

present craftsmanship is quite the equal of historic work. The process is extremely laborious, consisting of cutting out elaborate shapes from tiles fired on a relatively soft, usually yellow biscuit subsequently glazed with the required colour. The result is a flat and brilliantly coloured piece rarely more than 50 mm in any one direction and often merely 10–20 mm across, with minutely chipped edges. The work is carried out on a small anvil with a chisel sharp hammer. Rapid strokes eat away the tile until the desired shape is obtained but the rate of loss can be high. The technique is regarded as worthwhile because the finished surface is flat and the colour intense and pure having been fired at the temperature which best suits that particular pigment and glaze. Workmen carrying out designs in this technique are contemptuous of the multi-coloured, moulded mosaic pieces used by others, with their rounded edges, although the finished effect of these can be equally handsome. The secret of success in laying such mosaics is in the accuracy of the alignment of the surface and in very careful uniformly flat pointing in a sufficiently hard and durable but compatible lime mortar. It is a technique unsuited to severe weather conditions.

It was inevitable that such laborious and inflexible techniques as nibbling tiles to achieve the right profile would give way to moulding and modelling, particularly in the architecture of Islam which called for the extreme complexities of geometric shape in the form of muqarnas.

These arts flourished as far apart as India and Morocco in the late medieval period and a development of this technique involved the glazing of delicately carved brick. These bricks were themselves sometimes cut by the mosaic-making technique into elaborate geometric shapes. At other times they would be moulded and fired to a consistency suitable for carving. The complex low relief patterns which then resulted might then be highlighted by a repeat firing in which a glaze had been imparted to raised or other chosen areas. Thus a tile with a red body carved in bas relief might acquire a filigree pattern of ultramarine or cerulean glaze on its higher parts only. These techniques were particularly relished in Southern Central Asia – present day Afghanistan, Iran, Uzbekistan and Iraq – but they subsequently gave way to the use of patterned tiling in which multi-coloured designs were executed on tiles that were regularly shaped although not always rectangular. Sumptuous friezes and immense wall-covering patterns familiar from the fourteenth century were produced at centres such as Meshed, Rayy, Tabriz and Iznik. In emulation of carved stone and stucco calligraphy many of these tiles were modelled to heighten the patterns and emphasize the messages by highlights on the glaze. European potters made convincing imitations of such tiles in the

Figure 4.16 Natanz, Iran. Fifteenth century tile mosaic in a muqarnasat head of an entrance bay. The glazed tiles are formed to the required shape and embedded in a render carried on armatures.

nineteenth century while traditional manufacture continued, evolving in Iran into the distinctive Qajar style. Many close replicas are now made for repair and for the general market. As with Far Eastern ceramics, Italian maiolica and Dutch wall tiles, these products have become valued as art works in their own right and in consequence are prised off buildings. Their replacement by new materials, however good, invokes ethical arguments which have focused on restorations such as those surrounding the immense eighth century mosaics of the Great Mosque of Damascus and the more recent massive reconstruction of the great brick walls in the Timurid Mosque of Bibi Khanum at Samarkand.

European use of glazed tiling has been much less extensive and much later. Dutch patterns provided rectangular tin-glazed patterned tilework for interior use from the seventeenth century, and the most extensive use of glazed tile and brick subsequently has been for the reflectance in light wells and for cleanliness and sanitary purposes (Fig. 4.8), although there was also a late nineteenth century florescence of glazed decorative tiling.

4.9 Failures in tile and terracotta

The ceramics produced for facing buildings vary widely in quality and technique. Porcelain is rarely used, so that almost all tilework can be classified as being on an earthenware body Consequently detachment of the glazed surface from the body is one principal area of failure and concern. Physical damage and abrasion on the outer surface can be one cause. In hot climates salts drawn up through the structure and drawn into the porous body will detach glazed surfaces by destroying the integrity of the body at the junction of the two materials, the failure beginning at the edge of the glaze or along lines of fracture.

A problem perplexing potters throughout the ages has been the mingling of colours at the stage when the glaze becomes molten. A technique known to Spanish potters as *cuerda seca* (dry cord) employed lines of manganese pigment to separate colours, creating a dark line of delineation. The firing would be carried out at a compromise level of heat. Some colours would, therefore, be fired at less than their optimum temperature with a loss of intensity or quality and sometimes with imperfections of surface or even solubility in the glaze which could lead to its breakdown under weathering.

The conservator will be faced with two categories of failure and a clear difference in remedy. If the tile itself has failed due to a manufacturing technique or its inability to withstand the circumstance of its exposure its removal and possible replacement is the only

Figure 4.17 Failure of a late nineteenth century tile due to internal water pressure and crystallization with, for comparison, an unaffected tile placed higher in the same wall.

course of action. Its conservation as an object then becomes a separate question. If the failure is circumstantial and the fault lies in the structure itself, the conservator can look to the cause and having produced a remedy can leave the wall surfacing in situ.

Floors and walls are susceptible of structural movement and floors, in addition, are subject to wear and live loading. Any significant difference in load bearing capacity between the floor tile and its substrate and any significant difference in movement characteristics between a backing material and the surface material will be taken up in the bedding or transmitted to the tile. Where the background is stronger than the bedding, there is a likelihood of failure in the tile. It follows that a bonding material, softer than the substrate and the tile is desirable. Careful evaluation of the relative strengths of tile and substrate prior to or on first sign of failure may suggest the precautionary measure of lifting and suitably rebedding the tile.

The most common problem is the transmission of water, perhaps containing soluble salts, from the background material into the tile. The damaging effects can be several: the wet tile is weaker than a dry one because of the hydraulic pressures within the surface pores under imposed load and because of dilatation as moisture is drawn into the material. These transmitted pressures can break the internal bonding causing the surface to decrepitate. Further, evaporation from the tile can break down the surface material by the formation of saline crystals which exert internal pressure within the pores of the material, breaking down the bonds between the particles. This is

Figure 4.18 An Edwardian (early twentieth century) hopper head in terracotta. The long-term survival of this detail depends upon the regular maintenance of rainwater goods.

particularly damaging at the immediate edge of a glazed surface. Where the surface is a dense fireskin or is highly vitrified, the emergence of water will occur at the junction of the body and the vitrification if this is exposed. Deterioration is then progressive from the edge inwards, the softer underbody failing before the vitrified surface and removing support.

If the fireskin or glazed surface of the tile is crazed, the macrocracks in the surface which constitute the crazing permit the evaporation of water with possible consequential failure on the lines of the crazing. A deposited bloom may be little more than calcium carbonate leached out of the packing mortar but soil water is likely to contain sulphates, chlorides, nitrites and nitrates and other potentially damaging reagents. Ferrous compounds and others reacting with water may cause staining which can show up cracks in glazing and, getting under the glaze, can prove extremely difficult or impossible to eradicate.

Roof tiles are subject to more active decay and hazards than most other building components. They are always at risk of the failure of the bearing member, usually timber, and are subject to such fluctuating air pressures that they may lift and fall quite violently. Microbiological organisms are, perhaps, a more potent hazard on roof tiles than on floors and walls although they can be exceedingly damaging on both walls and floors, particularly by penetrating the adhesion layer, changing the localized pH value and, by calcium absorption, breaking down the bond. Roof tiles are subject to the greatest regular extremes of temperature change and to cycles of wetting and drying. The adequacy of their fixings is a key to their survival and many roof tiles are at risk of failures caused by the ferrous expansion of their pegs or the nails that hold the laths. Oak pegs are very durable if sapwood is avoided, but pegs of cheaper timbers are not durable in this usage. As a crude rule it may be said that roofs are more likely to fail due to the inadequacy of their substructure than because of intrinsic failure in the body of the tile. This is less true of wall tiles, which generally and wisely are bedded in position.

The failure of roof tiles due to frost damage frequently occurs on the underside where a capillary film of water feeds the tile undersurface to the point of saturation. Hence the camber on plain tiles and the drain grooves on the underside of large format clay tiles.

5

Behaviour of fired clay materials

5.1 Porosity

The size and nature of the pores in burnt clays depends ultimately upon the raw material and the temperature of its burning. Porcelain is the most fully vitrified of these materials and such pores as remain after its firing are few, small and effectively discontinuous, due to the extreme resistance to moisture movement of the very small intervening spaces. At near melt the surface tensions of the semi-liquefied silicates tend to reduce the surface area of the pore itself, thus driving out the gases through the continuous pore structure, leaving only occasional points at which the initial gases are trapped and entrained in near-spherical compartments. Thus, the material becomes virtually waterproof.

While porcelain is little used in building, its characteristics are those of the glazed surfaces of other fired materials used in historic building – glazed brick, glazed tile and even vitreous-enamelled iron. Some forms of engineering brick so closely approach the condition of being fully vitrified that they also are effectively waterproof, as most of their voids are sealed in this way. However, all fired earthenware retains the continuity of its initial pore structure but other factors contribute to internal change. Products containing a high proportion of naturally included or added sand tend to be very porous and permeable. Quartz sand on firing expands by inversion of the crystal. The inversion is sudden and occurs at a specific temperature creating a space or crack round each grain. This is a major source of continuous porosity. During high temperature firing the liquefied silicate can, with other fluxes, act as a lubricant between particles allowing them to move together and surface tension provides forces which causes the particles to rearrange themselves and change position. At most firing temperatures the silicates remain very viscous. Nevertheless, at a specific temperature significant shrinkage occurs as pores are reduced. The more fully vitrified the brick the greater the shrinkage and, therefore, the greater the change in the nature of the porosity. In the kiln the effects of loading upon the lower layers determines the height to which bricks can be stacked. Up to two and a half metres is usual. Greater heights and therefore higher superimposed loads cause physical deformation in the bottom layers at normal firing temperatures of 1000°C (and upwards).

Clays vary in their viscosity relationships. High lime clays are very rigid up to certain temperatures and bricks made from them can be set higher in the kiln. If the critical temperature is exceeded sudden fluidity arises and complete collapse occurs. Other clays have a much wider range of critical temperatures.

It will be apparent immediately that, at the lowest temperatures of firing (*c.*850°C), the structure of the brick will have changed relatively little from the arrangement of particles in the unfired material. The stability given to lightly baked earths is a consequence of the 'gluing' effect of the initial melt, although the arrangement of particles hardly changes: hence there is minimal shrinkage. As temperatures increase, the pores become more globular as the smallest spaces are filled with liquefied silica. At higher temperatures the particles tend to rearrange themselves due to capillary forces and recrystallization. Changes of this type and the transformations which similarly occur in hematite make it possible to determine the temperatures at which historic brick was fired.

A complicating factor is the creation of voids by the burning out of entrained organic material, some of which may have been introduced deliberately as powdered fuel and some of which occurs naturally. Many clays contain hydrocarbons or fossilized plant material – particularly those coming from carbonaceous measures. These have proved attractive to the brick manufacturer because of a reduced requirement for fuel in burning. Some clays (Oxford clays) are advantageous as they contain significant amounts of fossil oils which render the clays partially self-burning, the fuel being ideally situated in the heart of the material and, therefore, producing heat exactly where it is wanted.

A further factor in porosity is the formation of minute internal fractures during shrinkage, the material literally pulling itself apart to form minor fissures throughout the structure, and these minor fissures are themselves a part-characteristic of the porosity. However, when fired at temperatures from 800°C to 1200°C to produce burned brick and tile, the silicates, whether originating as one of the clays or as a mineral impurity, will soften and on cooling will take on some part of the characteristics of a super-cooled liquid. Its molecules do not rearrange themselves into regular lattice patterns but are linked in a semi-random arrangement with high internal strengths and stresses at the surfaces.

5.2 Mortar

Masonry in most historic structures contains one further porous element – mortar. Normally less than 20% by volume this may exceed 50% in some historic (Roman and Byzantine) brickwork. Mortars may be made of lime, hydraulic lime and other cements, gypsum or earth. On occasion bitumen has been used. Ignoring the latter, there are important differences between the types of mortar.

There is no precise boundary between the chemistry of cements and that of the hydraulically setting lime mortars. The difference between a lime mortar and a cement mortar must be thought of as a difference in degree of hydraulic set. At the one extreme the purest of lime with the purest of sands and clean rain water will depend for its strength upon the slowly acquired formation of crystals of calcium carbonate. At the other extreme, in a strong cement, the formation of the complex interlocking crystals based on hydrated calcium aluminate and calcium silicates becomes a rigid interlocking crystalline structure of far greater strength and much more rapid set than is achieved by the absorption of carbon dioxide. The initial part of this reaction is virtually instant although consolidation is normally a matter of hours. An instant set is described in discussion of cements in Chapter 12.

There is one further important difference between the behaviour of mortars at the lime-rich end of the scale and cements. The crystals of cement, while being stronger than those of calcite, lose their strength permanently if fractured. Vibration or strain can weaken a mortar by partial fracturing of the crystals and, unless a further cementing material is introduced, that weakening or failure will remain. In the case of calcium carbonate the presence of air and water at normal ambient temperatures can produce a slow regrowth of some of the crystalline structure, so that a lime mortar can, to a degree, be self-repairing. This regrowth will take place on the axial alignment of the crystals and hence is described as epitaxial growth. Consequently lime mortars in which a slow plastic deformation has taken place may develop a modest strength in their deformed condition as crystal 'repair' occurs. A maximum recovery of 10% of the original strength might occur. This self-repairing aspect of lime mortars, which can be stimulated by the introduction of calcium hydroxide in solution (lime water), becomes important in the conservation of lime-rich earths. Some mortars are of this type.

To the builder, the most important qualities of mortars have been the availability, workability and strength achieved, coupled with rapidity of 'set'. In gypsum the 'set' will occur within minutes and be virtually complete within an hour. With cements the set usually occurs in times measurable between four and thirty hours, but strengthening continues over a period of weeks. In limes the initial strength depends upon drying, followed by a period of absorption of carbon dioxide which can run into months and in massive dry structures can take years, or even centuries. Mortars contain sand, whose particle size is much greater than the particle size of the powders forming the cement. The crystals formed are very small by comparison with the size of the sands and they form a matrix around these larger elements. In producing concretes the further step is taken of introducing larger components – an incremental step in enlargement.

5.3 Water

Water is always required, both for the process of crystallization, where it is physically incorporated in the structure, and as a medium of transport. Initially, also, it acts as a lubricant, enabling particles to settle into position. The process, therefore, always involves the introduction of more water than is required chemically and this water must escape from the setting mortar. In consequence, there will be routes through the material capable of carrying water vapour. Water vapour will be deposited in these interstices in conditions of temperature and vapour pressure lower than in unrestricted spaces. It follows from the introduction of large volumes of sand that in mortars a greater proportion of larger particles are normally present than are present in earths and burned earths. The general conclusion to be drawn is that the nature of the voids in mortars differs from that in the adjoining brick or earth. Larger original particles mean larger voids but in richer mixes the high percentage of crystalline material produced means that within the interlocked crystals many extremely small voids exist. The nature of the pores left in a mortar therefore may differ considerably from those in earths and clays, burned or unburned. In a mortar there is a high proportion of extremely small spaces lying between crystals which have set in an interlocked pattern filling the spaces between the larger sand particles as well as coating the surfaces. In clays there is a very high proportion of even smaller spaces within the crystals themselves. Microscopically, therefore, there is little comparison between the voids in mortars, in burned earths and in unburned earths beyond a broad similarity of size range. Internally, the voids all have very different surface characteristics and therefore the behaviour of water within these voids varies and is a prime factor in determining the life and performance of porous building materials.

Chemically, water is extraordinary for its activity. In normal ambient conditions water is an active liquid reacting continuously with itself. The molecules conveniently classified in text books as H_2O form, break up and reform continually in a three-dimensional version of a process that has been likened to a reel on a dance floor where partners are continually changing. In reality water is in a continual process of breakdown and reassembly with hydroxyl ions (-OH) and hydrogen (-H) in continual ambivalence and disassembly. Some assembled H_2O moves off in a gaseous or vaporous condition due to molecular activity even at low temperature. The assembled molecule which has taken up energy in recombining then moves off, depriving the liquid of energy and thereby cooling it. At temperatures close to solidification (4°C) a proportion of the molecules in a less energetic condition move into a more regular and stable arrangement occupying a slightly larger space. In consequence the water density is reduced and this process continues into the stage of solidification (0°C). Due to this reduced density, liquid water between 4° and 0° rises above warmer water. There it forms a layer which can solidify as ice at 0°C at this same reduced density.

The behaviour of water in the pores of inert masonry is governed by two factors: the nature of the surfaces of the pores in terms of electrical attraction or repulsion of the hydroxyl ion and physical constraints due to the size and nature of the pores. As the presence of ions or molecules within the pores or in solution will change the behaviour of water, the circumstance is best considered initially in the absence of such complication, i.e. in deionized water. In visualizing the behaviour of porous masonry in relation to water it may be helpful to understand the several forces which govern its behaviour.

There is a chemical bond between hydrogen and oxygen which produces the hydroxyl ion – the relationship in which the hydrogen atom makes the single link of which it is capable, with an oxygen atom. The oxygen atom, being capable of two such links, seeks to form another bond with another hydrogen atom at an angular relationship of 105° in the 'spherical' surface. The consequence is that the combined molecule is imbalanced with both hydrogen atoms being held on one side of the oxygen atom. For reasons of the probability of the position of electrons, the molecule then has a weak electrical polarity with a positive field between the two hydrogen atoms and an equivalent negative field on the opposite side of the molecule. The structure is on the verge of instability and frequently breaks up, one hydrogen atom moving away first.

The hydroxyl ion and the hydrogen atom are immediately able to form one new bond each. If these bonds again form H_2O the same breakdown occurs. However, where a bond is with an ion – (an available radical) – a stable relationship can be effected; i.e. a compound is formed.

The behaviour of water in the conditions in which it affects brickwork is governed primarily by these weak electrical polarities and also by the stronger forces associated with the displacement of sub-atomic particles which make chemical bonds. The polarities control the location of the water in the porous structures of earths or brick. These polarities create relatively weak electrical fields. The result is that a natural spatial arrangement will occur in which, despite the constant internal change, the electrical forces tend towards regularity in the arrangement of the molecules. This regularity lies at the heart of the formation of crystals and the extreme complexity of the crystalline structures of water – as expressed in the snowflake – are a consequence of the incessantly irregular behaviour of the water molecule in breakdown and reassembly. The expansion which occurs as water solidifies, with a consequent reduction in available energy, is the most significant factor in its behaviour and has major consequences for the materials with which it is associated. Where water fills the pores of a rigid material the disruption resulting from such expansion can overcome the physical strength of the crystals and glass forming the ceramic bond. The expansion may be translated into hydraulic pressures within the pores when surfaces are sealed with ice and liquid water continues to expand inside the body. The consequent disruption of physical bonds by fracturing of the crystals may, in the case of burned brick or tile, be catastrophic for the material. In the case of unburned material reassembly can occur but an input of energy may be needed to re-establish the position of the particles. This may be achieved by the simple application of return forces such as gravity, recompressing material which has been forced apart or the cohesive effect of the surface tension of water held in the pores. Usually, however, there is no such force available to achieve recompression. The result is a permanent change in the structure of an unbonded material such as earth. Hence the need to tamp down a clay surface, often wetting it to restore its density.

The movement of water into microscopic spaces is governed by the phenomenon which causes the water molecule to be attracted to the surfaces to integrate with them. These surfaces are described as hydrophilic or water-attracting. At the junction of a hydrophilic surface and a liquid the molecules are drawn towards the surface and the surface tension of the liquid extends this attraction to the remainder of the liquid surface. The level of attraction between a hydrophilic surface and water determines the degree of curvature at their junction. This curvature is defined as a meniscus. The closer the surfaces, and the more continuous their enclosure, the more effective the attraction. This attraction will counter the force of gravity and cause water to rise in, for instance, a glass tube. As common experience indicates, the finer the tube the greater the rise. In porous masonry where the spaces are very small the effect is sufficient to draw water against the force of gravity to a height of as much as three metres.

This 'capillary' effect is reversed where the surface is repellent to water. Such surfaces are known as hydrophobic and are commonly associated with materials such as wax.

Two further mechanisms affect the movement of water in porous structures. The first is the relatively infrequent phenomenon of a sustained and consistent electrostatic field within the masonry. Such a field can be created deliberately or it can occur in circumstances where a metal has been introduced. Where such a polarity does occur the electrical imbalance across the water molecule will cause it to move in a direction determined by the field and, in consequence, there can be a slow rate of travel in accordance with the dictates of the field. This phenomenon is used in electrostatic damp-proofing, which is better described as damp inhibition. These phenomena are varied in degree by the presence of anions and cations of other elements in all water associated with built structures.

The more complex phenomenon is osmosis. The essential explanation of osmotic pressure in a liquid is the equivalent of the explanation of vapour pressure in a gas. In a situation where molecules can move relatively freely through a gas or a liquid, those of any particular type tend to achieve an even distribution simply because the mutual bombardment creates a higher pressure in the areas of greatest concentration and, therefore, a higher statistical probability of the molecules being projected outwards until an even distribution pattern is produced. The phenomenon may be reduced to a basic concept of pressures relating to each substance in solution. Where a membrane divides the solute, any substance which can pass the membrane will produce a uniform pressure on each side, that is to say, a state of equal bombardment. Where a substance is sufficiently large to pass the membrane only with difficulty the bombardment will be greater on the side of greater concentration and this is registered as a difference in pressure. In living tissue this pressure will lift solutions or create expansive forces. Osmosis is fundamental to plant life and to the disruptive effects of its pressures in the growth of roots in built structures. It is important in the mechanisms of damage to bricks and earths, but the molecular effects of solutes and solutions have wider consequences, particularly in relation to the distribution of salts in porous materials.

The phenomenon affects larger molecules or radicals – ions – the components of a solute having an available linkage which tends to move into even distribution throughout a continuous body of water. Therefore, in a material in which there is a concentration of ions water provides a medium which allows them to distribute themselves throughout the general body. If they are concentrated by the removal of water, conditions suitable for crystal development within the pore structure arise. This, in turn, can induce internal pressures in the porous masonry.

Where an ion is absorbed into such a crystal it becomes ineffective in terms of molecular bombardment. It is locked up. Thus a crystal in formation produces around it a lack of the relevant ions and osmotic pressure is set up which causes other ions to move in to fill the void. They in turn are absorbed, whereupon the process continues by replenishment. This is the process by which salt deposition occurs in masonry.

The capillary effect of water in masonry causes the liquid to seek out and be retained most tenaciously in the smallest spaces. However, complete water molecules in their energetic molecular movement continually break free of the liquid mass and pass into atmosphere within the pores and beyond. They may meet other liquid and be absorbed, or they may continue in vapour form, rebounding through the voids in the material, ultimately finding a lodgement in the very smallest spaces, or escaping into free atmosphere. This continual movement by evaporation or transfer into vapour, coupled with the continual capillary attraction, is a prime feature in the dynamics of the behaviour of water in porous structures and, in consequence, causes a difference in the lodgement of water depending whether the material is being wetted or dried. In a dry condition the introduction of water allows a slow percolation by capillary attraction, filling each pore or void successively. Thus a fully saturated condition can occur at the point of contact, while a virtually dry condition exists only a very short distance away. The gradual percolation of the water is preceded by a zone in which the water vapour pressure increases. At the forward edge of the penetrating liquid there is a zone in which the expansion caused by water taking up space between the particles moves forward as a frontier, leaving behind it a zone of swelling and so causing leverage upon the crystals at the point of transition and carrying forward a plane of stress which may inflict microscopic damage upon the adhesion mechanisms. The crystals themselves may weaken upon wetting. This damage can then be accelerated by water penetration into new cracks in material which was previously impervious, leaching out ions and dissolving away part of those surfaces. The wetting process can, therefore, be damaging to a porous structure.

Conversely, drying can, if sufficiently slow, be less stressful to the material. It proceeds by the emergence of water as a vapour. In a rigid porous structure the passage of water molecules through the pores will ensure an even gradation of water. In the drying process a vapour pressure gradient will form across the material, being at its lowest where the water molecules are free to escape into the atmosphere. An even vapour pressure gradient will form through the material as the water molecules migrate to the drier zone, ultimately to escape. This migration is entirely a consequence of statistical probabilities at molecular level.

It is generally true that greater internal stress can be set up in masonry by the process of wetting as opposed to drying due to the steeper gradients of water concentration which can occur at the wetting frontier. This may, however, not be true of the circumstances of rapid or uneven drying where similar frontiers of wetness are formed and it is a generalization which takes no account of the differences which are consequent upon the ability of mortars or brickwork to take up the strains of expansion and contraction. In a drying structure the liquid will be retained longest in the smallest of the pores, between which there will be a continuous pattern of larger voids. Consequently, there will be a much less precise frontier between wetness and dryness and, hence, much less localized stress on the rigid materials binding the structure together.

Damage at a microscopic level may be the prime consequence of expansion and contraction. When a material expands and it is confined or pressurized in one or two dimensions, it will deform in order to take up expansion in the second or third dimensions of the space available. In a rigid porous structure under load, wetting will cause an expansion, resisted by the loading, and the material, therefore, expands differentially – less in the direction of the applied load and more transversely. On drying, the equivalent contraction is taken up conversely. This means that each group of molecules is forced to distort with a bias due to restraint in the direction of loading. The very fact that the material is constrained and is, therefore, forced to distort during expansion may induce damage due to the stressing of the crystals or the glassy structures which cause it to cohere. Repetition of the process increases each successive strain perhaps to the breaking point of each crystal. The rate of repetition of the wetting–drying cycles therefore becomes a significant factor in the life of a brick or tile.

In summary, it may be said that the porosity of all brickwork is primarily a consequence of the involvement of water in the forming of mortars, earths and fired brick. Some water may be retained but most will escape leaving interconnected voids of micro-dimensions. The internal characteristics of these voids or pores are determined by the generally angular shape of particles of earths and the interlocking angularity of crystals. In burned brick these voids have been modified by the melting of the alumino-silicates and fluxes. This porosity is a distinguishing characteristic of the earths, fired clays and mortars used in building.

Finally, to consider micropores and the behaviour of water within them: all molecular volumes are, in terms of particle physics, largely space, but the perimeters defined by atomic structures (i.e. paths of electrons) are denied to other molecules, so the spaces occupied by molecular boundaries define the micropores within which water and its solutes can move. These pores, at their smallest, approach atomic dimensions. Consequently larger molecules, particularly of complex organic materials, which may comprise hundreds of thousands of atoms, are sieved out by the mesh of smaller spaces through which water itself can move. The nature of the solvent/solute relationship is thereby changed and carriage of solids by liquids is affected. The oxygen (-O) radical and the hydroxyl radical (-OH) of water themselves dissociate and can, therefore, penetrate molecular structures more intimately, their free behaviour being constrained by the surfaces of the micropores. Whereas, in a free state, interaction with other water molecules gives water a behaviour recognizable as temperature, pressure, condensation, viscosity, density and so on, in the constrained conditions of the micropore the performance of the fluid molecule and its radical is modified by interaction with the adjacent walls, and consequently its characteristic behaviour is changed in ways important to the conservator.

One directly visible modification is the locking up of water as a component of a crystal structure where it may be effectively fixed into one position in a molecule and removed from the interactive body of water until conditions change. There it becomes structural – water of crystallization. Another effect is constraint on the ability of the water to move through a continuous pore structure, its vibrations interacting with the rigid molecules of the pore walls to reduce its free movement. This important effect is described in Poiseuille's Law which shows that progressive reduction in pore size is reflected in resistance to fluid movement increasing resistance to the power of four inversely as the pore size diminishes. The extreme resistance of smectite clays to the passage of water is a consequence of this phenomenon. This resistance is not to be equated with waterproofing, as builders have sometimes mistakenly presumed. Clay is not waterproof but the extreme smallness of pore-size means that the rate of flow of water is exceedingly slow.

A further modification is in the behaviour of water vapour (and other gases) in contact with surfaces. This phenomenon becomes increasingly evident as the ratio between pore wall and volume changes with progressive diminution in pore size. These behavioural changes are known as Van der Waals effects, and they are particular cases of gaseous behaviour described by the Van der Waals equations which take account of molecular sizes and dipole effects of neighbouring molecules. Water vapour in micropores has characteristics apparently at variance with its performance in free air, one particular effect being condensation in conditions which would not cause vapour to condense outside the pore. The observable effect is a reduction in RH value within small pores. The effect is increasingly apparent with reduction in apparent pore diameter. The 'permanent' uptake of moisture in newly baked clays with consequent expansion is an associated phenomenon of peripheral interest in conservation.

More directly important are the pore characteristics in which the electrical effects of surface radicals and free radicals or ions cause the lodgement of metallic ions, particularly sodium, aluminium, potassium, calcium, iron and magnesium in the pore surfaces in such a way as to influence molecular structure. Ions moving into and out of solution on the perimeters of these spaces alter the structures of clays and their associated secondary minerals. These leaching effects are generally associated with specific electrical surface conditions and radical availability. In conservation terms they are visible as pH values, alkalinity (pH 7–12) and acidity (pH 3–7) and affect brickwork more by action in mortars than in well-fired clays.

5.4 Determining the presence of water

The simplest indicator of the presence of liquid water is the change in colour of most wetted brickwork. Visible signs, such as the growth of plants and microplants, are a clear indicator of the sustained presence of water, and, interpreted with experience, can indicate sustained water penetration, likely sources of entry, tracking through a structure and periodicity or length of time that penetration has persisted. Some inorganic signs can be interpreted similarly.

The solution of calcareous material in lime mortars will, over long periods, be evident as a white bloom or a semi-translucent white surface scale. Over very long periods this can result in the formation of stalactites (downward growing) and their converse, stalagmites. The phenomenon increases with time, because the lime dissolved out and then reforming as needles at the drip point is continuously being removed from the interior of the brickwork leaving ever-increasing aquifers allowing increasing water penetration.

The use of a moisture meter in brickwork is sometimes derided for its lack of precise accuracy. This is not justifiable. If appropriate sampling techniques are adopted moisture meters are a very important method of determining comparative moisture content at or near the surface. It must be remembered that the equipment measures electrical transmission between two probes a known distance apart, the transmission improving with the increase in available water.

Because transmission also depends upon penetration, salinity, porosity and other physical characteristics, a meter cannot give absolute values until it has been calibrated to the exact circumstances of use. However, this need not be done in most cases.

If the sampling is carried out consistently and at a sufficiently large number of points the comparative information will be critical in determining where moisture content is at its greatest. A careful meter survey will, therefore, involve measuring water content on a predetermined grid across wall surfaces. From the grid, contours of equal moisture content can be plotted which will provide important evidence of the nature of water entry and may also indicate hidden and inaccessible construction. It may sometimes be helpful to set up a wide-spaced grid which is then intensified in chosen areas to provide specific localized information.

Meters are cheap, simple and if carefully used, reliable.

Other more sophisticated methods of determining moisture content are available. Of these the fundamental method is to extract a sample, weigh it, dry it and weigh it again. Averaged over a number of samples this technique gives an absolute value which may be important in laboratory terms for calibration and testing. Such precision is unlikely to be justified in field work.

A variation on this method, involving less destruction, although providing slightly less accuracy, is to treat the powder from a drill probe in the same way. This technique has the merit of giving absolute values at known depths in brickwork without significant loss of strength or visual damage. The probe hole can be refilled with mortar or used to inject other materials, such as resins, and finally capped. The materials and equipment involved are simple and reliable. The method is sometimes called *gravimetric*. One defect is that a moisture gradient can imply misleading results.

The *neutron probe* method allows investigation of moisture content at depth in areas around a drilled hole. It has the advantage that it can be repeated at any time in a hole once drilled, and can, therefore, be used to measure change in moisture content in response to treatments or with the effluxion of time.

The neutron probe provides a measure of concentrations of hydrogen atoms, and therefore relates directly to water content including microporous water and water of crystallization. The values given may then be adjusted to disregard this locked up water. Expensive technology is involved, the process measuring the absorption of neutron scatter taking place in a volume some 100 mm in diameter.

A *radar* technique, using equipment designed for ground penetration (up to 1.5 m), can be calibrated to reveal moisture content in known types of brickwork. Here again costly technology is involved but the method can be repeated at will and at any time. It is non-destructive. However, salt pollution affects the readings significantly and the equipment cannot readily be adjusted to discard this factor.

It is also possible to use *microwave technology* to assess the moisture content of surface layers. Microwaves penetrate only 30 mm or so of brickwork so the technique is of help only in plasterwork and behind facings. It can be of considerable assistance in assessing relative changes in moisture content in sub-surface layers into which the probes of a moisture meter do not reach.

Finally, passing reference should be made to the use of *infra-red thermography*, which has been applied (see case study on San Marco, Chapter 13) to the comparative relative humidities of brick surfaces. The latent heat absorbed during evaporation will affect surface temperatures of brickwork, differentiating those surfaces which are evaporating most strongly. To be effective this non-destructive technique is usable only in circumstances where other factors such as insolation, air flow and similar external influences are not disruptively active.

5.5 The interaction of materials

Generally fired earths exist in buildings in isolation only as roof tiles. Otherwise they are normally used for walling or flooring and are essentially found in combination with mortars which are likewise porous materials of generally similar characteristics. Sometimes burned brick is bedded in an earth mortar and earth bricks are bonded with similar earths or with cementitious materials. Lime bricks and blocks are used with earth and lime or cement mortars, while crushed brick has been extensively used in the formation of lime mortars. Bricks and earths are frequently faced with renderings composed of sand, limes and cements, both internally and externally. Stones occur in combination as facing and as integral material so that the process of conservation must involve treatment of entire complex structural masses. These materials may be compatible in that they are physically inert, porous, of similar densities, similar compressive strengths and of similarly moderate tensile strengths. Their thermal capacities are similar. These behavioural characteristics group them together as materials. Their relationship to other common building materials – metal, wood and glass – crosses a divide which provides a convenient definition of a boundary but must not obscure the interaction of the materials when used in combination. In historic buildings an intimate relationship between porous building materials and glass or metal is rare and poses few practical problems but wood has consistently been used and misused as a component of such structures with consequent problems.

A short term view of building construction may justify the use of earth, brick, timber and mortar in close combination. The pioneers who built their stockades with timber driven into the earth banks hurriedly raised were entirely justified in using the materials, by the short life expectancy of the structure. Those who want to conserve the forts they produced have to make do without the perished timber, now vanished, or risk the censure of that part of the conservation movement which regards the reassembly of surviving components (anastylosis) as legitimate while disapproving of the replacement of known but vanished structural elements. The intimate mixture of timber with porous masonry produces many a problem of this sort. Of all the combinations of organic with inorganic materials, the presence of timber produces one of the greatest ethical and practical problems in conservation. The fundamentally practical aspect is simply the long term incompatibility of these materials due to differing rates of decay. While for practical purposes much brickwork and earth structure may be regarded as having an indefinite life this is much less true of timber and the anticipated rates of decay of the two groups of materials diverge appreciably. A strong case can be made for the removal of timber from an intimate relationship with moisture-bearing bricks where its failure can be anticipated, but there are many instances where the relationship is so interdependent that it must be maintained and some circumstances – such as earthquake protection – where its retention or replacement may be vital. Thus while it is practical and feasible to replace with solid brickwork a piece of timber built into a substantial brick wall it is much less possible to replace the wattle framework of a wattle and daub panel without destroying the panel in its entirety and thereby destroying the historic quality of the surfaces and structure. Elsewhere the timber may be an essential visual or structural component which cannot be discarded without historical loss.

In exploring these relationships and in making conservation judgements a balanced perspective and discretion are the essential guide.

5.6 Structural damage and failure

Types of failure may be grouped as follows:

1. **Impact**, which may include war damage from high explosives or the accidental impact of falling structures, vehicles, trees and like physical forces. In such circumstances inertial energy is absorbed immediately creating visible and invisible fractures in the crystal structures and the vitreous bondings.
2. **Deformation**, in which external forces are delivered powerfully and quite rapidly as in earthquake and storm conditions. These energies may include inertial effects but are primarily those of relatively rapid stressing which may exceed the capacities of brick and mortar, resulting in evident strain and fracture. Some rapid stressing is repetitive, e.g. traffic shock and vibration, resulting in cumulative damage by fatigue in the structure.
3. **Slow stressing**, in which excessive loading is applied gradually beyond the point at which the brickwork can sustain

it, as in the swelling or contraction of ground and movements associated with subsidence or land slip. Failures due to thermal movement fall into this category of slow stressing but are so inevitable, regular and recurrent that they take on a special importance due to their repetitive nature. Stresses tolerable singly may becomes strains when continually repeated.

4. **Loss of support**, in which an integral component (such as a timber lintel) ceases to be effective, producing changes in stressing. The effects are usually slow rather than immediate. The failed structure may be apparently unaffected until a sudden input of energy causes immediate collapse.
5. **Decay**, in which the brick or its supporting mortar fails by frost, wind erosion, solution, other chemical degradation or expansion of salts within the pores.

Every failure sets up a changed condition of internal stressing for which allowance may not have been made in the design. It therefore depends on circumstance whether there is sufficient strength in the structure to resist the new pattern of forces. Where there is such strength a new stability is achieved. An arch shaken by an earthquake may return to stability with broken mortar joints still providing compressive strength. The structure returns to

Figure 5.1 Sarajevo. The springing of an arch cut away by a projectile leaving a wider brickwork arch formed by corbelling within the pier.

Figure 5.2 Sarajevo. Impact by a projectile: the visible damage is extended by fissuring in the brick to a depth up to twice that of the impact crater.

Figure 5.3 A thermal movement crack in a garden wall, decreasing with height and being restricted to the lime/cement mortar (1:1:8) below the sixth course.

balance. Where there are insufficient forces of restraint to counter the new condition catastrophe follows. A wall gradually moving out of plumb due to salt crystallization on a side exposed to winds and sunlight will reach a point where the centre of gravity is located outside the line of support; sudden collapse follows, initiated perhaps by diurnal thermal expansion or wind pressure. On a calm day early this century the great campanile of San Marco in Venice gave warning of changes in its internal stressing by the development of cracks. The concierge and his family left, and within 24 hours the entire structure subsided vertically to become a heap of rubble. Only the cat, which had unwisely returned to its bed, was killed. In all likelihood the mortar, severely loaded on its lowest courses, gradually lost strength by the formation of salts with consequent changes in its crystalline structures – a common phenomenon in the city – and the declining strength of the material passed below the value of the superimposed deadload. If so this was a clear example of change in material strength producing an irreversible condition leading to catastrophic failure. Commonly such an event will be precipitated by additional loadings due to wind pressures or similar live phenomena.

Rapidity of loading is a significant factor in a composite structure. Failure will generally occur at lower values under rapid loading than under a slow sustained loading.

Figure 5.4 The effects of nitrates from cattle dung are apparent here well above ground level where crystallization has occurred beneath the surface of the brick. On the lower courses the brick has been more consistently saturated with crystallization occurring externally to less damaging effect. The damaged zone has been subject to subsurface crystallization.

Figure 5.5 Stress crack induced by a change in material. Above the brickwork the construction is a timber-frame. Differential movement loads the corner of the building. The induced crack moves diagonally through the structure.

Figure 5.6 Fracture of two terracotta components, the upper buff and the lower blue-grey. The continuity of the failure indicates a movement in the structure (probably thermally induced) rather than a failure of the terracotta itself.

5.7 Brickwork under stress

Brickwork can exist in many conditions of strength and weakness. The mortar may be stronger than the bricks in both tension, compression and elasticity. Conversely it may be weaker. Commonly, however, it may be stronger in some aspects and weaker in others. It may be stronger in compression but weaker in tension. It may be both stronger in tension and compression but not have the elasticity of brick, i.e. have a lesser bulk modulus of elasticity. The core of a wall may create one relationship between mortar and brick and the facing another. Massive and strongly bonded brickwork in a wall may support relatively weak arches of soft brick and conversely. Even individual bricks may vary widely in strength dependent upon type and firing. Physical linkages will vary. Ties may be of brick or of other materials. Walls often contain strengthening members, effectively sinews. Timber has been extensively used to provide lateral continuity and its strength and bond varies with its condition and age. Steel, wrought iron and reinforced concrete may be buried in wall thicknesses providing totally different conditions of strength to the general masses of masonry immediately adjacent. Stone may occur as binding courses, rubble fill, quoins or lintels. Some mortars will adhere to the brick surfaces holding them together: other mortars will have minimal adhesion and so will function more by holding the bricks apart.

Any serious attempt to understand the behaviour of wall masonry must take into account the patterns of the material strengths and connections in these terms. It is usually helpful, having deduced the precise materials and understanding their relative strengths, to set up detailed drawings through walls, vaults and other structural brickwork, superimposing strengths and forces on the drawings and relating these to the overall mass in terms of the cohesion its materials afford. In some circumstances such an analysis is essential to achieving a clear understanding of forces which always act in three dimensions. In others a specially formed model may simulate loadings that would otherwise not be calculable or understood intuitively. Model theory demands accurate knowledge and specialist construction before deductions can be made correctly.

5.8 Analysing failure

Rudimentary site testing can provide sufficient information for many evaluations. The strength of mortar close to the surface by comparison with the strength of mortar within the depth of the wall can be important and quite readily evaluated. In laboratory terms historic mortar will often vary in accordance with its position in the structure. In a practical sense the adhesion of mortar to the brick and the relative strengths of mortar and brick may be tested quite simply under uniform hammer blows, chisel cuts and abrasion. Such tests carefully carried out will be sufficient to tell the experienced operator the comparative characteristics of the materials and their adhesion. Likewise the comparative performance of different types of brick will be important to building up an understanding of the capacity of the structure to resist deformation. The drawn analysis is an aid to understanding the distribution of forces in a structure but it is no more than an aid. Conditions of the transmission of loads and cohesive strengths will depend upon continuity and discontinuities within the structure and most assessments of probable performance rely upon many presumptions of internal conditions which may not be verified even with the most sophisticated diagnostic techniques and certainly not by the individual conservator in the field. However, the weight and balance of knowledge which the careful practitioner can build up enables him or her to make broad judgements and to pinpoint behaviour by pursuing the analyses expressed in typical conditions as shown on drawings. The importance of the discipline of drawing cannot be overstressed. Only the simplest structural conditions cannot be more thoroughly understood by drawing.

Drawn analysis will suggest zones, planes and sections of weakness. In the simplest terms a section of wall carrying three vertically superimposed window openings presents an obviously weak sector in the same way as perforations between postage stamps. It may be, however, that an apparently weak part of the structure has not failed whereas another part of the building has failed for no reason apparent initially. The drawn analysis identifying the thicknesses of structure, the location of different components, the comparative strengths of materials and known loadings, shear planes, tensile strengths and typical elasticities can suggest the nature of the structure, the forces operating to cause failure and hence the reason for the apparently inexplicable observation. The actuality of movement can be important. Movement which has taken place and has ceased may be identified by earlier making good. The date of the movement together with the date of cessation may perhaps be identified by the evidence of earlier remedial action. The sequencing of movements, their current activity and repetition can all aid diagnosis. Once the cause of failure is ascertained the remedy can be considered. All too often the cause is presumed, perhaps wrongly, affecting the conclusion.

To work through an illustrative circumstance: a wall might show vertical cracking running through the building from top to bottom. The significant evidence to be gathered would be as follows:

- The position of the crack in relation to the overall mass of the structure.
- The age of the crack.
- Whether movement had ceased. If not whether the continuing process is accelerating or the reverse.
- The direction, nature and any periodicity of the relative movement of each section of brickwork.
- The comparative size of cracking at different levels from top to bottom of the structure.

Figure 5.7 The tomb of Yahya bin Kassim at Mosul. The entire right-hand section has first sheared away and has then moved outward and downward uniformly without the clockwise rotation which might have been anticipated. Comparison of the alignment of courses reveals a slight reverse rotation.

- Whether the crack passes through discontinuities in the structure.
- The homogeneity of the wall.
- The directions of advance of the crack.
- The nature of the crack, i.e. whether it passes consistently through both mortar and brick or whether it follows the line of mortar joints.
- The evidence of building sequence and subsequent workmanship.
- Comparative evidence of similar failure.
- The evidence of forces having been applied.

The conclusion will seek to identify the forces, their origins and significance.

Immediately accessible information having been obtained, the process of analysis itself must be undertaken with a careful discipline. Facts themselves are of no service if not properly marshalled. Some information takes a long time to gather: other information demands expensive or time-bound analysis. There is no point in spending large amounts of money nor waiting for a long time if the answers can be obtained from information already to hand.

The analytical process therefore requires assembly of all the *available* evidence, and its putting together in a time frame. The range of possible answers will then start to be apparent and it may be that further information is required to differentiate between them.

At this point the type of investigation can be decided. Decision-making affects the analysis. Calculation may show that the failure can be tolerated or stabilized and then tolerated: alternatively it may show that the original strength of the structure must be restored, and the next investigative phase will be designed accordingly.

The conservator must also be on the alert for a combination of causes of failure. Decaying mortar or failed pointing might be apparent although not in themselves the direct cause of structural failure. But water admitted to the wall as a result of that weakness might prove to have been the cause of failure in a load bearing timber which had allowed settlement in the masonry.

The age of failure may be indicated by subsequent fillings, remedial works, alterations and documentation such as earlier photography or even drawings and engravings. The effects of such discontinuity in the structure for a long period of time may also be apparent in other aspects of behaviour as the movement has been taken up in structure on either side of the defect. It may be possible to associate the fault with an event such as the adding or removal of a major load or the construction initially of two differently loaded structural elements on similar foundations without the introduction of a movement joint. The event might have been more dramatic – an earthquake, flood or subsidence due to erosion. If the cause has been identified and the movement is no longer taking place appropriate making good can be undertaken with confidence.

The activity of any movement can be determined simply and crudely, or by more sophisticated and more sensitive methods. A basic and reliable method is to fill sections of void with an inert, non-shrinking material (such as plaster) and to incise and protect fine lines engrained on the fill. Regular, sustained periodic measurement will determine whether movement is taking place and in simple terms what type of movement it may be. An annual periodicity will indicate sensitivity to thermal movement in the structure although not necessarily pinpointing thermal movement as the cause. Seasonal changes in humidity might equally, or in combination, be responsible. Rate of movement can be made apparent in the same way. The extent and direction of movement is less accurately determined by these basic methods but the broad information they yield will indicate whether more sensitive techniques should be employed. Comparison of the extent of visible movement at different levels will indicate whether sections of the building are rotating in relation one to the other, both along the plane of the structure and at right angles to it.

At the next level of sophistication a once popular method of monitoring was by the use of a glass tell-tale. Initially devised with metal mounts and screw fixings, it is used more simply with the introduction of strongly bonding adhesives such epoxy resins simply by setting a slip of glass in position across a discontinuity – crack or movement joint – cementing it in position and maintaining observation. Tell-tales are often fixed and forgotten or examined perfunctorily. If this is the case the poorest information is obtained. As a building moves the glass strip is stressed, perhaps to the point of fracture. The alignment of the fractures can then be used to interpret the strain and if the interpretation is skilfully carried out the direction and extent of the movement can be understood. However, much more information is to be gained by observing the surface of the glass prior to fracture. As movement takes place the surface is deformed and the deformation can be observed by reflected light. Very skilled interpretation is required but a much more accurate understanding of rate of change can be achieved. This can be particularly important in a building where movement may be accelerating. It may also be a valuable indicator of movements which are simply periodic. It should be borne in mind that glass has a sufficient elasticity to tolerate considerable torsional movement before it fails.

An acutely sensitive version of the same mechanism – the distortion of a plate fixed across a gap – is available with the technique of setting an adjustable reflector plate on to mounts fixed into the structure. By observing a rectified light beam – usually a low power laser – the angular distortion of the plate can be assessed using vernier screws by the adjustments necessary to bring it back into its original plane. This technique allows accurate measurement of the extent to which two sections of a structure may be moving out of alignment. The extent to which they may be moving apart may be sensitively determined by the use of strain gauges – devices designed to operate mechanically and to multiply the movement so that it can be read off a dial or a scale or monitored electronically by change in electrical properties of the strain gauge. By these latter methods and the regular plotting of observations the nature of the movements may be determined. A continuing acceleration or deceleration will indicate whether the structure is approaching stability. An accelerating change suggests otherwise. Reversal of acceleration and deceleration suggests periodicity where the structural gap is acting as a movement joint. Thermal cycling on an annual or even diurnal basis may provide this comfort to the conservator. Likewise fluctuations associated with regular changes in ground water due to tidal or climatic conditions may be indicated by a periodicity of movement which can be attached to a specific cause. Where movement is taking place in one part of the structure the pattern is likely to be repeated elsewhere. Movements in a vault are likely to be reflected in the supporting walls and piers and become apparent over openings.

The associated performance of discontinuities is, therefore, an important part of a general analysis. In interpreting the cause the position of any crack must be understood. The reason for its appearance at a particular point is a description of the zone or line of maximum weakness of structure in relation to stress. It may well be that in respect of a different stress the zone of weakness would not be the same and it therefore follows that the position and alignment of the crack will be a clear indicator of the nature and direction of the stress. Thus a simple vertical crack in a wall illustrates the presence of forces which are, or have been, exerted over a large part of the wall structure. They may represent a force moving one section of the wall longitudinally away from another. They may represent a force rotating one section of the wall away from a stable element. They may represent two forces acting wholly or partly in opposite directions. They represent at the point of failure a condition in which the tensile strength or shear strength of the wall is less than the stress which has been applied and the elasticity of the wall has proved incapable of accommodating the resultant forces over the period of its application.

Brickwork (and most other structural components of buildings other than plastics, timber and metals) has low elasticity (capacity to deform and regain its original shape) and is generally credited with poor plasticity (the capacity to deform without fracturing). Nevertheless, walls built in lime and earth mortars can exhibit surprising accommodation of stress or microcracking within the joints in response to forces applied over an extended period. Lime mortars, in particular, can be tolerant in this way and even be self-repairing in that they can achieve a restructuring by crystal growth which accommodates and stabilizes change. The measure of elasticity is 'Young's modulus' which defines the capacity of the material to suffer distortion in a linear sense and yet to recover and is extended as 'Bulk modulus' to describe the same effect in a volumetric sense. The measurement is of stress divided by strain, i.e. deforming force measured against actual deformation. The effect ceases at the elastic limit, beyond which the material will not recover. The usefulness of this measurement to the conservator is limited and relates to comparative values. Lime mortar with a high content of crushed brick has a higher elasticity than pure sand:lime mortar. A brick with a higher modulus will be more resilient – typically a better burned brick giving a ringing tone when struck. Beyond the elastic limit fracture will occur. This may take two forms, brittle and plastic, referring to narrow and broad zones in which fracturing occurs. There is, however, a continuity between the two types in that multiple brittle fractures may appear to be plastic. Soft mortars and some bricks may undergo irreversible distortion by multiple brittle fracture while apparently retaining coherence.

Sections taken through massive long-standing structures in which lime mortar is a substantial element, such as the great Byzantine church of Haghia (Aya) Sophia in Istanbul (AD 532–38) have shown that the masonry is capable of astonishing readjustments under sustained load and the variations in such loading consequent upon changed structural conditions can be related to the structural history of the building. Consequently a crack in a wall must be seen as the statement of an event in which the elastic and plastic limits were exceeded during the time the forces were applied and this in turn is an indication of the nature of the forces. A sustained stress, therefore, may be taken up in a 'plastic' deformation while an identical strain may be the result of greater forces acting over a much shorter period. The contrast is illustrated by the comparison between the strain represented by a vertical crack in a wall by comparison with the same elongation in the wall with minor multiple visible cracks expressed as a plastic accommodation and the result of a long sustained ground movement. In this latter case it may prove extremely difficult to identify the strain tolerated by the building because the effects may be virtually invisible. A wall in which such movement has taken place may appear to be entirely intact in the brickwork itself although the rendering or plasterwork with which it is faced may reveal the extent of the movement. In such a circumstance the bond between render or plaster and the load bearing brickwork will have been broken.

Plastic accommodation in brickwork will normally occur when a force acting (stress) is allowed to cause a deformation (strain) by a momentary pressure such as a tremor or vibration.

5.9 Non-destructive investigation

Measurements of the performance of brickwork offer some assistance in assessing internal and hidden conditions. Of the available techniques, magnetometry and deeply penetrating radar currently have minor apparent application, while ultrasound and thermography offer some, limited, assistance.

Ultrasound involves the emission and reception of a series of pulses at ultrasonic frequencies, the reception being translated into visual imagery on a screen. In essence the signal received differentiates the pulses reflected by material of variable density and identifies the distance of the reflection from the point of emission or the speed of transmission of the sound waves. Passed through the human body the return signals show bone, sinew and fluid differently, revealing an anatomical section. Passed into earth they differentiate between soil and various grades of rock. In brickwork they can reveal inconsistencies due to voids, metals, embedded timbers and fractures. The evidence is immediate but not normally recorded permanently, although electronic print-outs can be available. Important information to be gained falls into two categories: structural continuity/discontinuity and density variation.

In the first category lies the coherence of material. Ultrasonic testing is routinely used to check the coherence of stone in European quarries and increasingly elsewhere. It is less relevant in built masonry and, in clays, may be used to verify the coherence of large terracotta blocks, in theory, although not yet in practice. Potentially it is applicable to brick masonry but practical regular diagnostic use has not yet been established.

Thermography depends on the measurement of heat radiation. It, therefore, measures surface radiation, making an image which displays comparative temperatures (see Plate 5.2). These in turn depend upon the emissivity of a surface and upon the rate of delivery of heat to the surface. Weather conditions affect surface cooling and in consequence materials with a considerable thermal inertia require very specific conditions to provide a useful display. Voids, cavities, flues, detached rendering and mosaic, pipes, embedded timbers and the like can be detected in the right conditions but the technique is peripheral in the investigation of brickwork. It depends upon the use of germanium lenses which focus the received radiation on a detector cooled to nearly 200°C below zero. The large temperature gap is required to provide an accurate response to differentials as small as 0.1°C or less. In the words of Robert Demaus, specialist in remote sensing diagnostic techniques 'it is the thermographic camera's ability to identify these often minute differences [in temperature] that makes it such a powerful diagnostic and archaeological tool.' The sensitivity of the latest equipment allows information to be collected about the original construction, historical changes to the structure, structural failures and decay and damp problems hidden behind the outer surface, as well as variations that are not visible to the naked eye.

The cost of the equipment and its specialist operation makes it applicable only to circumstances where sufficient work of importance is to be surveyed and/or where ground level analysis saves the erection of scaffolding.

Application includes distinguishing between hidden materials, as in the case of timber embedded in brickwork, or in distinguishing between brick and other materials in a hidden structure. Thus the difference between wattle and daub and brickwork behind a rendered facing may be implied by the difference in thermal radiation. As the structure cools more heat is released to the outer surface from brickwork than from timber-framing or wattle and daub. Similarly the transmission and emission characteristics vary with the extent of damp contained in brickwork. A thermographic plot will, therefore, reveal the extent of contained moisture, allowing adjustments to be made to provide even drying out where (as in the case of the study of San Marco, Chapter 13) a uniform finished state is critical.

Magnetometry, referred to in the same case study can reveal the presence of embedded magnetic material – almost invariably iron although there are other magnetic elements. The presence of iron within brickwork masses can be very significant, and they can be reliably located within specific parameters. Known items, such as cramps in terracotta or cavity ties may be confirmed and positioned. Unknown embedded elements may be discovered by the same technique, which depends upon electromagnetic radiation exciting a magnetic field in – typically – an iron mass.

5.10 Forces within the structure

The course which a crack may trace through masonry structures is indicative of the nature of the forces involved, but will also be dependent upon the relative physical properties of the mortar, the bricks and the surfacing materials. A force drawing two sections of a wall apart on a line of weakness represented by vertical perforations, such as the contractive forces resultant from seasonal cooling or moisture change, or both, may be manifest as a crack or series of cracks following the mortar beds or perpends in the brickwork if the mortar is significantly weaker than the bricks or lacks adhesion to them. Where the mortar is strong, with a powerful bond, a different pattern of brick fracture will take place, one which may readily be confused with shear cracking caused by differential vertical movement between two sections of wall. Moisture and thermal movements are the two most important sources of vertical cracks in long continuous masses of masonry and are particularly prevalent where strong mortars have been used.

It is only by a careful analytical understanding of the wall structure, of the timing of stresses, the form of the strains of their subsequent movements, their direction and their pattern of activity that there can be a clear understanding of the causes of failure. To understand the mechanism of failure is important in some aspects of diagnosis. A physical break can only exist in brickwork where the structure has been subject to opposing forces, each requiring it to move differentially. A force may be visualized as passing through a near rigid structure such as brickwork with diminishing intensity. This may be illustrated by conceiving a long iron bar being pulled mechanically from one end. Virtually the whole 'pull' is received at the other end of the bar. A massive masonry wall, however, of the same length will transmit that same 'pull' less effectively, transferring much of the stress to its foundation.

In the case of the wall the stress could be plotted as a series of lines of equal or isometric stress at values reducing with distance from the point of 'pull' and independent of fixings or foundations. Failure will occur where the stress exceeds the tensile strength of the brickwork. Concentrations of stress will occur due to the presence of openings or discontinuities in the wall structure and the initial failure will be associated with some defect in the material, however minor. At this point the material will be strained beyond its elastic limit to the point where it breaks apart. The pattern of the lines of stress changes with the fracturing, their concentration moving to the ends of the fracture. At this point brickwork is subject to maximum stress so it fails and the fracture moves forward repeating the pattern until the point is reached when the intensity of the stresses has diminished or the fracture has reached the edge of the continuous structure.

These conditions can be visualized in alternative circumstances by conceiving of a very long wall subject to two different overwhelmingly destructive forces – an earthquake and penetration by a non-explosive projectile.

In the case of the earthquake the ground movement may be imagined as a pulse (or series of pulses) moving forward causing lift, heave and rapid expansion and contraction in the ground. Simplistically imagined the fracture will have been initiated in the foundations and at the base of the wall and will have been carried up the structure by the process, which may be visualized as concentrations of lines of equal stress at the end of the fracture producing a strain beyond the tolerance of the material, the effect moving progressively up the wall. The damage is therefore expressed as cracking.

The projectile, whose energies are equally capable of destroying the brickwork and creating a vertical rift, does not in fact do so. It simply punches a hole into or through the wall. In corresponding terms the action can be imagined as follows: the lines of equal stress take up a circumferential pattern as the shock wave moves outwards from the point of impact. The stresses on opposing sides to the impact act in opposite directions. Radial fractures therefore form. Stress concentrations cause the end point of the fracture to progress outwards until the diminishing lines of stress are no longer greater than the cohesive strength of the material. At this point the fracturing ceases and the consequential visible damage is, thereby, restricted. Exploring this phenomenon a little further it is clear that the stresses within the wall are set up by the

Figure 5.8 A brick structure moving coherently, Port Royal, Jamaica. The weight and cohesion of the reinforced concrete roof of this late nineteenth century ammunition store has restrained the brickwork during very severe settlement. Internal cross-walls provide bracing in compression.

forces represented by the momentum of the projectile being resisted by the pattern of compressive and tensile forces in the brickwork. Therefore, the stronger the brickwork the greater the stresses at the point when fracture is initiated. It follows that in an earth brick wall the weaker structure sets up weaker lines of equal stress so that the projectile passes through with less visible effect – except for extensive consequential damage reflecting the weakness of the material.

The contrast between the effects of projectile and seismic shock in these very simplistic terms illustrates the assistance that can be offered in an understanding of structural performance by visualization or estimation of the pattern of lines of stress set up by energy sources in a static structure. These imaginary lines can have the same usefulness as thermal contours (isotherms) or lines of equal air pressure (isobars). It is helpful to visualize them as wave fronts, comparing the earthquake effect with the patterns made by parallel waves running before the wind and the projectile effect with waves emanating from a single impact point where water surface is disturbed.

Where investigation makes it apparent that lines of equal stress emanate from a broad base the origin of the force must be sought in a widespread zone. Where it can be seen that they are focused, a localized internal pressure may be suspected and pinpointed.

In diagnosis of structural failure it is easy to overlook the need for a careful analytical method in which the principal tools are diagrams and detailed drawings, or even models. At the most sophisticated level, laminated transparent plastic models may be made to be loaded and photographed using polarized light under stressed conditions to build up an analogous picture of the forces within the structure. The model must incorporate the discontinuities and planes of weakness in the structure if the results are to be of use and not misleading.

Since building defects arise in many instances from combinations of forces sufficient to induce failure it is dangerous to omit from diagnosis the analytical step in which forces are identified. The simple and perhaps easy identification of a prime cause of failure can obscure the existence of secondary forces acting within the structure. In consequence other possible causes of failure may go unchecked and unremedied and, on occasion, a counterbalancing force may not be recognized until its withdrawal has altered the balance with unforeseen consequences.

6

Response to failure

6.1 Methodology

In any complex situation of failure a careful written analysis is, if not essential, extremely helpful. The analysis is likely to take in other components of a structure, but in some circumstances the breakdown will relate entirely to masonry.

A regularly structured format may readily be compiled under heads such as are set out below. They are personal to the conservator and will serve as an analytical discipline.

Visual analysis – material – area by area

Description of area – and relevant drawn information with dimensions

Type of masonry:
- face material
- core–cavity
- thickness
- bonding
- lengths and heights

Type of brick:
- manufacture
- distinguishing marks
- consistency
- comparative material
- dimensions of bricks

Nature of mortar:
- lime
- cement
- (gypsum)
- sands
- application – pointing – dimensions

Nature of plasters and renders:
- materials (as preceding)
- hairing
- finishes
- porosity
- number coats

Figure 6.1 A multiple cause analysis should be used to determine the interaction of factors leading to decay – quality of brickwork, age, rising water, salt deposition, pollution due to animals or storage, wind effects and others.

Extent of failure: – dimensions
– directionality
Visual analysis: – damage – area by area
Materials involved: – nature of failure
– form of decay

Locations of failure
Evident causes: – stress – directions
– loads – impacts
– biological – e.g. root
– physical – e.g. frost
– chemical – e.g. salt

Current activity of defect: evidence of progression or stability
Surface changes: – discoloration – fading
– soiling – patina
– encrustation
– efflorescence
– staining/paint
Disintegration: failure by: – delamination
– blistering
– flaking/scaling
– crumbling/chalking
– erosion
Deformation: – bulging
– mortar loss
– continuous hair cracks
– crazing
– fracture
Crystallization: – surface
– sub-surface
– uniformity
– localization characteristics

Biological growth
Area restrictions of: – fungi/algae
– lichens
– mosses
– plants
Defective detailing: – adjacent materials
– pointing
– applied materials
hydrophobic
paints

Background information: exposure, prevailing wind
– exposure, wind eddying
– exposure, pollutants
sea air
– other
aerosols
– SO_2
– CO_2
– average rainfall
– average incidence of frost cycling
– temperature/relative humidity cycling

6.2 Records of damage

Site observations should be recorded with care and entered on to drawn or written documents. Accurate photographic records are important and the information should be enhanced before photography by indicators such as arrows or pointers temporarily applied to the building to show the position and termination points of cracks together with scales and ruled or scored alignments to provide accurate and instructive comparison with later photography. Failure may be occurring in different planes and changes along the axis of a camera lens will not readily be apparent so carefully placed indicators and transverse photography may be advisable. The function of records is not so much to provide an archive of past events as to provide a base from which predictions and extrapolations can be made. To this end suspect parts of the structure should be monitored.

Areas around cracked brickwork should be examined carefully for signs of distress and superficial coverings such as plaster and render may be removed to disclose additional information. Surface spoiling or flaking of mortar joints may be an important indication of overloading or of structural movement. Areas of failure should be marked on drawings and these should be updated regularly.

The nature and cause of movement must be identified if possible. Even where the apparent cause has been eliminated, a change in structural continuity may result in progressive movement. A discontinuity (crack) is a change in the building design. A choice may have to be made as between crude but reliable techniques such as filling a crack to observe any further movement, the fallible but more sensitive method of providing tell-tales securely bedded in resin and carefully monitored or the use of sophisticated techniques relying on small fixed reflectors and careful instrumentation.

Effects: Symptons.	Apparent Cause: frost action	salt crystal i/zation	biol. growth	effect of environ. poll	mech. stress.	thermal stress.	water pene- tration.	surface erosion	non. chem. surface desposit.	conden- sation.	metal corros- ion.	fire damage
fading												
soiling												
staining												
graffiti												
encrustation												
efflorescence												
crypto- efflorescence												
patination												
crusting												
spalling												
scaling												
blistering												
crumbling												
pulverization												
sanding												
powdering												
chalking												
physical erosion												
peeling												
blistering												
loss of bond												
hair cracking												
linear cracking												
star cracking												
crazing												
rupture												
bending												
bulging												
bulking												
scratches												
cuts/incisions												
splitting												
bacteria												
fungi												
algae												
lichens												
liverwort												
mosses												
higher plants												

Figure 6.2 Matrix of causes of failure. Comparison of completed matrices allows comparison of failures and analysis of sequential faults. (Modified and adapted with permission from a NATO CCMS publication.)

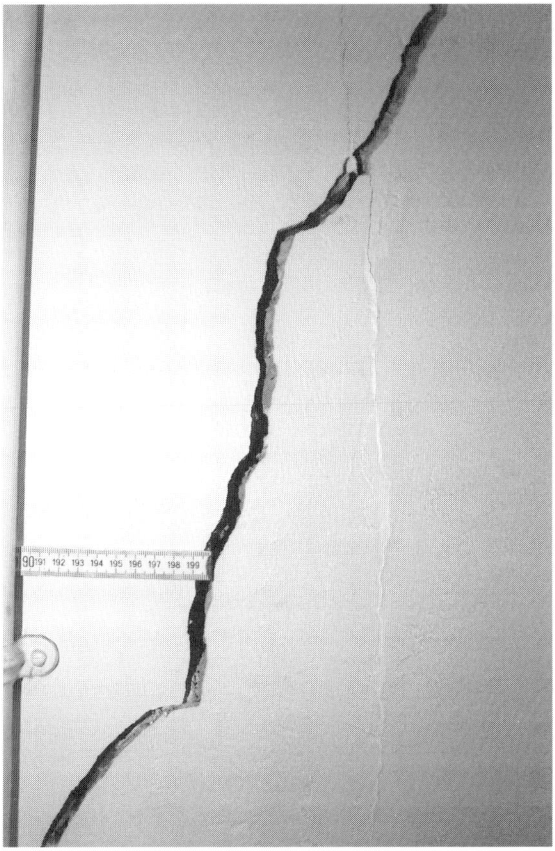

Figure 6.3 A near vertical crack in a brick spine-wall measurable as a 10 mm movement laterally and 2 mm vertically. Several different forces may be accountable and their effects may have been felt at different times. In this case the lateral movement may be primarily thermal and the vertical movement a consequence of adjacent rebuilding.

Every attempt should be made to trace the history of the deformation and the consequence of other happenings in the life of the building. Events such as periodic redecoration may well provide a clear indicator of the dates of formation and extension of cracks in brickwork. Movement cracks will normally be irregular and the irregularity on one side of the defect can be matched to the irregularity opposite. Careful marking, levelling and measurement, with a micrometer, if necessary can be used to determine the relevant movements, and to record the results.

In an important building it may be necessary to provide a three-dimensional modelling system to describe the effective nature of the movements. The next step is to determine, so far as possible, the probable lines of equal stress within the structure. Comparative stress diagrams as between the condition prior to failure and thereafter may prove helpful and if movement, failure or change in loading is a continuing process a prognostication of the stress effects expressed in similar terms will be valuable, both as an exercise in probabilities and an investigative process. Specialist engineering expertise may be important in all but the simplest of these operations, particularly where further failure may be predictable.

6.3 Secondary effects

Concern over structural failure must extend also to secondary effects. Bricks and mortar will have been affected by shock before the loss of strength is apparent. Brickwork which has received a severe impact will have suffered the passage of severe shock throughout the mass of the material and will have been weakened correspondingly. Such shock waves may take the form of repetitive minor tremors or vibration such as those due to traffic or they may be a single experience such as an exploding shell. Between these two lie the multiple tremors of earthquakes and the thrust of ground heave, swelling clay or rusting iron. Microscopically the effect is the transmission of a wave front of energy through the material. The elasticity of the rigid material carries the energy forward by the movement of each particle successively away from the impact point only to restore itself to its original position transmitting the energy to the next particle in sequence. Within the interlocking crystal structure of the mortar and the semi-vitrification of the brick this sudden movement stresses and may break the bonds between particles, the extent of the breakages being broadly proportional to the degree of stress. The higher the stress the greater the number of breakages. Below a certain level the effect is so small that it can be discounted.

In gauging the effects of stress the energy levels transmitted into the brickwork are the clearest indication of likely damage. While a bird flying into a brick wall will have a negligible effect a bullet of the same weight (having greater momentum and therefore greater impact energy due to its high speed) will produce high stresses at the point of impact destroying the coherence of the material over a small radius and causing a zone of weakness for a relatively small area beyond. The wave front moving out from the point of impact will rapidly diminish in intensity (by the square of the distance), whereas a broad wave front with a higher total energy value, emanating from a more distant point, such as a shock wave transmitted through the ground, will move forward across a near-parallel plane dissipating itself over a much longer distance of travel in the brickwork. The cumulative effects of energy transmission in this way are also significant. Loads placed on and then removed from the ground by traffic are delivered through the soil to the building structure as pulsations and are then carried through the building. While each shock wave may be so small as to be insignificant on its own, the repetitive effect can become cumulative and eventually sufficient fractures can be induced in the crystals or between the structural particles to weaken the material significantly. In addition, sympathetic vibrations may be set up either within the brickwork or within attaching structures such as floors and windows and as a result the energies can build up into damaging movements. These are often palpable within the building and they serve to illustrate the imperceptible effects which are carried into the structure. Actual structural damage within particles or at their boundaries will occur when other stresses are applied simultaneously. These may be thermal or due to moisture movement, setting up a condition of stress which is then heightened by the shock wave.

Stressing of a rigid material such as mortar and fired brick can be significantly damaging even though a point of failure has not been reached. This can be illustrated by the everyday circumstance of brick cutting. The bricklayer holds his brick in his hand and strikes it sharply with the edge of his trowel. The shock wave passes through the brick on the line of the impact and the brick fractures progressively along the line on which it has been struck, the line of failure deviating with inequalities in the material and quickly reducing in effect. At the point of impact the brick skin is completely shattered, initiating the crack which moves in one direction, the lateral stressing diminishing as it moves outwards in a spherical wave. The brick remains whole and complete. The blow is repeated in the same alignment on the opposite face. There may be no visible effect but after several blows at different points on the same plane the brick falls apart, the several shear planes having joined together. Although the fracture had been created to a substantial depth on first impact the brick had not apparently suffered significant damage. The plane of overstress had, however, been created in part, and, once associated with others similarly aligned the complete fracture was achieved, each crack once initiated moving forward at its point of cleavage. In this sense the behaviour is comparable to that of glass which is 'cut' by a surface incision. The initial fractures which had penetrated only partially are nevertheless present. If the surface of the cut brick is examined microscopically it can be seen to contain many small fractures. These are sometimes visible under a glass and can be exposed and seen under a microscope. The effect is clearly apparent on weathering. The absence of any surface coherence, even of a fireskin and the presence of a highly fractured zone invites the weathering effects of moisture penetration, biological accumulation, frost action and so on. Decay sets in far more rapidly after cutting than in the case of comparable whole brick. Therefore, where structural failure has occurred by fracturing, the nature of the stressing should be considered and zones of secondary damage and weakness should be allowed for.

6.4 Typical structural failures

Structural failure in brickwork will occur for a number of reasons, among which are:

- Loss of support such as the failure of a lintel.
- Slow stressing under adjacent distortion, such as a swelling tree root.
- Inherent distortion where changes occur within the structure itself, changes such as thermal expansion and contraction, freezing, irreversible moisture expansion or swelling due to salt crystallization.

- Excessive loading. (Relatively rare but occurring particularly where a failure deflects loads into a small area, as for instance in the toppling of a tall structure.)
- Shock – impact, vibration, earth tremors.

6.4.1 Loss of support

Where an element of brickwork is not sustained it will tend to drop away unless it is designed as an arch or takes up arching stresses. Adjoining brickwork which still enjoys support may provide successive cantilevers extending the zone of structural integrity. Where a lintel spanning an opening has failed or been removed the effect is to allow a triangular section of brickwork to drop away unless the tensile strength of the mortar is sufficient to hold the bricks in suspension, as may be the case with a strong cement.

Loss of support beneath a structure may allow a whole section to rotate, falling away as a coherent mass. Where impact has removed the edge of a structure the cantilevering effect of successive brick courses above will be apparent but the tensile restraint offered in these courses may be insufficient to keep them in position and they will drop away producing a successive corbelling.

Where loss of support has allowed part of a brickwork structure to settle before support is regained the effect will be seen as shear between that section which has retained support throughout and the section which has settled. The fissure will tend to the vertical and is likely to run through both brick and mortar.

Where one side of the structure has lost support the whole element may behave coherently and rotate. Here a vertical fissure will widen the further it is from the centre of rotation, and the fracture will tend to follow the mortar joints if the mortar is weaker than the brick. If stronger, a mixed pattern of fracture will develop with some bricks being sheared. Where support has been lost in a transverse section across brickwork, as for instance in the circumstance where a timber lacing member runs within a brick wall but on one side, and this timber has decayed, an eccentric loading will develop which may induce a shear fracture across the wall showing itself as a horizontal crack above the level of loss of support.

Figure 6.4 Brickwork accommodating thermal movement in the main body of a structure. In an ancient wall in Xian, China, movement has caused a fracture in the brick facing which, in the weak earth mortar, follows the line of bonding.

Alternatively an upper structure may rotate about the longitudinal axis of a wall introducing horizontal cracking, almost invariably within the mortar joints at about the level of the failure. The loss of the restraint of a buttress may provide a similar effect. The addition of buttresses with inadequate foundations can be to produce rotation in the upper wall structures. Historically many buttresses were provided with inadequately considered foundations. The effect of the additional deadload of a buttress immediately adjacent to the foundation of a wall is to distort the pressure bulb beneath the existing foundations by compressing the soil eccentrically under the structure. This effectively reduces support to the original foundations. The effect is dramat-

Figure 6.5 The Tomb of Yahya bin Kassim at Mosul. Left: After 600 years' erosion the River Tigris had undercut the foundations of the brick-built tomb. A massive square buttress was built in response against the cliff, topped with raking buttresses to restrain the movement. Top: A shear crack in the centre of the south wall, cut out and repaired when the buttress was built, is now a monitor of the continuing movement under the deadweight of the buttress itself.

ically seen in the crudely repaired mausoleum of Yahya bin Kassim. An initial rotation causing fracture was succeeded by direct downward and outward movement in response to the weight of the buttress. For this reason large and heavy buttresses on soft ground should be arched away from the foundations as flying buttresses.

6.4.2 Local stress

Fractures caused by swelling tend to be less radical than those caused by loss of support because the forces involved are significantly less. The point or area at which they are acting can usually be identified with some accuracy.

One typical contributor to the decay of buildings is rusting iron which can, in a free circumstance, expand to six or seven times its volume, particularly (as in wrought iron) where its structure is laminated. Wrought iron straps were much used in the eighteenth and nineteenth centuries to support flat or low rising arches over windows and doors. They were embedded in solid brickwork. Although wrought iron oxidizes more slowly than steel, examples may be found in which the expansive forces of the rusting iron have been sufficient to lift the brick courses between which they are embedded, in the upper levels of a structure, whereas the same construction decaying in the same way at lower levels and beneath a greater superimposed load does not exhibit the same horizontal cracking because the mass of brickwork has a sufficient deadweight to counter the expansive forces.

The swelling of tree roots beneath walls has the effect of lifting zones of brickwork so that courses once horizontal take an upward arc. The courses move one against the other. As their path lengthens the perpends separate and fractures develop vertically above the point where the lifting occurs. Every horizontal and vertical mortar joint will be broken over the whole area. These forces, the result of osmotic pressure and consequent growth, can be sufficient to lift as much as four or five metres of brickwork. The much weaker and more widely distributed forces of the expansion of clays are apparent only in very lightly loaded zones such as tiled floors. The loads of parallel walls will produce an upward pressure on paving between them which may be the effect of pure plasticity in the subsoil activated by the effects of changing moisture content, the walls settling during periods of dryness but being so heavy that they are not lifted back on the corresponding increase in moisture content. However, the increasing moisture content does create sufficient pressure to cause a swelling which is exhibited as a camber in brick paving between the walls with corresponding zones of cracking running parallel to the walls.

Earth movements can also effect buildings on a residual basis. Where a loading is removed the pressure bulb in the ground is amended and dimensional readjustments can occur over an extended period. In many soils the movement will be stimulated by changes in water content. Such changes can be particularly significant where neighbouring structural elements have been closely associated prior to the removal of one part. Thus if a historic building is built up to or close to another structure which is then removed, the subsoil, particularly if it is deep loam or ground that was once marshy, may recover over an extended period of perhaps several decades. This phenomenon can be particularly mystifying because there is no apparent reason for any ground movement whatever. It can be damaging, particularly if the readjustments provide forces acting at right angles to the main axis of a historic structure. The resultant deflections produce transverse misalignments in the walls for which the building was not designed.

Swelling due to crystallization by salts will be noticed, particularly where the differential effect of exposure causes greater evaporation on one face than another. Again the effect is countered by gravity. Chimneys are particularly susceptible as the sulphates emanating from flue gases are generally soluble (sodium sulphate in particular), and are drawn to the external surface where their deposition causes swelling of mortar and weakness. In the lower courses of a tall chimney stack the deadload restrains the tendency to expand but as the deadload diminishes towards the top the swelling effect is increasingly apparent with the result that the chimney adopts a curved profile which increases with height. This type of deformation is less likely to be accompanied by distinctive cracking except at points of particular stress such as the junction of two walls. However, significant internal cracking

Figure 6.6 Martello Tower, failure of external skin. The external facing of Napoleonic fortifications fails with the circumferential expansions and contractions consequent on thermal and humidity variation.

Figure 6.7 Looking up under the soffit of a polychrome brick arch. Soot encrustation has retreated under the erosive effects of salt crystallization and wind eddy leading to cavitation. Replacement bricks are distinguished by their sharp arrises.

may occur within the horizontal coursing as an outer skin forces the structure out of alignment and this cracking produces both structural weakness and exposure to waterborne decay.

6.4.3 Spalling, frost damage and salts

Spalling on brick surfaces may be due both to frost action and salt crystallization. The effects may be difficult to differentiate but as the remedies differ the problem must be looked at carefully. If a single expansion point can be identified (as in the blowing of lime) or if salt traces are visible as a white deposit the fault can be given a chemical origin and appropriate measures taken; perhaps poulticing, perhaps repointing deeply with a sulphate resisting mix and in the case of lime blow replacement of defective brick – or no action. Failure of a brick due to salt deposition is indicative of a weak or unduly porous brick. Such an effect may arise after a long period of stability if the internal temperature regime changes (e.g. improved heating) with consequent increase in vapour transmission.

Other forms of inherent failure (caused by forces acting within the structure itself) include frost damage and thermal movement. Where whole courses of brickwork have been lifted by frost action it can be presumed that other forces have produced horizontal cracking or failure of mortar, resulting in the saturation of

Figure 6.8 Settlement. Mining activity and below ground settlement will be taken up at the weak points in a wall by the formation of diagonal cracks such as that beside this doorway where lateral and vertical movement has occurred simultaneously.

Figure 6.9 A structural failure due to thermal movement in walls at right angles. Once the mortar bond has been broken on expansion there is no mechanism for recovery of the structure on contraction. The effect is to push out a wedge of material.

joints between the bricks and the cause is to be sought in the condition of the wall prior to frost action. Individual brick failures due to frost are readily identified and replacement is straightforward, although it demands considerable care. Frost action in mortar joints argues that the structure has been saturated and remedial action must eliminate the cause of wetting as well as replacing the mortar. Bricks and terracotta are frost resistant if they have an appropriate porosity and are adequately fired. These criteria vary with the type of clay. In an Etruria marl a porosity of between 4% and 7% may be appropriate whereas in a superficial clay, such as the Wealden, a porosity of 15% may be satisfactory. Firing can be related to hardness and colour in each specific clay. The degree of firing must have carried the clay to semi-vitrification to provide a sufficient bond between particles. This temperature, and the time over which it is maintained to be effective varies with the brick earth. Thermal movement and inadequate expansion joints can produce lateral movement in joints and slippage, particularly across damp proof membranes, often visible as projections at quoins. The effect of outward expansion followed by failure to return on contraction is sometimes known as a ratchet effect, visible as breaks at perpends.

Mortars rich in ordinary Portland cement are often the cause of joint failure, particularly where sulphates are present.

Temperature reduction also causes shrinkage in masonry structures. Long walls will contract and when temperatures rise, conversely, they will expand. This effect may be partially countered by swelling which occurs as moisture is taken up and contraction as it is lost. Where the mortars are soft the walling itself will take up the dimensional changes. In the process crystal fracture may occur and in the long term mortar may break down in consequence. Where the mortars are strong the wall may behave coherently, particularly if the construction below ground is in weak mortars. In consequence regular vertical cracking may occur at preferential points

Figure 6.10 The outer skin of a Martello Tower which has peeled away in consequence of fractures in the bonding bricks due to differential movement between the external leaf and the massive core.

where the wall is perforated, where the construction changes or where other walls are joined to it. The cracks are likely to be greater towards the top and the bottom due to vertical temperature differentials and ground restraint. With weak mortars the cracking will follow beds and perpends and with strong mortars is likely to follow a line through perpends and bricks continuously.

Ground instability may follow slippage of land on slopes over unstable sands or layered clays and sands. Due to the plasticity of soils, movement in an adjacent area may be expressed as a horizontal or primarily horizontal movement in the soil. Consequential cracking is usually diagonal. The cause of such movement may sometimes be no more than an adjacent applied load, perhaps a new building. The construction in Venice by the architect Sansovino of the great raking buttress to the Church of San Marco is the apparent cause of a series of fractures in an adjacent earlier brick structure, its pressure bulb extending beneath the earlier building. These fractures run approximately parallel courses diagonally up the earlier wall.

Earthquake damage, consequent upon a shock wave travelling horizontally, may result in the vertical or lateral movement of part of a building although during the period of shock each particle of the structure was lifted and moved successively, perhaps in an approximately elliptical path. A typical resultant effect will be the creation of a fracture in the building running at an angle close to the vertical. The shock wave is unlikely to have been travelling along the line of the walls and therefore a lateral displacement is likely to have taken place also.

Excessive loading will cause compression of a structure from which it may not recover. In normal conditions loadings remain within the elastic limits of brick and mortar and the structure will recover when the loading is removed. Excessive loading will cause the permanent deformation due to the fracturing of the crystalline particles. Excess loadings of this type can produce shear fractures in the brickwork. Additions and alterations to buildings may well have created higher loads than envisaged by the original builders on specific sections of a historic fabric. In these instances shear cracks may develop at some considerable distance from the point of conjunction of the differentially loaded sections. These cracks are likely to run diagonally in the wall unless patterns of weakness in the construction provide a path of vertical failure.

The creation of shear cracks in walls redistributes the stresses within the structure, perhaps reducing loads on some parts and accentuating them on others. Particularly vulnerable sections are window mullions and columns. Overloads may increase to the point where the crushing strength of seating of columns is exceeded. Further failure ensues which may then transfer loads elsewhere extending the damage.

Walls in which skins are bonded by header bricks may fail because the bonding bricks shear under the differential behaviour of the external skin and the internal mass. The result will be that the structural skins are disassociated and the effect can be quite invisible from either face of the wall. With the failure of each successive bonding brick the differential movements are effective over greater distances and the stresses build up causing further bonding bricks to fail. Ultimately the slenderness ratio of the detached skin causes it to buckle under load, particularly if the effects of salt crystallization cause unequal pressures across the mortar bed. Whole sections of brick facing can split or peel away in consequence. At a different scale the same effect is commonly observed in a rigid stucco or render. The failure of the external brick skins of Martello Towers is to be attributed to this phenomenon. Detachment is enhanced by the wedging effect of particles of loose mortar which drop to the bottom of a widening cavity providing a wedge against which further movement simply extends the breakdown.

Internal detachment of sections of wall can lead to a critical overload. Composite walls in particular are subject to this weakness. Skins of stone or flint built against brickwork may be stable and adequate to their loading while the structure remains coherent. Detachment of one skin from another can mean the transfer of the total load to, for instance, a brick inner skin which, being subject to increased loading and having a much increased slenderness ratio moves to a point of failure without visible indication of prior distress. The failure of cavity ties produces this condition.

A whole range of problems arise with defective original design. Historic buildings of every age contain details which can, in the light of experience, be improved upon or rejected out of hand. Missing damp-proof courses, inadequately protected steel, over-stressed flashings and lack of drip moulds are common causes of failure in historic buildings of recent date. With problems of this type the conservator faces decisions which involve improving upon the original work and sometimes altering its appearance.

6.5 Remedial action

In principle, the repair of structural failure can be divided into alternative types: those in which the repair accommodates the defect by countering the forces which caused it and those where the repair is designed to accept these forces and their effects. The initial question to be asked is, therefore, whether the original condition is to be restored or whether the defect can be accommodated and accepted as a new circumstance. A failed lintel is a simple case. Both structurally and aesthetically a new lintel must be provided to support the brickwork and the repair will be designed around this decision. The only question will be whether the lintel shall be of the original material or of that material in some way modified, or whether a new material is to be substituted – timber, proofed timber or reinforced concrete for example. Extensive cracking caused by expansion and contraction in the structure may present the opposite case. To provide such strengthening as would resist these movements could require powerful reinforcement in the knowledge that the containment of these movements at one point would result in their expression elsewhere. In such a case the naturally developed crack may be accepted and treated as a movement joint. Remedial work would be designed to provide the practical cohesion of the structure with adequate weathering.

In some instances the forces which caused the damage may be disregarded. Brickwork which has suffered war damage may be restored to its original appearance without a qualm although patriotic sentiment may ordain otherwise. Brickwork which has suffered damage due to subsidence may be confidently repaired on a like for like basis provided that the cause of subsidence has been eliminated. However, brickwork developing shear cracks which if unrestrained will lead to catastrophic failure may demand powerful temporary restraint and permanent reinforcement to the point when the coherence of the brickwork is dependent upon prestressed cables grouted into borings through the structure.

The decision as to whether it is practical or possible to counter the forces which have caused structural damage is likely to be reached on practical engineering grounds irrespective of aesthetics and historical integrity. It will be a presumption that so far as possible the historic structure will be saved and maintained as a coherent building. The extent to which the repair counters destructive forces or accepts them will be based on practical and engineering considerations. Within these broad guide lines there is a range of options and often acceptable alternatives. Where the cause of failure persists as a defect or a threat its removal is either a prerequisite or is prudent. Parallel precautions may well be called for, or be suggested by the event. Embedded rusting iron breaking up mortar joints, failure of wall ties, embedded decaying timbers and similar items of damage will be repaired at the same time. Their replacement by a material designed to do the same job – and perhaps to look the same – will be a natural and reasonable response. The replacement material will be chosen to be more durable than the original. It may be chosen deliberately to simulate the material it replaces with the objective of making a sympathetic historical statement. If so the material should distinguish itself as new on careful examination.

Brickwork may be bedded in a powerful cement mortar or at the other extreme it may be bedded in nothing more than earth. Between these lies a range of materials of varying coherence which require a careful understanding of their intrinsic properties when support is called for. Structural cracking defines a line across which tensile stress cannot be carried and in consequence stresses may be concentrated elsewhere in the structure, possibly to the point of overload. As stress increases towards the point of failure an increasing number of movements take place within it and many of these are virtually imperceptible. They are often movements which occur due to changes in condition; vibration, temperature, humidity and so on, from which the brickwork may not recover fully. A slow movement, however, may accelerate, ultimately with disastrous effect. The conservator will ensure that progressive movement is always countered by an increasing force in a stable structure.

Where the mortars in brickwork are weak the individual bricks may maintain a bond by little more than their weight and frictional force, but where mortars are strong, whole elements of brickwork will behave as one coherent mass, adhesion between brick and mortar being sufficient to have tensile effect. Support delivered as a thrust at a specific point will be effective only if the mass of brickwork is capable of distributing the stresses uniformly. If the brickwork does not have this intrinsic strength it is essential that the thrust is spread before it is delivered into the brickwork. In the provision of temporary support spreaders which have a surface softness and some elasticity should be used against brickwork to control and balance the restraints offered by temporary shores, struts or corseting.

6.5.1 Temporary support

The normal resource in the building industry is a screw jack or prop which can be hand loaded with considerable sensitivity and may be levered or even driven up to offer greater support. Hydraulically actuated props are available, being used extensively in the mining industry, and these are capable of far higher loadings. More importantly, their actual stressing can be accurately measured and varied. Equipment of this type is the principal recourse when major loads have to be lifted. 'Flat jacks' – pancake diaphragms of steel, filled with hydraulic fluid oil or water – may be used where the lifting pressures have to be generated in a confined space. Filled with concrete they may be used as a permanent support.

Arches must always be supported by formers bearing uniformly beneath them, never by point support. Leaning or bulging walls may be supported by raking shores or by directly acting shores or binders bearing against planked or boarded surfaces. A building in danger of collapse by outward structural failure – bursting apart – may be restrained by containing its outward thrusts within itself. A corset or circumferential arrangement of scaffolding can be designed to contain the reactions by balancing thrusts from opposite sides. Steel scaffolding can have an elasticity which can be sympathetic to distressed brickwork restrained by screw jacks. Timber interposed between the scaffolding and the jack provides additional softness. In tender circumstances a sandwich panel of plastic foam between plywood slabs may be used. In circumstances of great fragility support can be offered by a fluid (water or air) bed.

Structural deadload will normally be carried down through the brickwork but occasionally it will be necessary to support sections of the structure onto scaffolding, particularly where it is necessary to rebuild or remake sections of walling or to renew foundations. Such underpinning must be preceded by supporting the brickwork. For this purpose it is normal to make use of needles – bearers threaded horizontally through the brickwork via existing openings or holes cut for the purpose. Needles will normally pass through brickwork with an intervening distance not exceeding two or three brick lengths and they must be pre-stressed to support the existing structure with a pressure equal to the deadload. The structure must not be allowed to drop (even fractionally) on to the needles because its coherence will thereby be broken. In cutting out the voids required by needles the potential damaging effects of percussive tools (even hand tools) must be considered. Rotary drilling may be used to minimize shock.

Successful support can be achieved over short spans by drilling holes in a mortar bed, introducing threaded bars into these holes and offering on to these bars a perforated steel angle to which they are bolted. The angle is then supported mechanically, providing restraint across a substantial width of brickwork. An alternative involves chasing a horizontal mortar bed to about 50 mm and introducing into the chase an adapted steel angle bracketed on to screw jacks. Both methods are suitable for small spans – up to 3 m – and deliver eccentric loads to the jacks.

Support under needles is traditionally achieved by empirical judgement in driving together opposed wedges (folding or lazy wedges) until an apparently sufficient pressure has been obtained. While such methods may be adequate for small and straightforward structures there are occasions when more precise methods are required. It may be damaging to cause lift in brickwork due to the breakage of mortar joints. Excessive up-thrust must be avoided as much as lack of support. Careful calculation of deadloads can provide an accurate assessment of the support required and this can be delivered precisely through flat hydraulic jacks capable of small movement but of carrying large loads suitably positioned on or beneath the needling. Similar methods can be used to relieve structure of an excessive deadload which may be imposed by other forms of failure. A significant advantage in their use is the ability to measure the applied load by hydraulic pressure. Where minimal pressures are required – as in the case of plasterwork or single skins of detached brickwork – air may suffice. Tyre inner tubes, airbags and air beds may be considered when a light but uniform pressure is needed, with the advantage of the ability to measure pressures by straightforward techniques using equipment which is readily available.

6.5.2 Repairs below ground

A structure may need underpinning due to failures in ground support or in the foundation construction. Where, due to mining subsidence, or adjacent land slip, a transverse movement is detected it may be necessary to isolate a historic structure from the ground to minimize damage to the upper levels. No such operation should be carried out without careful engineering advice which will be aimed at ensuring that support is coordinated. To this end a grillage of steel joists may be used to isolate the structure from differential movements in the ground as changes take place. Grillages may be used in bridging over or diverting an underground water course, controlling drainage, restoring water table levels, arresting the motion of ground or transferring loads through a shallow insecure layer into a reliably load bearing lower stratum. It is possible to create artificial slip planes in ground by trenching and inserting sheet material bedded in sand. Proprietary boarding is available for such purposes. It consists of a honeycombed structure designed to collapse under load (e.g. in excess of 3 kN/m^2) when wet.

Reinforcement *for* Brickwork

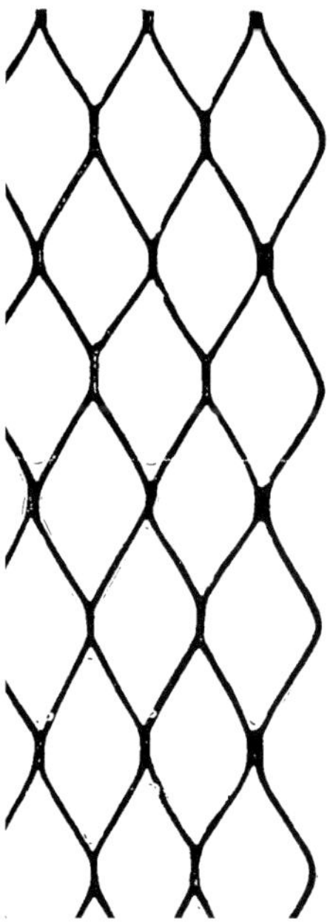

As a general rule, a strip of "   " 2½ inches wide should be used for every 4½ inch thickness of wall, and in most cases in every course. But the 7-inch and 12-inch strips will be found preferable for 9-inch and 14-inch walls respectively; and, obviously, combinations of the three widths of reinforcement may be used for walls of greater thickness.

In some cases it may suffice to stagger the reinforcement in bed-joints of walls 9 inches and over in thickness, i.e., in one bed-joint to lay a width 1 inch from one face of the brickwork, in the next bed-joint a width 1 inch from the other face, and so on, throughout the height of the wall.

Width of Material	Thickness of Steel	Approx. Weight	
		270 ft. Coil	75 ft. Coil
2½ inches	20 gauge	23 lbs.	6½ lbs.
,,	22 ,,	18 ,,	5 ,,
,,	24 ,,	14 ,,	4 ,,
4 inches	22 ,,	31½ ,,	8¾ ,,
,,	24 ,,	22½ ,,	6¼ ,,
4½ inches	22 ,,	32¾ ,,	9¼ ,,
,,	24 ,,	25¼ ,,	7¼ ,,
6 inches	24 ,,	37½ ,,	10½ ,,
7 inches	20 ,,	65 ,,	18 ,,
,,	22 ,,	51 ,,	14 ,,
,,	24 ,,	39 ,,	11 ,,
12 inches	20 ,,	112 ,,	31 ,,
,,	22 ,,	87 ,,	24 ,,
,,	24 ,,	67 ,,	18½ ,,

Coils of other lengths, or bundles of flat strips of any length up to and including 16 feet, can be supplied. Unless specially ordered otherwise, 270 feet coils are sent.

Figure 6.11 An advertisement in the architectural press from before the Second World War. Buildings, now historic, in which this type of material has been used may be subject to serious failure due to rust expansion in the joints. The specified 1" cover of the steel is inadequate in an external skin.

Piling may be used to solve problems of this nature. Sheet piling might be used to contain the supporting soils, hydraulically driven rigid (timber or precast concrete) piles might be used to compact and consolidate soils and bored piling might be used close to the structure to provide a method of transferring deadloads to load bearing strata. A historic brick building should never be subjected to the use of intermittently driven piles in its immediate vicinity. Repeated shock can cause internal damage which is beyond detection. Conservation must err on the side of caution in such matters.

Where steel or precast concrete sheet piling is introduced at some distance from the existing foundations it provides, in effect, a dam against which earth pressures will build up. Consolidated piling may be designed in serried ranks to arrest ground movement and to densify sub-soils while bored piling may be used to carry a ground beam to be linked to any grillage or underpinning intended to provide permanent support. The engineering work must not impose shock, vibration or sudden stress upon the building.

Bored micropiles have been used extensively in recent years, patented systems allowing pressure injection and pressure expansion of the tip of the pile to maximize bearing. Being bored these systems are particularly appropriate to brick structures, integrating readily with the existing foundations and working on the principle of supplementary support in which the pile accepts additional loading as the structure moves or loads increase. If immediate support is required flat jacks may be provided between the pile and the under-bearing. For permanence such jacks may be filled with concrete under pressure.

Exceptionally the conservator can have recourse to dewatering and freezing. In the former technique perforated tubes are driven into an over-wet stratum and are connected to pumps. Surplus free water is extracted. The negative pressures produced carry water out of the soil until the free water content is reduced to an acceptable level. This type of action is usually aimed at bringing a saturated soil below the fluid limit to stabilize it but is of no assistance in bringing a soil below the plastic limit. Shrinkage must take place with possible consequent settlement. Freezing on the other hand will stabilize wet soil and saturated ground without inducing shrinkage. In fully saturated ground expansion will occur although normally without adverse results. Pipe coils are introduced – sometimes sheathed in hollow piles. Brine rather than refrigerant may be the heat transfer medium. Ground can be made both solid and 'waterproof' by such methods (more readily so in a temperate climate) and the condition may be maintained, at expense, as long as desired. Neither technique is instantaneous, cheap or permanent but the objective of stabilizing a historic structure may be achieved initially by such means. Circumstances that might demand such methods could include subterranean operations such as the construction of large drains and the removal of water pressures to allow the insertion of permanent waterproofing material.

6.5.3 Reinforcement

Tensile reinforcement of brickwork takes several forms, namely:

- The introduction of mesh or other grillage within the courses.
- The introduction of rods within brick structures, perhaps tying to adjacent structures, usually to provide resistance to shear.
- The introduction of cables into borings in the brickwork, by which the structure may be prestressed by post tensioning.

Plasters and renders are reinforced by materials of a mesh or grillage type similar to those used in the reinforcement of brickwork.

Exotic forms of reinforcement employing plastics and alkali resisting glasses have been used as brick reinforcement set within mortar beds. However, some current research suggests that alkali resisting glass reinforcement has a limited life. Chopped strands or mats of such materials are employed to reinforce renders and plasters. Natural strand materials such as hair have long been used for such purposes, as have fibres of split cane and other plant forms. Materials which will decay organically are not to be recommended in remedial work, other than in plasters.

The extent of the employment of mild steel reinforcement in non-historic brickwork is often not recognized. It has been used in USA, Australia, New Zealand, Britain, India and Japan and European countries in this century. Impetus was given by savings that could be achieved by reducing the thicknesses of walling and advantages in resisting ground movement and shock. Model bylaws in the 1930s in Britain allowed the reduction of wall thicknesses by half a brick (i.e. 4½" [110 mm]) if the wall required was upwards of 1 brick thick, i.e. a 13½" (340 mm) wall could be reduced to 9" (230 mm). A contemporary manufacturer recommended the use of mesh in every course and in dimensions which gave the steel 1" (25 mm) cover of mortar of unspecified strength! Inevitably structures of this type have been prone to failure due to rusting of reinforcement and even where the mesh is galvanized an indefinite life is not to be expected. Many more failures will become apparent. Such hidden reinforcement may be detected by magnetometry. Conservation of such structures must begin with the rejection of the initial method of construction. Repair is likely to involve reconstruction to a different specification while retaining the built form.

Reinforcement laid in the beds of brickwork became popular in the wake of the all-consuming success of reinforced concrete. Mild steel has sufficiently similar coefficients of expansion and contraction to concrete made with Portland cement to be effectively coherent and the adhesion of cement-based concretes to clean steel is good. The actual differential in coefficients is not sufficient to break a well-made bond at normal temperatures. The bond forms an alkaline environment coating the steel and this provides some considerable resistance to oxidation, i.e. rusting. This resistance, however, is only effective provided that something between 30 and 60 mm of dense undamaged concrete shields the steel from the weather. The thickness depends on exposure, density of concrete, and the presence of fissures. Permeability to water and oxygen is the crucial factor. Prestressed concrete, being unlikely to develop fissures under normal conditions of dynamic loading, does not suffer from the defect of unstressed concrete which contains minute fissures consequent upon the movement whose development can be associated with

Figure 6.12 Buried steel is not protected by a skin of brickwork. Rusting in mild steel packing plates on top of a steel stanchion lifted the corner of this house inducing diagonal cracks from the steel to the window cill – made good (left) – and rust on the face of the stanchion forced away the brickwork of the quoin (right).

varying loads. Generally brickwork does not provide adequate protection to mild steel although the mortar between the beds of the bricks in historic buildings of this type may well be composed of a substantial part of Portland cement. The brick itself does not reliably provide the weatherproof cover which shields steel sufficiently from liquid water. Water can saturate up to 12" (300 mm) of brickwork and the thickness of a mortar bed is insufficient to give the additional protection required against saturated brick. Defects in laying the mortar accentuate the problem and ageing reduces the high alkalinity of the mortar, allowing oxidation (i.e. rusting) to increase progressively. A particular cause of failure in framed buildings is the use of brickwork as a skin or part of a curtain wall. By definition such brickwork is not under load. Therefore structural movements, particularly under wind load, tend to be higher. These can induce horizontal cracking which permits water-entry. Both mortar and concrete contain excess water at time of laying to produce flowability. As it dries out this water leaves continuous pores which provide a passage for further water entry.

Unless, therefore, reinforced brickwork is contained within a structure in a weather-shielded position, or unless the reinforcing steel is non-corroding there is a danger amounting in many cases to certainty that steel reinforcement will corrode, losing its strength by diminution of cross section and opening up cracks in brickwork by rust expansion which will allow increasingly rapid water entry. Failures of this type are likely to be associated with other similar defects in design and specification involving fixings, damp proof courses and flashings. The functions of many of these details and their best use was often ill-considered in early buildings of the 'modern' movement. Many of these buildings are historically important and if the same principles of integrity are to be applied to them as are applied to buildings of earlier periods the conservator will need to retain the quality and characteristics of the buildings without repeating defective methods of construction. Where rusting reinforcement, rusting cavity ties or fixings and ill placed or failed flashings have been detected and are the cause of defects there may be little alternative to the removal of substantial areas of brickwork prior to its rebuilding with improved materials and detailing. However, in some cases it has been shown that ties, located magnetically or by evidence of rust lifting the brick courses, may be drilled out and replaced with resin-bonded fibreglass rod embedded in a polyester resin. While there is insufficient evidence of longevity in this technique it may be appropriate in circumstances where the destruction of the wall will occur if no action is taken.

Ash mortars and walls permeated with sodium chloride are particularly inimical to embedded iron ties, whether or not they have been galvanized. In some industrial areas the use of ash to gauge lime mortars was common from the early nineteenth century. A high sulphate content may arise and a slow change from high alkalinity to mild acidity provides the circumstance for persistent corrosion. These mortars have been called 'grey' or 'black' or 'black ash' or 'black lime' mortars.

Brick reinforcement may be advocated as a method of providing additional strength to a historic structure to increase the rigidity, carrying capacity or coherence of areas which have failed or simply to provide additional tensile strength across proven zones of weakness. Where, after careful analysis, reinforcement is included in conservation work non-rusting materials should be employed. Zinc coated steel is not adequate in the long term.

Proprietary mesh and expanded metal products are available for such purposes and their use may reasonably be contemplated where the following criteria can be answered positively:

- Will the reinforcement be protected permanently from decay by its situation and physical protection?
- Will it transfer loads and stresses within the structure adequately and without causing them to create ruptures or breakdown at other points?
- Is there no alternative more consonant with the historic nature of the structure?

Where, taking these constraints into consideration, it is decided that brickwork should be reinforced for reasons of safety, security or simply historical accuracy (it might originally have been reinforced) a choice may be offered between placing in the brickwork a tensile material which induces compression or a material capable of acting as a beam or rigid member. Into the latter category fall those elements of brickwork, unattractive as often they are, where simulated arches are carried on metal angles or bearers. This is a technique not merely of recent construction but one which goes back beyond the eighteenth century in European brickwork. Wrought iron members were commonly used as formers and permanent shuttering for low pitched or flat brick arches and in many instances were placed too close to the external skin of the building. Although wrought iron has a lesser tendency to rust than mild steel, many such members now betray their presence by rust expansion.

6.6 Embedded ironwork

Wrought iron is a material of widely varying compositions depending upon its manufacture. Historic wrought iron may, by its nature, be laminated having been beaten out of iron billet. The working on the forge (hence wrought) reduces the available carbon and highly worked wrought iron may approach the reduced carbon content of steel. The alloy known as cast iron contains substantial amounts of carbon and in consequence has considerable durability in the presence of water. Other alloys with nickel, chromium, vanadium, manganese, etc., are similarly resistant to oxidation and are known as stainless steels. The process of forming wrought iron, which has the advantage of being more malleable than cast iron, drives off substantial amounts of carbon. An even greater proportion is taken out during the purification processes that produce molten steel. Rust inhibiting properties will depend upon other, usually metallic, elements with which it may be alloyed. All iron embedded in brickwork must, therefore, be treated with suspicion. It should be a presumption of all work in historic buildings that iron is not introduced to or embedded in brickwork. Where for positive historical or structural reasons the initial presumption is overcome the greatest care should be taken to avoid the formation of rust. Bedding in lead was the common answer historically, expansion being taken up in the lead or lead wool packing. Whatever the case, iron in the process of rusting should be removed from brickwork and mortar. By contrast rusting iron can (advantageously) take a very powerful grip on structurally compacted earths.

Methods of removal have been devised in which intermittent bricks are taken out allowing the ties to be cut off and extracted. The inherent dangers are that elements of rusting steel will be allowed to persist while the replacements and their fixings may themselves be inadequate. Stainless steel fasteners, plastics and epoxy resin fillers are offered for these purposes. Their successful and effective use depends a great deal upon the technical skill of application. In addition to the necessary expertise in fixing the new materials there is an intrinsic danger in the removal of the decayed iron. Where the original mortar contains a substantial amount of Portland Cement the brickwork will be hard and resilient. Fractured along the lines of stress caused by rusting iron, it may be difficult or impossible to remove, and percussive methods of removal or, worse, methods involving thrusts or pulling, can cause serious damage, much of which may not be visible. The greatest care must be taken in the removal of reinforcement from historic brickwork and in its renewal or replacement. Sawing out mortar beds is generally to be preferred to any form of hammering. Preparatory sawing will provide a discontinuity which reduces transmission of shock and accumulates minor strains as wedges or chisels are introduced.

An important and wide application of the use of metallic mesh as reinforcement is in the provision of a key to plasters and renders when applied to brickwork. Such methods should not be regarded as a panacea and there should be a presumption against the introduction of special and modern types of reinforcement in structures not designed to receive them. However, there may be circumstances in which their use is justified and the presumption may be overcome.

Such circumstances would include the existence in brickwork of a thermal movement joint which had formed and was to be allowed to remain or was likely to persist in a position which gave rise to a corresponding crack in the superimposed plaster or render. Such a crack might be visually unacceptable and have no relevance to the intentions of the original designers. By the use of sheets of expanded metal reinforcement extending across and well beyond the line of the crack the effects could be carried into a sufficiently wide zone of plaster or render for the movement to be tolerated without cracking. The mesh would be securely fixed to the wall at an adequate distance from the thermal movement joint and would then provide a secure background for the replacement plaster or render of a thickness calculated to tolerate movement. In some instances deliberate detachment of the plaster from the brick to a distance of up to 500 mm across the crack will serve to spread movements sufficiently for the plaster to tolerate them. Detachment may be achieved by paper or plastic film.

Other similar uses of reinforcements of this type include the conservation of decayed ceilings, where laths might be in poor condition and the nails holding laths to ceiling joists even worse. Removal of the ceiling and its loss can be avoided by supporting the plasterwork, cleaning out all decayed material by suction, laying down trays of (ribbed if necessary for rigidity) non-corrodable metal mesh or lath, on to which is poured a gypsum plaster. This plaster, bonding with the keys of the existing plaster and locking on to the mesh, is then capable of transferring the deadweight to the ceiling joists to which it will be fixed by non-corroding nails.

6.7 Tie rods and bars

Many historic buildings have been tied structurally by the use of rods with terminal plates (pattress plates) set on the face of brickwork. The function of these rods is to resist bursting pressures in walls due to loss of restraint in roofs or floor structures. Typically the structural problem will arise as a building ages, perhaps due to settlement, weakening of mortars, or timber failure. Timbers embedded in walls are frequently subject to sustained dampness with the result that the bearing surfaces fail, becoming friable and lacking adhesion to the mortar bed on to which they were once laid. These may be wall plates, running longitudinally in the wall or on top of it, or embedded joists. The loss of frictional cohesion allows other forces, usually dynamic loads associated with wind or thermal effects to induce minor movements in the structure. There are no counter-forces to assist recovery. As the building distorts it becomes apparent that restraint is required and rods may be threaded across the building perhaps at joist level (for concealment) to be secured against the outer face. Wrought iron straps formed into the shape of a cross, an S or left as a simple bar running vertically may suffice as a bearing surface, but the problem was so common that enterprising iron-founders produced circular and oval pattress plates which have also been used for the purpose, often with a date or name cast on them, providing valuable historic evidence. A traditional way of tensioning such rods was to provide a screw thread with a nut at one or both ends and to heat the rod. While it was at maximum elongation the nut would be brought up to full adjustment allowing it on cooling to draw the structure together. Heating would be by charcoal or coke in a trough around the rod. The process could be repeated a number of times and its success as a technique is testified by the many extant brick buildings whose relatively thin skins continue to serve their design function.

Failures were often a consequence of rot which affected the surfaces of timbers built into solid and therefore damp brickwork, causing loss of adhesion – particularly where wall plates fail causing tie beams to be ineffective. Where roof structures are the cause of the problem the design of the rafter couples or trusses may depend upon a collar (a tie between rafters fixed above the foot) intended to function also as a tie beam. Principal rafters are sometimes weakened by morticing to accommodate the tenons on such collars with the result that they deflect on the line of the collar and deliver an outward thrust at the eaves. A builder or designer anxious to save money might have used such an expedient in order to avoid building the eaves to the full height of the ceiling and the same financial constraints might have led to the use of undersized roof timbers. The ultimate effect is an outward thrust on a wall perhaps unduly thin for the same reasons.

These remedies have now become historic entities in their own right and whether or not essential are protected parts of the historic fabric. The same methods, however, are used today. Modern pattress plates are provided in cast steel and may be obtained in a self-aligning form to take account of variably oblique lines of restraint. The face nut is ball-finished. Tensioning is by nutting up from the outer face but bottle screws may also be used. The pattress plates are thickened and ribbed to distribute stress. Other sophisticated techniques and tensioning devices are available but the principle remains the same. It is common and acceptable practice (subject to safeguards) to anchor such members in concrete beams cast in situ and carefully keyed to the structural brickwork but waterproofed against corrosion. A painted d.p.m. on the concrete is a common precaution. Provided that such undertakings are carried out recognizing and guarding against the dangers of the leaching of salts from Portland cements (see below) there is no more effective general way of delivering a well-distributed restraint across the breadth of a brickwork structure.

The conservator will normally rely on specialist technical and engineering advice which will take into account the nature of the brickwork and mortar. The wide experience in the use of these devices relates as much to stone and reinforced concrete as to brickwork and much of the key experience has been gained in civil engineering applications.

6.8 Mechanical fixings

Other fixing devices used to secure brickwork are the ranges of fasteners and anchors developed for civil and mining engineering. There are some circumstances in which elements of historic brickwork may be tied back to native rock or other masonry by mechanical fixings.

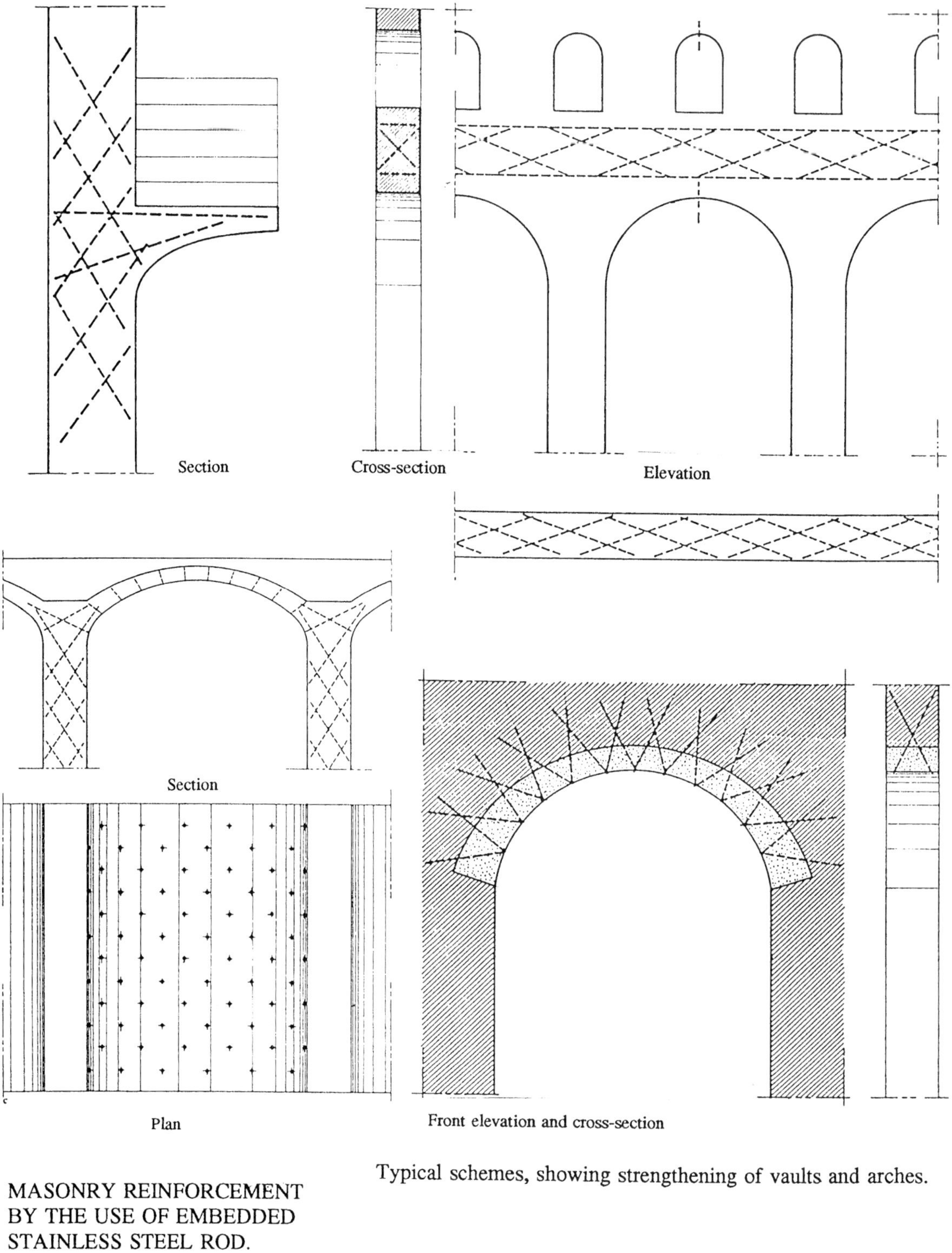

Figure 6.13 Illustrations of the methodology of introducing tie bars and tension cables into masonry by the Fondadile techniques. (Courtesy SAGEP Editrice, Genoa.)

The principal available types depend upon an initial drilling operation succeeded by the insertion of patent bolts which expand to grip the walls of the bore or are embedded in a resin cement which may be delivered in a capsule to mix and penetrate under pressure within the bore once the fixing bolt or fastener is in position.

On occasion fixings are hidden by the removal of part of the outer skin so that the fixture may be inserted within the thickness of the brickwork, the outer bricks being then replaced.

The specialist engineers must ensure that the drilling operations are kept well within the tolerances of the strength of the brickwork. Methods are available by which fixings are shot-fired into the structure. Normally there is no application for such methods in historic brickwork due to the shock wave produced.

The tensioning of rods in or across brickwork by screwing or by contraction is designed to restore the position and alignment of brickwork or to pre-stress it or perhaps a combination of both. The conservator should be clear which of these options is being pursued and the extent to which such tensioning operations are allowable. Such decisions may be as much a matter of judgement as calculation but the process of calculation should always be exploited to its maximum extent before any judgement is entered.

Measuring techniques of very great precision are now available and where historic brickwork is to be restrained, eased, repositioned before being stabilized and reinforced the work will be planned on the basis of the amount of movement thought to be tolerable, the rate at which such movement should be achieved, the acceptable stress in relation to the anticipated strain and the ultimate method of fixing – perhaps bonding in position by grout, perhaps bonding by additional brickwork, buttressing or permanent mechanical restraint. Some movements may be designed to take place over very extended periods. An operation which would have no chance of success if carried out in a matter of minutes might be entirely satisfactory if carried out over several months. It is also most important in any operation involving the rectification of a material with some inherent flexibility (as has brickwork) to pre-consider the monitoring process during the operation and to know what action is to be taken if the effects experienced depart from those predicted.

Metal has been used to provide tensile strength for, or to contain pressures in, brickwork for a very long time. Perhaps the most famous extant example is the immense chain which is built round the base of the great structural brick cone within the dome of St Paul's Cathedral in London. In such cases the tensioning is through the settlement of the brickwork being taken up by the metal over a period of time (post-tensioning). An entirely new technique, in which Italian contractors have played a leading role in developing and applying the expertise, is the reinforcement of brickwork and stone masonry by the introduction of tension cables passed through holes bored in the masonry mass and embedded in grout. These are then tensioned mechanically and then locked into position (pre-tensioned, i.e. they are stressed prior to taking up the natural load). The technique is perhaps the most dramatic and far-reaching application of modern technology to historic brickwork.

Various patents cover parts of the system, with specific attention to the type of drill head. This has to be controlled accurately while cutting at distances of up to 30 m from the entry point. By this means complex grillages of cables and wires can be carried through the masonry structure, a cementitious grout can be injected and on tensioning (commonly by hydraulic mechanisms), brickwork can be given a coherence previously not considered possible, without apparent external effect. One difficulty commonly arises, however, from the percolation into brickwork of the soluble salts derived from the grout. These eventually appear as a grey bloom externally, to the disfigurement of the monument. Their removal is virtually impossible because the source of the problem cannot be removed.

7

Physical repair

7.1 Cutting out and making good

Where a brick must be cut out percussively (by cold chisel or by powered equipment) care must be taken that the energy delivered in the impact is released laterally. In practical terms this means that every blow should be delivered at a point adjacent to a void so that the stress pressure resulting from the blow is released as the fracture occurs rather than being contained or pent up within the matrix. By this means excessive stress is avoided. The practical means by which this can be achieved in a flat surface of brickwork is by drilling, grinding out or chasing initially adjacent to the intended cut. The percussive blows should begin adjacent to the drilling(s), working outwards and always having one free side into which fractured material can move.

When the source of the defect is fully understood, corrected or is allowed for, consideration can be given to the repair of voids or cracks in brickwork. It may be necessary to lift a damaged or decayed element back into position, to secure it or support it and then to repair the damage. The decision may be to repair a crack in the knowledge that future movement will take place, the repair serving simply to reduce the size of the crack rather than to eliminate it. Where thermal movement is to be anticipated flexible pointing techniques may be involved and the further opening or closing of the crack may be allowed by applying a flexible mastic or plastic foam filler surfaced with a flexible polysulphide or similar sealant of appropriate colour. The merit of providing such a filling will generally be the exclusion of weather and air movement in general terms while accepting a minor crack with the weathering and replacement consequences that this implies. This principle will be followed only where the conservator recognizes the need for a permanent movement joint and usually only where the joint is concealed or is not visually prominent should this type of solution be employed.

Such a crack may well follow the perpends and half beds in pursuit of a vertical alignment. Where repointing or sealing is intended cutting out and remedial work should be not less than 20 mm deep and preferably substantially deeper if the joint is wide. The principle is that the depth of pointing should exceed its width. Each perpend will be treated individually where a crack of this type is being repaired.

A long-lasting weatherproof repair may be achieved, however, using one of the many ranges of mastic sealants. Colour matching may be a problem since factory batched materials will be limited in their range of colours. By special order matching colours can be obtained. Surfaces must be dry and firm and the anticipated movement should never exceed 30% of the width of the mastic pointing. Moderate longevity is now assured in the case of the reputable blends of flexible sealants – elastomers of various types. In appropriate situations with adequate protection and good workmanship a sound and weatherproof joint can be achieved.

No such solution, however, transfers significant additional strength across the joint. Discontinuity is the correct approach to a crack where movement is required or where a powerful bonding will transfer unacceptable loads to other parts of the structure. Where it is required that the structure shall be continuous the strength required must be determined. Where substantial shear or tensile stresses are

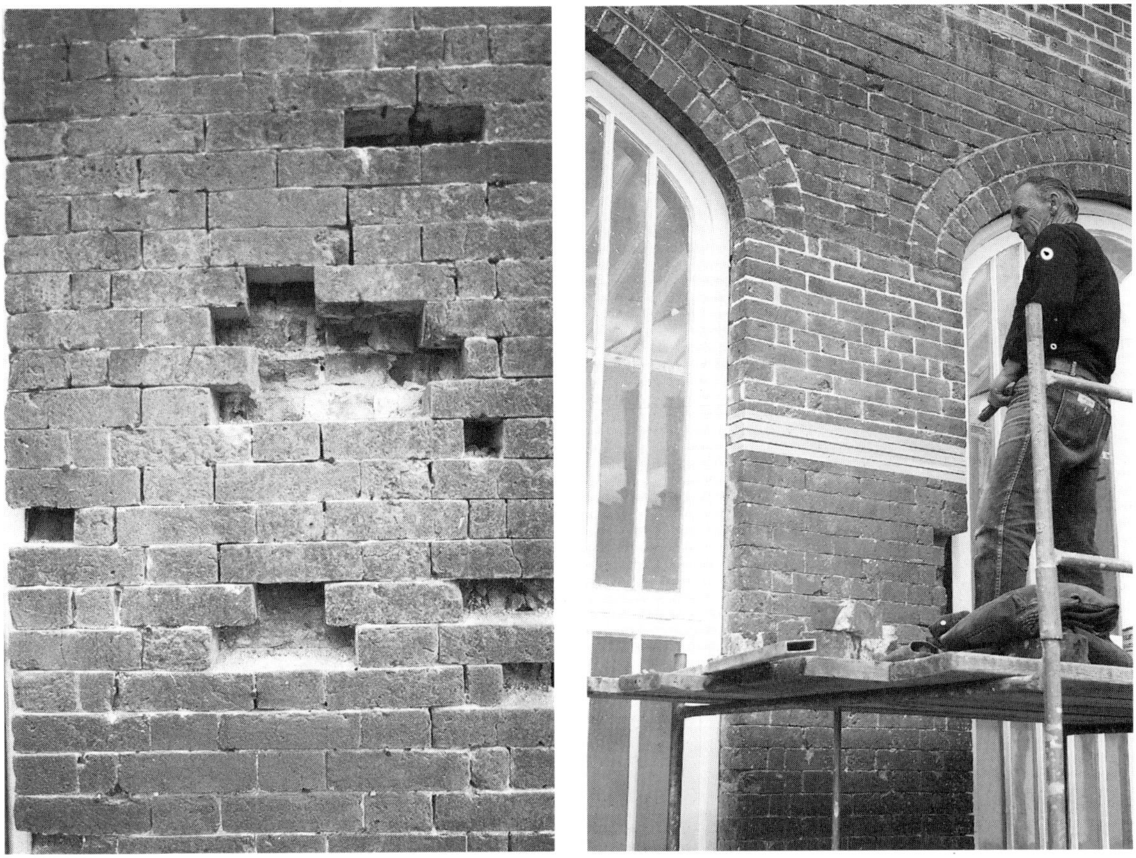

Figure 7.1 Gauged brickwork with very fine joints prepared for the insertion of matching material (left) and in the process of repair, joints being raked with short lengths of hacksaw blade and pointed between protective tape with lime putty (right).

to be anticipated metallic elements may be included as connectors. Stainless steel or bronze dowels are commonly used, particularly where shear forces are to be resisted. Where tensile strength is required connectors with ragged or butterfly terminations may be chosen. Dowels may be secured using resin adhesives in drill holes subject always to the difficulty of manoeuvring and aligning the drilling. Ragged or butterfly ended connectors will be built into mortar. Metal connectors of this type will always be stronger than the brickwork into which they are fixed or embedded and in calculating maximum capability the critical strength is that of the brick and mortar at the fixings. Careful calculation will generally result in the specification of units in standard wire gauge dimensions placed at frequent intervals rather than heavy connectors which tend to be widely spaced. Shear forces and tensile stresses should be considered separately and in some circumstances it may be desirable to provide reinforcement against shear where differential loadings may be involved without necessarily providing tensile strength. This can be done by setting dowels into pockets or even more thoroughly by setting dowels into embedded sleeves. Situations justifying this approach will be those in which it is desirable to equate bearing pressures across differently loaded or differently supported sections of wall while at the same time permitting flexibility in the lateral sense to tolerate thermal movement.

Where cracks are to be filled by inserting or making good the brickwork itself as a continuous structure attention must be given to the extent of the bonding throughout the depth of

Figure 7.2 A repair of impact damage executed in a cementitious render struck to receive a matching pointing. This technique may be appropriate where the existing brickwork must not be disturbed or where the damage is retained as a visible statement. A suitable choice of mortar is essential. In this case a mix unduly rich in Portland cement was evidently chosen, causing a 'bloom' visible on the right and of unsatisfactory colour. A pigmented mix with an aggregate of crushed brick would have been more successful.

the wall. Bonding may be repaired either by cutting out and inserting tying bricks with suitable broken jointing so that there is a positive brick bond throughout the renewed zone or alternatively, at much less cost of excavation, by making good the facework and filling with an appropriate grout. The nature of the fractured faces will normally provide a very sound surface for grouting and where shear or torsion is to be anticipated in the joint filling or grouting may be a better solution than the use of bonding bricks. Where tension may occur across the joint brick bonding is more calculable and more reliable. The strengths achieved are probably higher because the bonding surfaces will be greater due to the depth of the stitching bricks. The strength of the bricks in use will be known from laboratory tests and mortars must be matched to the existing. Under compression both methods of making good are equally satisfactory.

In placing new brickwork to reconstruct a section of wall previously fractured an opening must be made sufficient to allow the adequate bonding of new bricks. The total dimension to be spanned will be greater than the original dimension by the width of the fracture. Since bricks vary in size and the opening will be two or more bricks in length, careful choice of the replacement brick should yield perpends of widths similar to the original wall.

Failure to select brick lengths with care will leave a pattern of over-wide perpends. The objective will be to reinstate bricks of similar colour and texture in the pattern and bonding previously used. In some circumstances this may not be possible because the adjoining sections of wall will have moved vertically so that any replacement brickwork will be out of line. This effect may be minimized by providing an adequate width of new work or accentuated, perhaps as a historical statement, by amending the bedding at the junction so that the mortar thickness accommodates the new dimension and remains permanently visible.

7.2 Salvage and infilling

The art of salvage in brickwork repair is a subtle one. Recovery of brick can save both time and money and, more importantly, can improve the quality of the finished product. It can also prove deceptive and where the repair could be mistaken for original work care should be taken to ensure that the repair is identifiable. A date marked on the brick with a greasy crayon on the concealed face is sufficient. In most structures bricks suitable for facework are lost to view as they are built into the mass of walling. Sometimes they disappear below ground. In careful conservation where the original work was in lime mortar there may be manifold opportunities for recovery of bricks with an adequate or excellent full face or header and these can be used to replace the damaged or fractured bricks to complete a full wall. Occasionally bricks may be reversed. Where a surface failure has not been due to the quality of the brick itself but due to impact, stain or maltreatment the originally concealed face of the brick may be recovered by reversal. Whichever way the effect is achieved it is usually advantageous to make good use of original material in the knowledge that the newly placed bricks will weather in the same way as the surrounding wall. New work may be differentiated from old by detailing or material used in the mortar joint. New work should be designed to weather towards the character of the existing structure even if its initial colour is at variance. There is always a temptation to pursue an immediate match in both mortar and brick surface which may give initial satisfaction but in the longer term will unintentionally weather to contrast with the original structure.

On occasion the conservator may be aiming at a visible difference between the repair and the original work in which case this may be achieved quite deliberately. Both material and technique can be varied. There have been periods when integrity in conservation has been sufficiently fashionable and dominant to cause repairers to make good cracks with visible repairs. This technique usually amounts to the introduction of tiles in multiple courses in place of previous bricks or stone, the finished effect being to show sections of making good (stitches) between three and eight tiles deep interspersed with filling. On occasion the effect can be picturesque while elsewhere it can smack of pedantry. It does, however, provide a positive indication of the event which the repair represents.

Where bricks are salvaged and reused they will frequently have mortar on the face. Most lime mortar is susceptible of physical cleaning which can be carried out by gentle tapping and trowel cutting of the mortar, not the brick, followed by brushing with non-ferrous wire brushes or bristle. In stubborn cases soaking and rebrushing will complete the work. Only in the case of cement gauged mortars is it likely to be necessary to use acid-based cleaners. Where this is necessary the bricks may be placed face down on a pad, soaked in the material, or a tray containing a shallow layer. Alternatively brush application may be used. The period of time over which the bricks should be in the bath is to be carefully controlled and is to be followed immediately by repeated washing off with sufficient bristle brushing to ensure that all acids are removed from within the brick. It is good practice to wet the bricks before acid cleaning. This ensures a boundary of dilution which controls penetration and facilitates removal of acid.

Figure 7.3 Severely eroded sixteenth century terracotta quoins with relatively modern replacements.

In laying, the normal rules in good practice should be carefully followed, particularly as to wetting of the adjoining brickwork sufficiently to ensure that any intended setting of mortars is not inhibited by lack of moisture due to suction from the adjacent structure.

In use care should be taken of the context. No reused brick should be laid in such a manner that it can be thought to have been originally in that position. Where doubt could exist a date on a concealed face will assist future historians.

7.3 Shaping and cutting of bricks

The normal image is of a burly bricklayer delivering, on to a hand-held, bright, well-burned brick, a ringing blow with the edge of his trowel. The brick is reversed and the blow repeated whereupon it obediently separates into two clean-cut pieces of precisely the size required. There is very much more to brick-cutting than this particular expertise which, successful and practical though it is, causes the loss of substantial numbers of good bricks and produces a cut which may wander considerably from the intended line as well as leaving a surface full of minute fractures.

Anyone who needs convincing of the disadvantages of this method of work has only to examine bricks cut in this way and used in paving or in other circumstances where they are subject to heavy or abnormal weathering. By comparison with uncut material the cut brick will always fail and often very quickly. Impact cutting of this type must also include guillotine cutting. In both cases the application of a powerful force rapidly applied sends a shock wave on the intended line through the fired and fairly homogenous brick with the effect that the stresses created in the centre of the wave front exceed the tensile capacity of the material. The effect may be visualized, if somewhat inaccurately, by thinking of the lines of equal stress as an expanding membrane on to which an object impacts. As the expanding surface moves forward there is a high level of stress in the centre and dispersal towards the sides. The central stress is sufficient to exceed the tensile strength of the material at the point of stress-concentration at the end of the crack. Seen in this way it can be visualized that the surfaces adjacent to the line of the actual fracture will also have been highly stressed by the shock wave passing through leaving a finished cut face which is weakened by the event and in any case does not have the advantage of the fireskin or cooled superficial layer which protects the remainder of the brick. As a surface it is weak and unsatisfactory. That said, it has to be understood that brick cutting with the trowel edge is a common, normal and recognized technique of craftsmanship, acceptable where the cut face is set away from the weathering surface. It is, therefore, acceptable for the provision of 'snap' headers, the cut face lying within the wall, but is not acceptable for the production of closers – short lengths of brick lying laterally. A variation on the technique offering more control is to make a series of successive strikes with a broad cold chisel (bolster) along the intended line of fracture, with the brick resting on a soft bed. On completion of the circuit a suitably fired brick will normally come apart.

There are other methods of cutting bricks – sawing, grinding and guillotining. These are mechanical – flame cutting involving excessive thermal shock and laser cutting techniques not yet being available to the building industry.

Some terracotta and faience may be cut in the same way as glass. A material harder on Mohr's scale than quartz can be used to scratch through the fireskin of a tile or paviour brick. The scratch can be accurately positioned and repeated on as many faces as is desirable. The principle as with glass is that scratching the material cuts through the superficial layer, releasing the stresses in the surfaces. In the fireskin of terracotta these stresses are much less than in glass and in some forms of terracotta, those closest to normal brick, they may be minimal. In glazed tiles they are at a maximum for this type of material. Glass in its cooling solidifies with surface forces within the skin. Surface solidification takes place while the core is molten with the ultimate effect that the solidified material contains high surface stresses – normally compressive. On their release minor additional stress, such as a shock wave, will initiate a rupture at the bottom of the scratch and this will travel through the material. The differential stressing of the surface and core of solidified glass is

Figure 7.4 Traditionally cut and shaped glazed tile mosaic seen from the front and the back. The chamfered face is to allow the cut edges to meet with a minimum of exposed mortar. This technique focuses on Fez in Morocco.

the reason for the difference in behaviour between a glass filament and glass sheet, the ratio of surface to core being different. A glaze on a ceramic may, however, be in compressive or tensile stress depending on its coefficient of expansion. In the case of tiles the effect is less dramatic and the results less reliable. Dependent upon the material, however, scratch cutting followed by stressing is employed, although guillotining and grinding are more certain.

Bricks are frequently cut using high-speed resin-bonded fibre-reinforced discs of silicon carbide (carborundum) or coarsely crystalline aluminium oxide (corundum). Discs are now also frequently edged with segments of diamond in sintered metal powder. In essence this is a method of sawing, each successive abrading particle forming a minute scratch or point of impact in the depth of the cut. The great advantage of sawing as opposed to rupture is that the energy contained in each successive shock is very much smaller and it is, therefore, quickly dissipated (as heat) without putting excessive strain on the neighbouring layers of brick. From a practical viewpoint the greatest advantage is the precision of the cut in materials which may show rough arrises and faces by impact-cutting and where in a non-uniform body the impact cut may run off line. Alternatively, soft bricks may be sawn in much the same way as wood in carpentry. traditionally, saw marks were removed by rubbing with a harder brick, often using sand as abrasive, hence 'rubbers'. Half way between these two techniques stands the art of fine-adzed – or fine-axed – work. Bricks of a suitable degree of softness may be shaped delicately by hand tools delivering small successive blows. An adze or chisel may be used and fine finishes may be achieved by abrasion with materials such as silicon carbide. Quartz (as sharp sand) may also be used, dislodging rather than cutting other quartz particles in the brick. In work of this type much skill is needed. Crude and clumsy work is always visible. With hand tools, patience and precision shapes of considerable delicacy can be achieved in material of some softness. Because bricks used for this type of work are not highly vitrified their weathering qualities may be less than other brick in adjacent positions. However, there is a zone of happy compromise between degree of firing and sufficient porosity to allow cutting by this method. Techniques vary from the extreme refinement of the sculptors bench with drills and other small mechanical tools to the hand held brick trimmed with the comb (scutch) end of the bricklayer's hammer by multiple light blows. Throughout the whole of this technique the essence is the delivery of a blow sufficiently small to abrade only the area immediately under the impact. If not held in the hand the brick will be bedded on a soft surface. Fine-axed work is commonly succeeded by shaping up with manual abrasive methods.

In good modern practice carborundum stones should be used replacing the hard brick and sand abrasive; iron files should not. The use of iron cutting tools in this work, however, is traditional and may continue to be acceptable provided that fragments of iron are washed away or brushed off at the completion of the work. Iron wire brushes should not be used if the risks of staining are to be avoided.

For a long time bricks have been shaped using saws. The bowstring saw in which a wire or fine blade was tensioned across a bow and was then worked in a groove into which sand and water were introduced was an established method of marble cutting in the ancient world. No doubt some enterprising Roman used a device of this sort to cut brick at times – perhaps to make terracotta tesserae for inlay and in the Islamic world the same techniques have been followed. The bow-saw and drill, the chisel and the fine adze have been used to cut decorative bricks, to carve their surfaces, to shape the spirals of columns, the bulbs of their bases and the richly coloured geometric shapes of ceramic tiles. Bas-relief carving of bricks, sometimes prior to their further firing and glazing, and their shaping into frets and grills constitutes some of the most laborious and refined work in the material, justifying laboratory repair techniques appropriate to conservation as artefacts rather than as building components.

The shaping of bricks by abrasion has been widely practised on historic buildings, particularly in those periods when high precision was sought in both line and form.

If an even finely finished (gauged) surface was required bricks which had been previously rubbed on an abrasive surface – perhaps a flat rubbing stone or the flat side of a wetted revolving sandstone – would be laid as true to face as possible and finished finally by rubbing down with pumice or sandstone blocks, or similar bricks. Bricks sufficiently soft to be shaped in this way would be known as 'rubbing' or 'rubber' bricks although for gauged work quite hard bricks would sometimes be used. Such work is preponderantly eighteenth century in origin and appears to have originated in the attempts of bricklayers to produce work as finely finished as the most highly wrought stones. It is a product of an age before mechanical and mass production when a precision finish was indicative of a very high quality hand produced material. The advent of very precisely formed machine-made materials diminished the esteem in which such accuracy was held and confused aesthetic judgement so that rougher material was again prized for its own qualities. However, where gauged brickwork is the subject of conservation the same quality of finish will be required even though it may be obtained by less laborious means. The use of suitable abrasives on belt sanding machines can achieve a gauged finish, although the effect is not a match for the hand technique. Its use will be considered with care in relation to the circumstance.

'Rubber' bricks were commonly required for formation of arches – flat, low-pitched and fully rounded – and they constitute one of the peaks of the bricklayer's art. Such bricks were often deliberately made with more carefully refined clay and brick earths. A higher clay content and moderate firing was often favoured for this purpose. Rubber bricks would be shaped finally to the precise dimensions required by rubbing in the wet, typically with a circular motion on a flat piece of suitable sandstone. Stone such as millstone grit is too hard for this purpose and a typical stone selected would have come from paving quality beds.

Under some circumstances bricks are rubbed against each other and again water may be used as a lubricant. An unduly soft brick may fail under this treatment. When fully finished the arrises will be sharp and the brick precisely shaped to within a millimetre. Such bricks will be positioned with care. A good craftsman will never hold the brick by the arrises and the lime putty will be applied as though it were butter. Each brick is accurately formed on a template and it will be positioned on a carefully carpentered former. The initial shaping may be done with a fine axe, but it is more likely to be sawn and in modern work cut with a grinding disc prior to the final rubbing. The high quality of craftsmanship and the careful labours involved are reflected in the costs. Work of this calibre is particularly applicable to dressings and arches where adequate keying is vital.

Sometimes a groove will be formed in the beds of the voussoirs. This can be done with a file, the mortar filling opposing grooves acts as a key. In Islamic work keys cut into the bed may be carried through in the form of joggles and these form a common and important part of the design vocabulary.

It will be readily appreciated that the special shapes to be achieved in highly finished work of this type demand skill and care in the cutting which extends to the use of special packing boxes of timber into which the brick can be firmly set, so that it is supported on all faces before the crucial finishing saw cuts are made. Rubber bricks may be cut from oversize blocks of material fired to the right degree. Alternatively they may be moulded initially to dimensions which after firing yield the intended shape with a shrinkage tolerance of –0% to +10% ready for rubbing to the finished size.

7.4 The support and construction of minor arches

The repair of any arch depends upon proper support while the structure is extant – a more difficult process than building *ab initio*. All voussoirs must be given uniform support during repair and this is usually achieved by centring. Voussoirs are normally set with their long axis across the line of the arch, that is to say parallel to the alignment of the through passage beneath. The exceptions to this rule are structures with 'ring arch' (pitched) construction in which slab-like voussoirs are laid in successive rings, each slab being placed crosswise to the span, i.e. its greatest width being aligned between the springings. This type of construction of is of very ancient origin being traceable to the Nile Valley and Mesopotamia well before Christian times. The advantage of the system is that it allows the flat face of the voussoir to obtain sufficient adhesion with rapidly setting mortars to gain support from the preceding arch or part arch, so allowing the mason to span a structural opening without employing centring, other than as a template. In the repair of such arches the procedure is not reversible and centring is used.

All centring must deliver a uniform pressure to the intrados of the arch. Whether the centring takes its bearing from seatings at the springing or from cill or ground level, it must be constructed so that it is possible to increase the pressure of its support under the voussoirs as a matter both of adjustment and balancing

Figure 7.5 Venice: permanent horizontal shoring in brick. The techniques of arch construction are here applied to flying shores.

the load. It must also be possible to reverse the process so that when the mortar has set the centring can be removed. Traditionally this variation in loading is achieved by the use of double or 'lazy' wedges, but with modern equipment, screw jacks may be employed either to lift the entire centring or to give selective support. They have the important advantage that no shock is involved in their removal. Where delicate brickwork is involved foam packing or air inflated packers may be used to distribute the loads. Hydraulic jacks may be used in circumstances of major loading where elements of a broken arch are to be repositioned.

Fractured arches may consist of two or more segments which it may be possible to reposition if the remaining structure is given temporary support and detritus above the segment is cleared. Misaligned segments should not be lifted by the upthrust of centring formed to the line of the original or intended arc but should be supported on skeleton centring, formed within the finished profile. The centring, having been positioned, should be used to provide the base on which each segment is supported and on which it is differentially lifted. If in a damaged or failed structure a significant deadload is imposed on a broken arch this must be relieved before the arch is repositioned. For this purpose needling may be necessary. Percussive methods of driving holes through walls for the insertion of needles are inappropriate. Drilling is the acceptable alternative followed by sawing and the needles chosen will be in the form of bars supported and eased up on a separate structure prior to the reinstatement of the arch. Operations of this type are delicate and justifiable only where relatively small movements are involved. Ultimately they are likely to be secured by grouting and the arch itself will be reconstructed beneath the supported section of wall. Alternatively substantial superimposed wall sections may be removed. This may involve significant disturbance to other aspects of a delicate decorative and complex fabric. Many fractured arches which have reached a new stability have been repaired by grouting up fractured joints, leaving the visible deformation as a historical statement.

In the event of significant damage to the refined rubber brick arches of seventeenth, eighteenth and nineteenth century building, some voussoirs are likely to have been overstressed and to have fractured beyond reasonable repair. Their replacements, carefully shaped, will be set in beds of pure lime putty or putty containing an equal admixture of fine sharp sand. In eighteenth century British and European practice some arches were set in mixtures of white lead, whiting (ground chalk) and linseed oil. Both lime and whiting mixtures take a considerable time to set and the centring will be left in position for some weeks. Even three months may be an appropriate period although the superficial hardening (oxidation) of linseed oils takes no more than a day. Where really fine joints are required (down to 1.5 mm) a bedding of Shellac and white lead has been traditional. Shellac is a natural resin produced from the secretions of a South East Asian tree (*Ficus resinosa*) following attack by a parasitic insect. It is soluble in alcohol and water and is thermoplastic. Mixed in equal parts with white lead it was a reliable setting putty, the finish being indistinguishable from neat lime putty. Although Shellac is still readily available the use of white lead is prohibited in many countries, lead being absorbed into the body system as a cumulative poison. In the UK and the European Community the use of lead compounds is governed by stringent regulations and their use is sanctioned only with appropriate precautions in special circumstances where it is necessary to simulate historic conditions of work.

7.5 Voids in brickwork

Voids may have been designed in brickwork or they may have occurred as a result of leaching, decay, structural movement or even insect attack, particularly in tropical regions. It is important to consider the purpose of filling any void before deciding on the methods and material to be used. In some circumstances there may be merit in leaving a void, provided that the permanent security and stability of the structure is not in doubt. A void designed in the original structure – such as a cavity wall construction or a flue – will be filled only if there is a special structural reason for so doing. The haunches of arches were often left void or partially filled with brickwork, and in principle these voids should be retained unless structural conditions have changed.

The materials available for such purposes are almost all fluid or plastic in form, consisting of loose fill, grouts, putties or cements and expanding synthetic materials such as flexible sealants and polyurethane foam. Compatibility or flexibility in the material to be introduced will be a significant factor. Methods of introduction will vary from direct placement of brick or broken brick in mortar to the injection of refined grouts through fine nozzles under pressure. Direct placement generally necessitates a sufficient void to allow hand working although in some situations such a material can be poured. Grouts are best placed in historic brickwork through plastic tubes under modest hydraulic pressure. A head of a metre is normally quite sufficient for the purpose.

The introduction of liquid materials can normally be achieved under their own hydrostatic pressure although pumped pressures are sometimes necessary to ensure full penetration and unlike earth structures brickwork is often sufficiently strong to accept the considerable hydrostatic pressures achievable with pumping equipment. Where hydrostatic pressures must be avoided or where the water of the mix might damage or endanger the structure an immediately setting mortar may be specified. This may be pressure-injected using a gun with an additional supply line through which a solution of sodium silicate is forced, at up to 2% of the mix by volume. Setting is immediate – within a second with full curing within 24 hours. Immediate support may be gained without pressure, seepage or weeping. There are problems of control and finish. Circumstances in which this immediately setting mortar is required are rare and the application is by specialists so that the 'on-costs' are high. However, there are occasions when its use meets a need which cannot be satisfactorily met in any other way.

Lightweight materials can be injected by airborne methods. These are normally required for insulation purposes alone and consequently are materials easily carried on an air stream such as rock fibre or glass fibre, expanded vermiculite, expanded perlite or expanded beads of thermosetting resins, such as polystyrene. They have little application in historic structures. Occasionally the introduction of an inert poured fill is justifiable. Fine sands at one extreme and sintered clay nodules at the other define the range and the most commonly used material of this type is pulverized fuel ash (PFA, see above) which will set off in the presence of moisture to provide a weak, solid material.

In fine voids the introduction of flexible or plastic materials can be achieved with great accuracy. These can be as simple as lime putty introduced through a hypodermic syringe but frequently comprise sealants which are to remain permanently or semi-permanently flexible or setting compounds whose function is usually to provide a bond between surfaces. Sealant technology with pressure injection techniques has now reached a level of performance capable of delivering specified amounts of material into voids at considerable depths. Many such compounds are capable of being made self-curing, or of curing by interaction with air or water so that they can be considered as setting or semi-setting materials. In addition, gaseous generation within viscous liquids provides the opportunity for the injection of expanding foams, of which the principal groups are isocyanates, ureaformaldehydes and polyurethanes. Their delivery can be achieved via nozzles and pipework within specific zones or cavities to produce a material which, under pressure, will fill irregular voids and will set off to provide a rigid material firmly adhering to all sides of an irregular cavity.

Pressure injection implies some sophistication in instrumentation and equipment and such processes generally move the application into the field of specialist operations although increasingly direct methods of applying both foams and setting compounds such as epoxy resins are available to the building industry. These range from hand packs to gun applicators into which precharged disposable cartridges can be fitted. The judgement of the conservator will be required not only as to the suitability of the materials in any given situation but also as to the need for special skills in application. Visual prominence, compatibility, protection from ultra-violet light and other damaging agents, guarantees of materials and performance and above all the function to be performed, will be the decisive factors in decision-making. Any judgement as to the materials, methods and merits of filling voids in brickwork must relate to the intended function of the fill. This can be broken down into:

1. Elimination of air movement.
2. Inhibition of water movement.
3. Physical consolidation by structural attachment of opposite sides of the cavity.
4. Strengthening of the structure to improve compressive loading.
5. Integration of divergent sections of material.
6. Amendment of the distribution of forces – perhaps prior to some other building operation.
7. Avoiding infestation.

A wide range of requirements, a wide range of available materials and a significant variation in techniques of application offer the conservator a significant range of options to be evaluated against criteria of acceptable performance and cost.

The location of voids may itself be a task requiring careful definition or, at the other extreme, may be a problem offering its own solution. Where voids are such that grout or expanding foam will fill every irregularity the extent and nature of the void need be known only in general terms. A wall of high porosity due to failure of the lime mortar in rubble fill may be stabilized by the introduction of grout through a pattern of holes drilled at regular intervals. The grout finds its way through voids whose shape remains unknown. At the opposite extreme the size and position of a void may be critical and physical methods of investigation may be destructive. The adhesion of a delicate mosaic to a surface such as the hemispherical brickwork in the head of an apse presents problems of great delicacy where the introduction of a suitable adhesive by hypodermic injection will require exact placing and minimal generation of pressure coupled with skilful methods of uniform support. Such a case was the securing of the very important sixth century mosaic in the apse of the church of St Catherine on Mount Sinai, where the tesserae and their backing had become detached from the brickwork and in effect formed an extremely thin shell of very fragile material, the collapse of any one part of which would have brought down large sections, so destroying the mosaic as an original artefact. In such cases air inflated supports can provide the low even pressures necessary while grouts or low pressure expanding adhesives are introduced.

The simplest method of injection is by pouring. A hydrostatic pressure is generated, being directly related to fluid pressure at the maximum depth of liquid. A grout may be introduced through a plastic tube fed from a funnel whose height can be varied in accordance with the pressure sought. Mechanical pressure injection techniques can vary from the simple diaphragm pump to pressurized vessels with systems of control involving agitators and pressure-equalizing devices. Serious damage can be done where high pressures are maintained in attempts to force grout into zones where it would not otherwise penetrate. Skins of brickwork can be forced off by excessive hydraulic pressure and essential bonds can be broken. Generally hydrostatic pressures of a metre are sufficient to propel a satisfactorily fluid grout into voids in brickwork, and these pressures can be achieved readily by the simple tube and funnel technique. Leakage points sealed with clay will resist pressures of this order adequately. Pre-wetting of cavities to be filled is desirable, avoiding the coagulation in small voids of drying grout. Exceptions to low pressure methods are voids and semi-voids under foundations, behind retaining walls and under floors perhaps already filled with water. These applications are of an engineering nature in most cases and will require the appropriate skills.

Inert filling materials must be compatible with the structure into which they are introduced. They must be able to perform their function without damaging their host material by the introduction of salts, by excessive expansion under changes of humidity or temperature and they must not fail by excessive contraction, pulling away or weakening. A lime-based grout with a crushed brick aggregate will be acceptable in filling voids in ancient brickwork whereas a strong cementitious concrete will not. Its different thermal expansion properties and the introduction of deleterious salts make the use of such material unacceptable in ancient brickwork. However, it may be necessary to fill a cavity, allowing greater tolerance of movement than lime mortar would accept in which case a grout might be contrived using PFA and/or expanded polystyrene beads. For even greater flexibility a suitably formulated foam might be used, particularly where other properties, such as adhesion were thought necessary, but great care must be taken to understand ageing characteristics and achieve reversibility. It is helpful to remember that the need for thermal and hygroscopic compatibility reduces as the size of cavity reduces. A material which might be unacceptable in large volume such as an epoxy resin/sand mixture might be suitable in a relatively thin cavity where it is necessary to achieve a strong bond from face to face. Reversibility, however, might still be a problem.

The principal characteristics defining utility of filling materials can be tabulated as a guide:

- Dry sand: no adhesion, flexible, no wetting effect, good strength where contained. Wedging effect as material infiltrates fine spaces; exerts some 'fluid' pressure; dense. Flow characteristics in soft fine sands will allow escape through small voids which may be present or may develop.
- Lime-based grouts: wetting effect on application, relatively high density, compatible with ancient and most modern mortars, rate of setting can be determined, stable and of moderate strength, low strength, significant compressive strength, low to moderate adhesion. Clogs in fractures below 3 mm depending on sands used; moderately dense.
- PFA grouts: may be modified with cements and limes to produce greater strength, but alone sets off with moderate compressive strength: wetting effect on application and capable of supplying sulphates and other damaging salts if not carefully selected: modest compressive strength and low tensile strength with some adhesion; with pre-wetting, capable of penetrating fine fractures down to 1 mm; medium density. Dry, it has a very small (30°) angle of repose but wet, it will stand with a face approaching 90°.
- Earths: introduced as liquid slurries have wetting effect and shrink on drying. Earth containing calcitic binders can produce a very modest set and others mature into a self-stable condition. Adequate compressive strength when confined, capable of nourishing plant life and being broken down by insects, compatible with earth mortars where used for bedding fired bricks. Penetration poor unless finely sieved; medium density.
- Cementitious mortars: have wetting effect on application, very modest shrinkage on setting and drying out, high rigidity and strength, modest tensile strength, moderate adhesion. Capable of introducing deleterious salts into brickwork with consequent external staining and grey deposits. Brittle, but can be modified with limes and gypsums to be softer. Significant strengths partly a consequence of types of sand used. Penetration less good than PFA; medium density.
- Gypsum plasters and mortars: capable of liquid application and wetting with controlled but relatively rapid setting. Material expands on hydration with gap filling effect. Can be adapted to carry lightweight aggregates such as vermiculite. Moderate compressive strength, modest tensile strength and moderate adhesion. Should not be used in situations where wetting will be persistent; medium density.
- Polymers and surfactants increase flowability at reduced water content in most of the foregoing.
- Polymer additives, e.g. emulsions of polyvinyl acetate, polyvinyl chloride, etc.: can be added as mixed emulsions to all water-borne fillers improving their properties of adhesion and tenacity, being carriable by water into all pores normally accessible to fillers but not capable of being carried into the smallest pores penetrated by water due to the size of their long chain molecules.
- Resin fillers – acrylic, polyester resins etc.: non-aqueous and generally non-wetting, capable of being introduced as high, medium or low viscosity materials with controlled curing times and capable of carrying fillers such as sand or other 'inert' materials to form solid blocks, generally of

substantially higher compressive tensile and torsional stresses than brickwork or mortar and with high adhesion. Modulus of elasticity and thermal coefficients distinctly higher than those of brickwork. Can provide effective connecting zones, waterproof. Resistant in many forms to biological and animal attack. High modulus of elasticity. Capable of being reinforced and of bonding reinforcement into brickwork. Generally requiring skilled or specialist introduction and careful preparation. Will discolour the brickwork.

- Expanding foams – polyurethanes, ureaformaldehydes and others: non-aqueous delivery, capable of producing substantial internal pressures within cavities and of complete filling of voids and large associated pores. Not generally capable of penetrating small pores, e.g. below 0.5 mm. Powerful bonding and adhesion, moderate tensile and compression strengths, generally closed cell structures, waterproof and sometimes vapourproof. High flexibility and shock tolerance. Not fireproof. Requiring sophisticated equipment and application by trained personnel except in the case of small scale application by disposable units. Of short life by comparison with historic building materials. Properties may vary with ageing, particularly as regards loss of flexibility. Not capable of being modified by the addition of other fillers or available site materials; lightweight.

The above guide sets out some of the principal characteristics of the general range of available fillers which may be used in conjunction with voids in masonry. Others such as expanded polystyrene beads and sawdust can have special relevance to situations where organic materials are acceptable. Both materials can effectively be mixed in a sand/lime/cement matrix where their particles form isolated pockets which produce a lighter and more insulating material in thermal terms with substantially reduced compressive strength but with improvements in elasticity which allow the differential in thermal expansion coefficients between brickwork and the infill to be taken up in the introduced material. The tendency of both materials to float up out of a liquid mix requires that they are treated as the bulk constituent of any infill, the lime-cement acting as a binder rather than as a true matrix. Fillers using organic materials so radically different from the general patterns of limes and cements can have special characteristics. The voids due to the unusual aggregate may vary from 2 mm to 3 mm up to 20 mm and the binder will in itself tend to contain large pores resulting from the nature of the mix. Such special mixtures will be required only in unusual situations. Even more exotic combinations are possible – in particular the use of an aggregate consisting of crumbs of rubber derivable from waste motor tyres whose special qualities include the dampening of vibration and are applicable to forms of underpinning where such problems occur. Engineering advice is essential.

7.6 Strengthening brickwork

While grouting may improve the strength of brickwork containing voids, it is of no assistance where a solidly built wall suffers from weakening mortars. Ageing, water penetration, vibration or simply poor quality mortars may make it essential that the brickwork is strengthened without rebuilding. A typical problem may arise from the deflections in an eccentrically loaded plate in the wall, perhaps carrying joists with fluctuating live loads. Public safety and increasing usage may be involved and the wall may perhaps carry important plasterwork or be of a significant texture or historic construction. In circumstances of this sort strengthening or stiffening may be achieved by lattice reinforcement.

Holes may be drilled diagonally in a vertical plane, i.e. at 45° downwards nearby to the full thickness of the wall, so that, working from opposite sides, their alignments cross although they are displaced laterally on a regular pattern. Lengths of ribbed reinforcing rod are fitted within these borings, stainless steel being used where there is any danger of water penetration.

Borings of 20 mm and rods of 6 mm diameter are usually sufficient and a pattern of three to four holes per square metre may be adequate. It is important to plan the work so that there is adequate overlap or scissors effect. If necessary, particularly in thin walls, the angle of drilling will be much closer to

Figure 7.6 A structural crack of some age and confirmed as being stable being 'stitched' with stainless steel rods returned into holes drilled into the mortar beds. Above: Joints raked out and drilled. Middle: Rods before placing. Below: Rod inserted and being bedded. (courtesy of Martin Welch, craftsman.)

the vertical than 45°, and holes may be as much as three times the wall thickness in length. Drilling may be started with advantage at the low point in a perpend. A rotary non-percussive drill is best with a compressed air line or water lance for clearing the hole of debris. The grout should be compatible with the mortar.

It may in some circumstances be necessary to drill horizontally, but normally the objective will be to improve strength vertically, when a slenderness ratio problem is apparent. Sometimes, where only single-sided access is possible, upward drilling is also necessary. In sophisticated work the holes may be drilled to conform to a curvature in the wall.

On occasion, as illustrated (Fig. 7.6), stitching with stainless steel rods is adopted. In principle this repair is a simplified variant of the method described.

7.7.1 Densities and strengths

Density and compressive strength tend to run in parallel. There are extreme examples (very rare in historic brickwork) of bricks with a density below unity (i.e. lighter than water, being made from special clays with particular manufacturing techniques), but the general bulk densities of brick range between 2.5 and 2.9 g/cc, (that is between two and a half and three times as heavy as water at 0°C). The specific gravity of fired clay is higher, but the meaningful figure is the bulk density which takes account of porosity. The crushing strengths of historic bricks in sound condition vary from 5500 lb/in^2 (38 N/mm^2) to below half this value which is still in excess of that of a good lime mortar. Blue bricks are often to be rated in excess of 10000 lb/in^2, (69 N/mm^2) and some Flettons are rated at less than 2500 lb/in^2 (17 N/mm^2). In conservation, consistency of materials is the crucial factor and, therefore, the important criteria are not the achievable strengths but the compatibility of strengths. While modern bricks can have strengths ranging from 1500 to more than 16000 lb/in^2, materials at the extreme ends of the range are unlikely to be called for because they do not relate closely to the materials being conserved.

7.7.2 Porosity and adhesion

Crucial qualities in the durability of brick and its cohesion to mortar derive from the nature of the porosity both on the surface and within the brick. Surface porosity is indicated by the speed at which water is taken into the brick surface. The fireskin has little effect on suction rate (but not on water absorption). The ideal relationship is that the mortar should be drawn in plastic form into the surface cavities of the brick without there being excessive suction to dewater the mortar before penetration and setting. Suction rate has been used to measure the effectiveness of the bond between mortar and brick. High suction rate dewaters mortar and tends to cause microcracks at the interface resulting from rapid mortar shrinkage: hence recommendations for dunking or docking (water dip) of high suction bricks. The objective is that the mortar paste should penetrate the pores and set with irregularities corresponding to the irregularities in the brick itself. The undulations and porosity of the surface should therefore be sufficiently large to allow the adhesion which can result from the filling of the irregularly faced openings and indentations. This mechanical interlocking is the prime bonding mechanism between brick and mortar. Consequently a smoothed or vitreous surfaced brick has a very much lesser bonding capacity, as will be known to those who have 'cleaned' various types of recovered brick.

Mortar creates a bond by filling surface irregularities. While the cement paste does not enter surface micropores the needle crystals formed as cementicious hydrates may do so. There is a ratio of approximately 10 times in improvement between the typical smooth-surfaced brick and its rough equivalent. The smoothest surfaces are not without bonding capabilities and can provide tensile bonding strengths of 10 lb to the square inch upwards but a well-bonded rough-surfaced brick may achieve 10 times this value. The lowest values in the range are the result of the direct bonding of calcitic materials to a vitreous surface. The interaction between the crystals of calcite forming at the surface and the silicate skin does provide initial strength.

7.7.3 Rates of absorption

The absorption capacity of a brick can be tested by measuring the uptake of water under standard conditions. This practical test is in the nature of field testing because the results are dependent on circumstances and conditions – temperature, wetness of brick and methods of handling. A frozen brick or one which has been fully saturated has no effective absorption. The essence of the test is that it provides an indication of the ability of the brick in a defined condition to produce an appropriate bond. Codes of Practice which related water absorption to adhesion have been shown to be unreliable to the point of being grossly misleading. They are now replaced by British Standards using tables of available masonry panel strengths to which reference should be made.

However, the IRA – Initial Rate of Absorption – may be determined by a standard test, which has fallen into disfavour in the manufacturing industry as being imprecise. A dry brick immersed to a depth of 3.18 mm for one minute and having a face area of 194 cm^2 should absorb between 10 and 40 g of liquid if it is to provide good bonding with a normal mortar mix. The ability of the mortar to retain its liquid content is a factor in the relationship and where the retentivity is high (with more absorptive brick) good bond strengths are generally developed. Adhesion can be enhanced by the use of bonding agents in which long chain organic polymers provide continuity between the surfaces. Good practice usually dictates priming with a dilute mix of the emulsified polymer.

8

Destructive processes in brickwork

8.1 Freezing and thawing

Pore structure and pore volume are important in determining the behaviour of brick and the stresses of freezing and thawing. Water expands by about 9% of its volume when it moves from above +4°C to –1°C. Loss of heat from an external surface of brickwork causes a plane of reducing temperature to move progressively through it and as the temperature falls to freezing the expansion occurring during the liquid phase moves ahead of the plane of freezing, forcing water through the pores, being trapped by dead ends and resisted by very small voids. A zone of hydraulic pressure thus moves ahead of the ice. Where water is trapped ahead of this, front rupture can occur in order to release hydrostatic pressures. The failure will be in the vitreous bonds and crystal structures.

Porosity capable of providing routes permitting a sufficiently free movement of gas and liquid is therefore an important protection against failures caused by freezing of contained water. Isolated failures will produce a gradual softening of the brick in successive cycles but coherent patterns of pressure will cause delamination or breakage along internal planes of weakness. Freezing from two sides simultaneously is a cause of rapid failure in bricks of low frost resistance – probably because the internal hydrostatic pressure cannot be relieved on either face.

A further factor in resistance to freezing in brick is the quality of its burning which produces hardness and strength due to the formation of more and better vitreous bonds. A further factor is water content on freezing. Highly porous bricks containing a greater amount of water due to their higher absorbency will tend to be weaker under the freeze–thaw cycling than well-burned bricks of lower porosity. Small pores containing gases are the effective protection against destructive internal pressures. Most good bricks will withstand these stresses without undue damage, but poorly burned and excessively porous material will fail quite rapidly. The ratio of internal strength to porosity has an obvious correlation: improved porosity reduces the internal pressures but if that porosity is accompanied by weaker internal strengths, even the reduced internal pressures will be enough to cause failure. Given adequate strength in the brick the key factors are the nature and size of pores. A test which has been used to provide a guide is the absorption coefficient, which seeks to determine the difference in the absorption of the brick under normal conditions and its ultimate absorption capacity. The coefficient is now in disrepute in the manufacturing industry due to the many exceptions. However it remains a feasible and usable field test. It is normally measured by comparing the amount of water absorbed by a brick reaching saturation in normal ambient conditions and a brick which has been placed in boiling water for 5 hours and then cooled in water. On cooling, steam in the voids condenses and – other gases having been almost entirely displaced by water vapour – even the smallest pores fill with liquid water. In most brick no further absorption will take place if boiling is continued beyond 5 hours. The absorption coefficient is expressed as the weight of water in a saturated brick expressed as a percentage of dry weight. Its particular utility is in comparison of bricks from the same type of clay, but of different burning or manufacture.

Figure 8.1 The deleterious effect of lime mortar on the softer yellow brick of a mixed wall construction. The better-fired outer surfaces of the brick have survived. Cavitation of the core weakened by calcium carbonate crystallization leaves voids in the wall structure.

An alternative, quicker but less preferred method of determining the coefficient is to place the brick in a vacuum chamber into which water is released to cover the brick while returning to normal pressure. A saturation coefficient below 0.7 suggests a well-structured material with sufficient additional voids to relieve unusual pressures whereas significantly higher values suggest that the voids are excessive and may be a cause of weakness. The vacuum method is commonly used for quality control in manufacturing.

Figure 8.2 An appropriately weak (½:1:8) cement:lime:sand mortar eroding back from the original finish leaving the brick surfaces intact.

Figure 8.3 A carefully repointed wall in which the strong cement mortar focuses erosion on the central area of each brick.

The porosity of a brick also affects its permanent dilatation or expansion on removal from the kiln and exposure to water vapour. Anhydration of clay molecules occurs, reversibly at 350°C and above. The levels of firing and calcitic content also have some effect but as the linear effect does not exceed 0.05% and is usually less than 0.04% dilatation is a relatively unimportant factor in determining durability. The effects of external surfaces being wetted and dried while the internal faces remain relatively constant in terms of moisture content have yet to be evaluated in relation to the stressing of thin leaves and some aspects of delamination.

The absorption coefficient has been used as an indicator of frost resistance. However other factors such as the distribution of pores and

Plate 1.1 Wood-fired hand-made stock bricks of early nineteenth century date laid in a Flemish bond in a solid brick-thick wall with 'penny struck' joint. The stretcher faces are bright orange and the headers a silvery blue-grey.

Plate 2.1 Hand-made, clamp-fired Wealden bricks, burned with coke breeze. The mixed brindle colours and dark spots of coke burn are typical of this process.

Plate 2.2 Brick structural wall faces with small hard chalk and with occasional brick inserts, the windows being dressed in brick (nineteenth century).

Plate 3.1 Early post-medieval brickwork in East Anglia. A mature style had emerged, using features of contemporary stone detailing, but expressing the quality of brickwork in the dark diaper patterning.

Plate 4.1 Late nineteenth century glazed terracotta architectural wares achieving a high standard of perfection. Among others Doulton and Lambeth potteries in London, with the advantages of rail transport, imported both clays and fuel to achieve products of very high quality (reproduced with permission of Lloyds Bank Ltd.).

Plate 4.2 The Temple of Heaven, Beijing. An extreme elaboration of roof tiling: cresting, finials, terminals and verges are sculpted and glazed with motifs of symbolic importance. Individual items are minor artworks in their own right. Originals may be replaced even before failure.

Plate 4.3 The Temple of Heaven, Beijing. Eaves terminals to roof tiling have been replaced regularly and faithfully during the life of the building. The green glaze on a central tile is now failing after perhaps 150 years.

Plate 5.1 The thirteenth century tomb known as Yahya bin Kassim at Mosul. Erosion of the bank of the Tigris led to the building of a massive brick buttress to support the tomb built originally a considerable distance from the river. The weight of the buttress caused a settlement which sheared the structure from top to bottom, visible as a 'stitching' in the centre of the façade and a failure above in the drum.

Plate 5.2 Non-destructive techniques such as thermography can provide information on hidden structure. Here an apparently uniform gable wall is shown to have brick noggin (panels) in a timber frame on the basis of differences in infra-red radiation. (Courtesy: the Demaus Partnership – Building Diagnostics Service.)

Plate 6.1 Failure of brick due to physical damage. These paviors, laid at the same time, weathered satisfactorily except where they had been cut manually to accommodate the boundary. After 25 years the cut and damaged bricks decrepitated under frost and microbiological attack. A purpose-made shaped brick would have survived.

Plate 8.1 While brickwork below the waterline has remained sound for centuries lack of a damp-proof course allows a constant (and polluted) water supply to replace evaporation losses with the continual destruction of brick surfaces by inorganic and biological agencies. Some deterioration is attributable to the Istrian limestone used at lower levels. The resultant continuous replacement of surface layers is evident.

Plate 8.2 Flue tars. The deposition of emulsified wood or coal tars within a flue results in their very visible transfer to the outer surface of the brick. They are carried by vapour condensing in the flue to the outer face. In this case later pointing separates the stained brick which will also contain sulphates.

Plate 8.3 A panel of tile mosaic in which damp penetration typically has destroyed the dark blue and white glaze. The light blue glaze is more resistant (*lower*) but also fails (*centre*). Physical damage has been repaired in mortar with irrelevant tile embedment. (Samarkand, fifteenth century.)

Plate 8.4 An identical brick above and below a course of Portland stone. The bricks above the stone have retained their faces. The bricks below the stone, washed by lime-rich surface water, have lost their facings entirely, and are being eroded at several times the rate of the upper bricks.

Plate 9.1 Liverwort well established in decayed brick paving and superseding the initial colonization of mosses.

Plate 9.2 Roof tiles of different type and manufacture will promote different types of plant life. While in this instance clay plain tiles carry a thin coating of a small grey lichen the adjoining concrete tiles have built up a heavy growth of moss edged with a bulkier yellow lichen.

Plate 9.3 Fifteen years growth of a russet/orange lichen obscuring the natural red of kiln-fired brick.

Plate 10.1 Damage caused by the failure of soil drainage. A combination of plant growth, mortar erosion and loss of brick surface is evident.

Plate 11.1 Heavy-duty iron fittings embedded in lead packing surviving for seventy years in the massive brickwork of a nineteenth century military building without causing damage.

Plate 12.1 A repair to terracotta in a render of matching density. The life of the original terracotta can be extended in this way. In the process, however, the terracotta has been cleaned down and its fireskin damaged as can be seen from the surface texture. The colour of the repair is a poor match.

Plate 12.2 Ill-advised cleaning of tuck-pointing. The bricks were already sooted when the tuck pointing was redone. The base mix was, therefore, made dark. On cleaning the poor shape of the base bricks is evident as the base mix reads more strongly than the lime tucking.

Plate 13.1 Laughton Place. The surviving sixteenth century tower after restoration. Until the middle of this century the tower was enwrapped in an eighteenth century mansion leaving only the central part rising above the surrounding roofs.

Plate 13.2 Laughton Place. New terracotta to matching profile in situ. The difference between new and original is clear both by colour and from the embossed designs on the sixteenth century work.

Plate 13.3 St Pancras Station. The Great Central Tower cleaned, repaired and stripped of its scaffolding.

(a)

(b)

Plate 13.4 Restoration work in Baghdad. Prefabrications of complete assemblies. Cut brick carved, mortared and assembled in a site workshop before placing in position (a), and placed in position (b) on a repaired mosque.

the nature of bonding within the brick structure consequent on firing render the coefficient unreliable on its own.

A major factor in frost damage is the freeze–thaw cycle. Resistance – i.e. longevity – may be expressed this way. Each type of brick will be able (typically) to stand up to an identifiable number of cycles. Therefore a climate which subjects brickwork to a high frequency of cycles will be more damaging than one, apparently more severe, with less rapid change. A northern Continental climate is likely to settle into a permanently frozen condition after a short period of cycling whereas in a more temperate climate the fluctuations may persist all winter. In Britain as many as ten freeze–thaw cycles may be recorded in a 24 hour period. The annual rate of cycling is therefore a better guide to the probable life of brickwork than the extremes of temperature.

8.2 Chemical attack

Brick is usually regarded as being chemically inert, a view which in practical terms is effectively true. Only hydrofluoric acid is a serious threat to the structure of a brick and this acid is very rarely present, even in the weakest concentrations in the atmosphere, other than in circumstances of abnormal chemical pollution. Sulphuric acid which is commonly present in very weak solution as a pollutant can, however, initiate attack on brickwork and may ultimately dissolve up to about 1% of crushed brick left in a strong solution. At that point the level of attack appears to be halted by the lack of further materials available with which it can form a bond. Normal rainwater, which even in unpolluted atmospheres is slightly acidic due to the presence of dissolved carbon dioxide is capable of dissolving some brick, to the extent of about 0.01% by weight, beyond which the process seems to go no further. However, these dissolutions do break down a part of the brick's structure throughout the body, and to these must be added other forms of attack. By such very slow chemical processes water will partially dissolve brickwork, the rate of decay depending greatly on the nature of the brick and of the contents of the water. A well-burned brick is more resistant to decay than one which is lightly burned. The specific chemistry of a brick determines its surface susceptibility. The amount of water, its pH value, the presence of other ions, of active biological organisms, temperature, pollutants and many other factors will determine the behaviour of the brick. Essentially two mechanisms are responsible – solution and the formation of sulphates.

Some forms of glass are soluble in weakly acidic water. Although pure silica melts at 1720°C, well above the temperatures at which brick is made, the presence of fluxes and other materials produces various forms of silicate glass at much lower temperatures and well within the range of the firing of brick. Consequently vitreous materials – semi-glasses – are a normal constituent of well-fired brick. The nature of the glass varies from type to type producing critical differences in response to the availability of water but in all cases a glass surface is an asymmetrical or random molecular arrangement not typical of the body beneath and chemists are not yet satisfied that the processes of change caused by water interacting with the glass surface have been adequately understood. Some types of glass undergo rapid change in the nature of the surfaces on exposure to water, becoming alkali deficient, the residual silica rich zones changing in physical character to allow the further passage of water to secondary layers. The variety of vitrifications in glazing and fired ceramics offers a range of possible weathering mechanisms.

In many instances the acceleration of reactions is a function of temperature coupled with the availability of ions which are capable of entering the glass structure because of the abnormal interatomic distances typical of surface conditions. By passing through the surface layer the ions enter the semi-porous sub-surface. Ion exchange takes place and the nature of the glass surface is modified. The effect is particularly evident where a small drop of solute evaporates leaving a potassium rich deposit which acts as a focus for condensate, on which the process can repeat. The nature of the vitreous surfaces within bricks and their behaviour awaits recognized conclusions but knowledge of the behaviour of glass in similar conditions and particularly recognition of the highly variable nature of attack on

Figure 8.4 A very evident difference in the weathering of brickwork above and below a limestone arch.

glass surfaces suggests that there will be a wide variation in its effects upon brickwork.

Soluble silicas can arise through the presence of some vitreous combinations, particularly with potassium fluxes and acidity. These solutions, leaching persistently to the external face of brickwork, may produce silica staining, a hard white deposit almost like a glaze, occasionally found on historic industrial structures. This is unsightly but not damaging, except insofar as it seals the surface of the brickwork. Its removal can be achieved, with care and with full safety precautions, using hydrofluoric acid.

The third mechanism of chemical decay identified relates to the formation of gypsum within brick structures. The surface layers of 300-year-old historic brick studied using X-ray diffraction spectrometry (Franke et al.) showed the presence in the outermost layers of the brick of substantial quantities of calcium sulphate, concentrations reducing with distance from the surface. Calcite within the brick or deriving from mortars in the presence of carbonic acid and sulphur can produce calcium sulphate utilizing naturally available sulphur of biological origin, or airborne sulphur present as a pollutant in which case the reaction proceeds more rapidly. The effect is to produce gypsum deposits within the outer pores of the brick. Their expansion is contributory to spalling or failed face work. A similar process can affect mortars. The activities of sulphur-fixing bacteria are responsible for much sub-surface sulphate deposition.

8.3 Porosity and water content

Water in any substantial quantity can be the cause of quite rapid deterioration in mortars and less rapid deterioration in the brick itself and by permitting salt deposition, biological growth, frost action, staining, softening and erosion water is the prime contributor to slow degradation in brickwork.

Brickwork dilates slightly, as do most other substances on taking up water. This effect is seen most distinctly in the first few weeks after firing and this first dilatation is irreversible in brickwork. Expansion is more evident in underfired brick which has more 'unsatisfied' clay molecules to hydrate. The effect of initial dilatation on brickwork is

Figure 8.5 Surface spalling induced by deposits of industrial fumes. The pollutants are drawn into the brick surfaces differently in accordance with the quality of the brick itself. The rate of decay varies in consequence.

negligible but internal stresses are caused by the pressures generated through recurrent dilatation and these act more effectively upon wet material which is slightly softer. This softness is generally thought to be a product of the lubricating effect of water within the interstices, but it is probably also to be associated with the effect of dilatation in that the internal pressures generated by water movement into the pores minutely move the frictional surfaces apart reducing the pressures between them and, therefore, reducing friction as well as providing lubrication. That this phenomenon of softening is more perceptible in under-burned brick supports the theory. There are fewer bonds of fusion within the under-burned bricks and more dependence upon frictional and interlocking effects. Similar broad generalizations are true of mortars, although there are no bonds of vitreous fusion. Brickwork, therefore, behaves as a coherent mass, the extent of softening and dilatation depending upon the exact type of material in both brick and mortar but with effects moving in parallel and not being seriously divergent. Water enters brickwork impelled by one or more of four forces, gravity, capillary attraction coupled with surface attraction (Van der Waals forces) and hydrostatic pressure. These forces may work together or against each other, their net effect being the 'driving force' at any particular point within the masonry. The relationship between the driving force and the porosity of the masonry determines the extent, rate and form of penetration.

Gravity is generally of minor importance in the behaviour of water in masonry. It determines the flow of water in large cavities as, for instance, water on the inner surface of the leaf of cavity brickwork, and it is the force which ultimately limits the rise of water being drawn up through the capillaries. Hydrostatic pressure, likewise, is of limited significance and is effective only where there is a continuous body of water contained within rigid boundaries. Thus it adds to the driving force when water enters brickwork set against saturated soil. The prime force operating is the capillary attraction of water entering voids. It is unduly simplistic to say that the smaller the void the greater the capillary attraction. Over the range of porosity between a few millimetres and substantial fractions of a millimetre this is, however, true. Capillary attraction depends upon the surface tension and the angle made at the container wall of the liquid due to surface attraction of the liquid molecules against the solid. Where this attraction is positive the material is known as hydrophilic and where the effect is the reverse – repulsion – it is known as hydrophobic. The capillary attraction can in this circumstance be measured as negative and this is the effect produced by the coating of surfaces with silicone resins.

As pore sizes are reduced other factors become critical. Water drawn through a porous solid by capillarity is responding to the interfacial tension at the free surface of the liquid. A dry brick having been placed on a wet surface is permeated by water lifted by a driving force drawing the liquid upward against the force of gravity. Energy is required for this process to provide the potential energy in the molecules raised above their former position and this energy can only come from the thermal energy available in the liquid. If there is a surplus of thermal energy available in the brick itself (i.e. it is at a higher temperature), this will be transferred to the liquid in compensation. In a porous substance the flow of a liquid must take account of viscosity and the surface effects of the liquid at its junction with the solid. Two key equations express the phenomenon. 'La Place's' equation provides a measure of the equilibrium pressure difference at the liquid surface in a circular pore. This formula gives an interfacial tension. Where pore sizes are relatively consistent and in the medium range a measure of the driving force can be obtained. However, the flow of liquid through a capillary network is governed by Poiseuille's Law. The significant comparison is that while the pressure difference is a function of the inverse of the radius of the pores (La Place) the resistance under Poiseuille's Law is an inverse function of the radius, raised to the fourth power. Consequentially the rate of flow falls off very rapidly as the average radius diminishes. Put empirically the 'frictional' resistance to viscous flow apparently increases with disproportionate rapidity as pore size diminishes. Therefore in brickwork it is only the large pores which make an important contribution to the internal flow of water, even though the very small pores contribute considerably to the effective driving force. (The term pore must be confined to interparticular spaces and must not be used at the intermolecular level where internal polarities predominate.) The pressure effect of the small pores is felt throughout the whole liquid body although the movement through the small pores is greatly inhibited. Within porous masonry the liquid will rise generally to the point where the driving forces of capillary attraction are balanced by the contrary pull of gravity. The rate at which the liquid actually flows will be determined by pore size. The smaller the pores the longer it will take the liquid to reach the equilibrium level although ultimately, whatever the pore size, the same levels will be reached. The height, therefore, determines the levels to which salt contamination can be lifted while the pore size determines the rate of delivery of water and solutes to the surface. It follows that the lesser the distance the solute has to travel and the larger the average pore size the greater the volume that can be delivered to the surface.

8.3.1 Deposition and movement of moisture in brickwork

Gravity, capillary surface attraction and hydrostatic pressures are the static forces which serve to move liquid water through brickwork. An important further force is air pressure differential between two surfaces of brickwork. This frequently rises to levels that contribute significantly to the movement of liquid water. Where cracks, lesions and substantial capillaries occur air will move through the brickwork. It will be particularly prone to move through joints at material junctions and similar discontinuities. Where there is liquid water at the point of entry this water will be given an additional pressure which will physically drive it through openings, sometimes bridging gaps which would otherwise remain waterproof, or being lifted over barriers which would normally impede or inhibit its further progress. Irregular penetration of masonry by water is often attributable to fluctuations in air pressure which will, of course, act in different directions depending on the direction of winds, and the opening or closing of apertures in buildings.

Figure 8.6 Under the relatively light loading of a metre or two of brickwork the expansion of mortar due to differential crystallization can produce a pronounced deformation of walls, piers and chimneys.

Moisture also enters masonry in the gaseous state and where that masonry is sufficiently cool for the air to reach dew point condensation occurs. When this happens the latent heat of evaporation is given up, warming the material slightly and slowing further condensation. A further complication is that in very small pores the local dew point rises in relation to the size of space. Expressed more accurately the relative humidity at which dew point is reached reflects pore size and may in a very small pore be as little as 80% of the value in a free condition due to the increased molecular containment, i.e. the effects of the rebound of water molecules from the containing surfaces. Water can, by the means of air movement and deposition, appear as a liquid deep in brickwork more rapidly and over a very much wider zone than would be possible by capillary action. It can form a layer

Figure 8.7 Blistering in brickwork due to surface absorption of pollutants, in this case, coal tars and sulphates in an urban environment. The pollutants have been absorbed in solution and have returned to crystallize below the surface.

within brick as a boundary along a transition zone. It may occur because warm moist air within a heated building is being cooled as it moves under high vapour pressure outwards through brickwork. Alternatively a building being artificially cooled in a warm humid climate may suffer the deposition of moisture for similar reasons as the outside air is condensed within the brick wall.

Except when fresh from the kiln, all brick harbours a minimal water content within the smallest internal spaces. By virtue of internal condensation of vapour a further amount of water is normally retained within the pores. Size of pores, salt content and ambient conditions have much to do with the amount. It is effectively discontinuous in the sense that no water pressure is transmitted through the brick by this liquid. It is, however, available as a medium of chemical activity and is responsive to and responsible for a water vapour pressure within the interstices. The variation does not cause commensurate dimension change. Water drawn or deposited within the pores in volume may expel almost all the air as saturation is reached. This is an abnormal condition for brick in other than permanent underwater (below water table) circumstances. It is not necessarily a deleterious condition. Saturated brick, however, is likely to be physically damaged by hydraulic pressures consequent upon frost action.

The significance of substantial amounts of water within brickwork is perhaps more to do with its support of biological activity and its functions as a solvent/ion transfer medium than its hydraulic effects.

8.3.2 Aerosols and airborne contaminants/pollutants

In the sense that a dandelion is not a weed until it takes root in the garden, an aerosol is not a pollutant until it arrives where it may cause damage: and, like the dandelion seed, it probably arrives by air. Aerosols are near molecules – particles so fine that they remain suspended in air due to the continual impact of gaseous molecules. Commercial 'aerosols' are not necessarily aerosols in the scientific sense. True aerosols may be solid or liquid particles but they are not truly dissolved in air in the same way as a vapour such as water in its gaseous form. Floating, they are in a perpetual state of agitation, due to molecular buffeting, but can become lodged in or fixed to liquid or rigid surfaces as a result of impact. Thus the nature of the surfaces has an important bearing on the degree of absorption.

Brick and mortar surfaces are rough to the naked eye and exceedingly rich in cavities at a near molecular level. Whether these cavities contain liquid and the extent to which even dry surfaces are receptive to airborne particles then become determinants of the rate of take-up of these floating materials, many of which are contaminants or pollutants. The air close to the surface of a rough material such as brick is intimately related to the surface irregularities and these interfere with its normal behaviour. The complex movement patterns consequent upon the rebounding of molecules from the many sharply angled facets modify gaseous behaviour and within small voids the frequency of impact has the effect of lowering the dewpoint progressively as pore size is reduced. Where pores are exceedingly small, as are the spaces between clay platelets, lodgement of water molecules is maximized and molecular surface polarities fix the water molecules. In large part it is this phenomenon that accounts for the take up of water in newly manufactured brick to between 2% and 4% by weight, with consequent dilatation.

Molecular movement in air is the prime determinant of the behaviour of water in gaseous form in relation to earth and brick. The scale of the phenomenon can be described by considering a litre of air at standard temperature and pressure (0°C, 760 mm mercury) as containing some 30 million, million, million molecules travelling at an average speed of 450 m per second, every molecule being involved in some 5000 million impacts every second. These impacts are not totally elastic, their deviation being governed by energy exchanges expressed in Van der Waal's laws, the net effect being to maintain an energy constant defined as temperature and pressure. Water vapour is a normal although variable constituent of the air mass, the extent of its presence being measured as absolute humidity and its readiness to condense out of the air mass being measured by relative humidity. As it condenses it will first form minute droplets – an aerosol – normally seen as a mist or fog. It may also be drawn out of the air by absorption into a solid or liquid in which case the phenomenon is known as hygroscopy, by which the water vapour will lodge within the molecular interstices or even within large molecular structures.

Measurement of the degree to which a solid is hygroscopic is determined by taking a sample of the material in a totally dry state and then determining the amount of water taken up under known conditions of temperature and relative humidity. This can be expressed graphically for a brick containing various percentages of sodium chloride compared with the curve for brick which is salt free, illustrating very clearly the extent to which the presence of a pollutant can affect the hygroscopicity of a cellular material.

Calcium chloride is a material so hygroscopic that it will continue to absorb moisture to the point when it has become a solution within the moisture absorbed. This phenomenon is known as deliquescence. Calcium chloride has been added to mortars to improve frost resistance during laying and is sometimes formed within brickwork due to the presence of calcium ions from calcite and available chlorides. Historic brickwork will normally contain a range of salts absorbed from its surroundings and it is, therefore, inadvisable to add to these. Calcium chloride is no longer recommended for promoting frost resistance in fresh mortars and concretes in BS Codes of Practice. Its deliquescence and chloride ions are disadvantageous in terms of damp resistance and in their tendency to promote rust in ferrous metals.

By virtue of some limited hygroscopy brick takes up a certain amount of water vapour and liquid water as an aerosol, in addition to liquid water, which may be delivered by precipitation, capillary action, etc. Thus all historic brickwork contains some water lodged within its structure in a condition of dynamic balance. The rate at which water is taken up from the air depends upon the vapour pressures within the air and brick, the nature of the surfaces and the extent of other materials such as salts already lodged in the brick. At a molecular level the particular circumstances of the laminae of air against the porous surface of the brick and their interaction are the governing factor in determining how many water molecules move across the interface and in which direction. In terms of deposition, the wetness of the brick, the wetness of the air, temperature, air flow, etc. are the other principally governing factors. They can be categorized usefully as follows.

1. Constituents of the air, consisting of the available water coupled with air-borne particles, available sulphates, nitrates, fluorides, nitrites, chlorides, carbonates

and rarer pollutants, such as phosphates will have a significant effect on the surface condition of the brick and hence on water transfer (these ions may be expressed as acids, e.g. carbonic, nitric, etc.).

2. Temperature of the air and temperature of brick surface, which, as it varies, alters the speed of chemical reactions, the formation of water droplets and the solidification (freezing) of water. Fluctuations of temperature with changes of insolation can have a marked micro-effect upon surface conditions.
3. Air speed, which determines the impact of droplets, aerosols, and particles on to the brick surface, coupled with size of particles or droplets.
4. Relative wetness of the brick and humidity of the air. In conditions of condensation where the surface of brickwork already carries moisture water transfer is more rapid than in conditions when the brick surface is dry.

The application of experimental work in wind tunnels on deposition in dry and wet conditions is beginning to expose the behavioural patterns of aerosol-size particles (0.11–1000 microns) in the laminar or quasi-laminar layers (the boundary layers in aerodynamic terms) adjacent to the surface. The depth of these surface air layers through which particles have to pass between free air and brick surfaces and which controls their behaviour is in the order of 500 microns in conditions of moderate air speed and is largely determined by viscosity and frictional resistance (Hicks). Where condensation is occurring and vapour is being taken out of the air the replacement air flow increases the deposition velocity of all the available pollutants. This can be expressed at standard temperature and pressure as an increase in deposition velocity of about 0.005 m^3/s for every 100 W/m^2 of latent heat produced by the condensation. This 'Stefan flow' affects gases and aerosol particles similarly. Other related mechanisms act primarily on aerosols. Temperature and humidity have the effect of driving the particles towards cold surfaces or those where evaporation is occurring. This is the Stefan flow. Its effect is to accentuate deposition of pollutants in specific areas.

The phenomenon is sustained and persistent when air is turbulent and when the surface is damp.

Wet deposition occurs in precipitation or when airborne water as aerosol-sized particles is carried at speed against a surface. The inertial effects are then sufficient to cause some water to be deposited by impact. These periods of deposition are on average, and in total, shorter than the dry depositions. Not only is the physical texture of the surface important; the chemical nature and the biological circumstance can also be significant both by modifying the roughness and by affecting the chemical nature and condition of the material against which the pollutants are deposited.

A further and closely related factor is what might be called the local topography of the surfaces in the historic structure. This refers to the physical form of the surfaces themselves, the massing of the building and the wind directions. A slow air stream can be responsible for sustained high velocities as it moves over or round a building. The effect of obstruction in the air stream causes compression which produces acceleration in the air mass. This localized acceleration over ridges and around obstructions or changed angles can then be accentuated by eddying and diminished by protection. High air speeds reduce pressures on surfaces across which they travel – the Venturi effect. Air passing down a roof will eddy under the eaves leaving a zone of relatively still air immediately beneath the eaves but causing more rapid surface movement below the still zone. Downstream of ridges there are zones of reduced air pressure. Thus tiles are sucked off roofs by air pressure beneath them – a wing effect. Likewise air passing round a corner will cause accelerated air movement, perhaps with an intermittent pattern due to buffeting. Where there are prevailing patterns of air movement over long periods of time there will be variations – perhaps very wide variations – in the level of deposition of moisture and pollutants with the result that bricks of similar calibre will weather very differently as a result of their placing in the structure. This phenomenon is, of course, very commonly described in the single word **exposure**, which then becomes a catch-all for many different effects – for differential wetting and drying, for temperature variation due to loss of latent heat, for varying forms of biological climate sustaining or denying sustenance to different organisms, for different levels of deposition of pollutants and, of course, for abrasion by the impact of particles carried where air speeds are sufficiently high.

Careful study of the shape of the building and consequent wind behaviour can be of considerable assistance in reviewing the extent of wear, weathering, erosion and decay which has been experienced and is likely to be experienced in a structure. This evaluation can be used to determine the quality of brick to be used in remedial work and the need for any special precautions. Sometimes relatively simple solutions can provide satisfactory answers. Wind speeds can be reduced by the simple act of planting: either adjacent to the building or as a surface cover. Espalier fruit trees or wall climbing plants can change the surface air speeds on a wall significantly, whatever their other advantages or demerits.

8.3.3 Liquid water in brickwork

Some historic brickwork was designed to be below the water level permanently and in this condition will normally remain saturated and stable virtually indefinitely. Even in saline solution well-burned brick set in a powerful hydraulic mortar has a high durability and in the event of accident can be replaced with like material in confidence. As in so many situations, it is the availability of oxygen which promotes the active mechanisms of decay.

Failure of building services systems – sewage and water pipes and rain water disposal systems – account for the entry of much water into brickwork with consequential effects of plant growth, decay of mortars and structural erosion. Some water will contain contaminants, leaking foul water drainage systems providing an obvious source of nitrates, ammonia and organic acids. Trade, agricultural and commercial effluents may well contain ranges of damaging chemicals requiring specialist advice and action. Ineffective sewage disposal may produce ground water contaminated with the products of the breakdown of organic matter – particularly sulphates and nitrates – and the extensive use of artificial fertilizers is widely recognized as also being the source of significant concentrations of nitrates. The problems of water entry into masonry should always take account of the possibility of the failure of pipework, conduits, channels and drains. Hidden, small, persistent leakages can have serious long term effects on the durability of brickwork and associated structure. On a major scale this can be the loss of water pumped into piped water supplies raising the ground water levels to unanticipated heights; on a minor scale it can be as simple as a pint of water, daily discharged into a piscina to drain apparently innocently into the lower levels of brickwork but culminating in fungal decay penetrating the mortar joints and the destruction of significant timber. Due to temperature change pipework embedded in masonry may be subject to significant movements and stresses causing partial failures which may be as difficult to detect as they are persistently damaging. Water delivered via pipework is unlikely to be chemically pure and, therefore, the nature of its contamination must be recognized if extended leakage has occurred.

The absorption of water from the ground through foundations is one of the most persistent and potentially troublesome failures in brickwork. In areas of high rainfall the concentration of salts carried in ground water is generally less than in more arid zones, but there is always a risk of salts being deposited in brickwork when water evaporates. Historic brickwork can be protected against rising moisture by the introduction of completely or partially impermeable membranes, by increasing the evaporation at low levels and by inhibiting mechanisms. Such changes should be considered carefully in relation to the effects on the fabric. Most ancient structures had little or no defence against rising damp and were designed or intended to survive by allowing evaporation through above-ground wall surfaces. Brickwork was specifically designed to perform this particular function under the sole plates of timber framed buildings. Early medieval timber framed buildings in Europe generally had minimal foundations and base walls but later were built on stone or brick plinths which increased in height to the point where the timber was positioned above the general level of moisture rise in the

Figure 8.8 Lack of rainwater goods and a damp-proof course produces a combination of ground-splash and rising damp which disfigures the wall and accelerates weathering.

brickwork or masonry. With improved building techniques damp inhibiting courses of semi-impermeable materials have been introduced, such as slate laid with broken bond in a dense cement matrix, blue brick used similarly or alternatively layers of impermeable material such as bitumen, plastic and lead. Sometimes a very dense brick laid in a strong mortar has been used. Structures intended to be free of rising water were created and provided that these barriers can be maintained the originally intended protection will be preserved and probably will be sufficient.

As a matter of general principle the first consideration should be to allow brickwork to function as originally designed. Only where the original work is patently defective or where conditions have changed and new uses require different levels of moisture restraint should a changed form of construction be adopted. It is always desirable that a historic structure is returned to the condition in which it originally functioned satisfactorily. On this principle brickwork which performed historically by allowing the natural capillary rise of moisture through brickwork with evaporation in the lowest zones of the wall should be restored in such a way as to continue the same function rather than having impermeable damp proof courses cut into the brickwork or impermeable linings imposed internally or externally, unless it is clear that the method of construction was defective from the outset. A decision of this sort may have a bearing on the use of the building, suggesting for instance that a cellar should remain a cellar rather than become a living room waterproofed with bituminous compounds which drive moisture further up the walls.

Damp-proofing and damp-coursing

The height of the potential rise of moisture in brickwork is determined by the capillarity of the wall and the rate of rise by its porosity. The level which the rising water reaches is dependent upon the rate of evaporation as against the rate of rise. Thus the effect of providing a tanking or impermeable lining internally or externally on the wall can be to raise the hydrostatic base above which the rise occurs and certainly will be to lift the surface area of the wall through which moisture can escape by evaporation. It follows, therefore, that moisture will rise internally to higher levels following the formation of tankings which can, therefore, be effective only if pursued to the full feasible height of water rise. However, an internal tanking can be effective if it is sufficiently high to provide externally a sufficient area of evaporative surface dispose of the rising damp. It is on this principle that waterproof dados are successful. Internal tanking may consist of a multiple application of asphalt – in three or more layers up to 30 mm thick – followed by a coat of rendering, a much thinner skin of bitumen, perhaps rubberized and applied as an emulsion in three coats, or of a plastic-based ribbed sheet fixed to the wall and keyed to carry a plaster finish. All these methods are governed by established codes of practice and manufacturer's advice. Correct assessment of the conditions and careful control are essential to success. Of the application of bituminous emulsions it should be noted that they are prone to blistering under the pressures of salt deposition on the inner faces of the mortar courses, the blisters ultimately failing unless a solid rendering has been applied to sandwich the bituminous membrane and hold it in position.

The introduction of an impermeable membrane or layer through the wall can, in brickwork, be the most effective and satisfactory method of eliminating rising damp. An inert material, such as sheet lead, or a bituminous or plastic membrane or a combination, may be introduced in short lengths as the wall is progressively undercut usually by a combination of drilling and sawing in short sections alternated along its length. Alternatively the brickwork may be made effectively impermeable by injecting resin into its lower courses. The effectiveness of such a damp-proof course usually depends upon the nature of the detailing at the internal face. Failures in bonding the short sections of impermeable material may be relatively insignificant because a very high proportion of the wall will have been rendered impermeable and the small amount of moisture then capable of passing by capillary action will disperse with minimal effect unless sensitive areas of plaster or timber are nearby. On the inner face of the wall such a damp-proof course may be continued downwards behind a skirting to meet a damp-proof course or impermeable layer inserted in the floor and if this work is well done an effective water barrier is achieved. It is particularly important to avoid any form of construction which allows the impermeable damp-proof membrane to be bypassed internally. The principal disadvantage of this approach to the problem is that two distinctly different zones are created within the building – one above the damp-proof course and the other below it in which totally different regimes of humidity subsist. The damp-proof course, however, may act as a slip joint allowing some movement between the upper and lower structure dependent upon the deadloads. In situations where the internal surfaces of the masonry are faced with timberwork, significant plasterwork, painting or other artefacts the arguments in favour of an impermeable damp-proof course may be sustained.

In much ancient brickwork, however, the upward migration of damp through the walls was anticipated originally by the builders and a sensitive conservation approach will be to ensure that the amount of water entering and the surfaces available for evaporation are consistent with the continued good health of the brickwork. To do this it may be desirable to reduce ground water levels by discharge to agricultural drains, french drains and aprons. Commonly it is important to reduce soil and detritus which has been piled against the walls and so to minimize areas introducing moisture and to increase the evaporation zones. If such measures are insufficient it may be prudent to reduce the upward migration of moisture by reducing the permeability of the lowest zones of the wall. This can be done by resin injection. At the time of injection the resins have very low viscosities and they are introduced usually as two part materials mixed at the pressure gun. They set chemically to fill the pores of brickwork over limited zones which can be made to intercept. Generally, however, the effect is to substantially reduce the available routes through which the capillary effect can lift ground water. Less water rises and consequently the upper walls become very much drier since evaporation proceeds at the same rate and on the same surfaces as previously.

Other methods of reducing the amount of moisture passing through the lowest sections of a wall involve the use of electrostatic pressures by which electrodes within the wall are charged similarly to those in the rising water with the result that the opposing electrostatic potentials produce a resistance countering the pressures of the capillary effects in the pores. The electrostatic potential can be set up actively by means of batteries or passively by the introduction of metallic anodes and cathodes buried in the ground and the wall. In theory and inconsistently in practice such methods reduce the capillary rise of moisture in the walls sufficiently to allow the upper sections to have a reduced moisture content.

Figure 8.9 Rising damp showing through limewash to a height of about 2 metres, above which the moisture content reaches an equilibrium.

One further method of reducing the effective moisture rise in such wall surfaces is to increase the rate of evaporation from the wall at a low level. This can be done by the introduction of porous earthenware pipes at regular and quite frequent intervals along the wall surface, the pipes being set into the wall in a near horizontal position sloping slightly downward and outward. In theory internal convection currents are set up within the pipes allowing the air which has absorbed moisture by evaporation and therefore having cooled to fall to the bottom of the pipe and so discharge down the slope carrying moisture with it. Replacement air is drawn in to the top of the pipe to continue the cycle. This still air condition is generally inapplicable, the outward movement being more a product of turbulence at the outer end of the pipe and in windy conditions of the Venturi effect by which the lateral pressure of a moving air stream is less than the still air pressure. However the air exchange has the same desired effect. Proprietary systems of this type with decorative end covers may be found in a number of historic buildings. The effectiveness of the evaporative technique depends upon the filling behind the pipes linking pipe and brickwork. This must be complete to be effective and at the same time must be porous. Poor workmanship in this regard frequently mars the effect of the system. Ageing tends to fill the pores of the pipes with salts and dirt reducing their effectiveness. Their principal defect, however, is the visual appearance which is often inappropriate in historic brickwork and, therefore, restricts the use of this technique to areas in which the wall is not seen but is at the same time accessible to the air.

Evaporation on a brick surface proceeds at a constant rate while a continuous water film is maintained. As the brickwork dries the rate falls off to be an equilibrium controlled by the rate of delivery of water through the brick. Porosity then enters the equation.

The most effective method of improving the evaporation of surfaces of brickwork is to provide good lime mortar-based pointing in the tradition of ancient brickwork. It is possible to return the wall structure of historic buildings to evaporative balance by undertaking repointing where necessary and clearing accumulated debris from walls. The effects of this work can then be improved by appropriate reductions in ground water levels (taking care not to cause subsoil shrinkage) and then to further reduce the amount of rising moisture by inhibiting or reducing the areas of permeability in the wall until the point is reached in which the general rate of evaporation is sufficient to maintain a satisfactory balance between the moisture rise plus absorption on the one hand and evaporation on the other. Such a policy demands sensitive management, monitoring and investment over a period of years with careful recording of actions taken and effects observed. By their nature many historic buildings are susceptible to this approach – churches and monuments in care in particular.

A factor which has some bearing on evaporation from wall surfaces is run-off at the upper levels. The absorption of surfaces and the discharge of run-off are the factors to be considered. A wall that has been rendered less absorbent by renewal of its external coating – a decorative coating of tallow and limewash perhaps – will provide increased resistance to moisture penetration therefore, and will deliver more surface moisture to the lowest zones of the wall. A wall whose upper levels are impermeably coated – with, for instance, weatherboarding – will discharge large volumes of water downwards when rain is directed by wind against its surface. The careful arrangement of drip moulds or secret gutters can quite dramatically reduce the amount of water being delivered to brickwork below it, to the extent that no further damp inhibition measures are necessary. The conservator should exploit every practical method by which the rates of run-off can be controlled, water rise be reduced and the rate of evaporation increased, his ultimate objective being a balance compatible with the long term health of the building.

8.3.4 Precipitation

Much of the water delivered to brickwork comes as precipitation – principally rain – but also mist, snow and direct condensates such as dew and frost.

Rainwater is generally regarded as being pure, particularly in the sense of containing no calcium and dissolved salts. It does, however, contain active anions in the form of carbon dioxide and some sulphur dioxide in solution – principally the former, except in areas of industrial pollution. These are gained by the absorption of gases from the atmosphere both being naturally present and augmented by man's activities. Although the carbon dioxide component is usually larger, sulphur dioxide forms a more corrosive acid in solution. Water in a rain cloud preferentially condenses on available solid aerosols and, therefore, raindrops commonly contain particles of dust, smoke or similar products which have been dispersed in the atmosphere and deliver these to the tile and brickwork surfaces which they meet. In consequence rainwater can initiate chemical activity and must be regarded as a mild, but active, chemical. It is also a solvent and hence a cleaning agent. The major protuberances on a building's surface demonstrate the effectiveness of cleaning upon exposed surfaces and the reduced effects on concealed under-surfaces. At worst material cleaned from above may be redeposited below. The face of a brick arch may be bright orange with near white pointing, but the soffit is unlikely to reflect the same pristine colours. Visually this may have some advantage: usually not. The cleaning process, however, is unlikely to be simply a matter of washing away the deposit. In the rough surfaces of brick and mortar dirt accumulates within the voids and frequently it is associated with growth, with waxy particles and with processes of adhesion which relate to microbiology or physical chemistry. The cleaning effect of rainwater, therefore, reflects more than simple washing away: it can usually be taken as indicative of surface erosion. The much loved patination of brick and mortar is part of this long process of weathering, of wearing away, of lodgement of dirt and of the removal of brick surfaces.

Where lower walls are thicker or drier than surfaces above them free running water can be absorbed as it progresses down the face, depositing superficial dirt from upper levels in long vertical streaks below. The adroit inclusion of a drip may overcome the problem at least to the extent of providing a clear line of demarcation. Calcareous stones – particularly limestone but also some types of lime-cemented sandstone – will deposit calcium-rich run-off on brickwork leaving white or grey stains which can be redissolved with hydrochloric acid. If this type of cleaning is employed careful washing off is essential using small amounts of water at frequent intervals.

8.4 Weathering details

By experience builders have evolved details which improve the resistance of buildings to the weather. These are often applied unthinkingly, or even misapplied, and they rarely allow for the special conditions of individual circumstance. Often they have become part of a stylistic dress-coat and can be omitted or amended only with visual offence. This is particularly true of the 'classical orders', rarely executed in brickwork. Common, however, are projecting string courses and drip moulds whose use has historical connotations.

Figure 8.10 Surface erosion weathering back from the joints. Lack of a maintained pointing has contributed to the failure which now requires replacement of the bricks.

Figure 8.11 Surfaces where rendering has protected the brickwork by acting as a sacrificial layer, in which crystallization has occurred. The brickwork above cill height has suffered surface loss due to the rising water, partially accentuated by the lower rendering despite its porosity.

Some walls have two sides exposed to the weather either as adjacent faces at an angle or as opposing parallel faces and this produces a variation in water absorption which, coupled with exposure to wind movement, orientation and therefore insolation and the position on a wall face, will substantially alter the nature and effect of weathering. A historic building, therefore, carries a series of microclimates of exposure which will determine the levels of acceptance of pollutants and the maintenance of life forms. These will be evident to the practised eye. Bricks within a very short distance of each other may for these reasons have different moisture contents. Such variations allow different forms of life to flourish on and in them and to present surfaces which may be resistant to the deposition of pollutants or alternatively may (as with wet and chemically active surfaces), be virtually perfect sinks for atmospheric pollutants. Careful conservation judgements will take into account the exposures of which existing weathering gives evidence and which the restored and repaired surfaces must accommodate. In many instances it may be necessary to introduce minor adjustments in the light of the visible evidence of weathering and decay. The following are typical of such adjustments.

- Almost invisible drips of lead, zinc, malleable stainless steel or phosphor bronze introduced to throw moisture away from a feature and eliminate surface runnels and staining. An ineffective drip mould may be made effective by providing a secondary drip, perhaps under a capping to a parapet wall or verge. The end of an otherwise perfectly satisfactory weathering may require the addition of a return drip or flashing to complete its usefulness. Where calcareous stone, concrete or reconstituted stone is staining brickwork a careful detail throwing the water clear of the wall face can change the pattern of movement so that the stain may be removed without fear of its reappearance. Drips from the ends of badly detailed or worn cills can be dealt with in this way.

- Half hard copper may be introduced as a drip mould or control where superficial fungi and lichens are a problem. Rainwater running over the metallic copper will absorb it to provide a solution which will act as a biological control in the courses below it and provided that the exposed metallic surfaces are small, copper salts should not be deposited. If they are too large a 'verdigris' stain can result. Adequate maintenance of pointing in copings will maintain water flow to the intended pattern avoiding unsightly stains below eroded joints.
- The introduction of secret gutters, perhaps with small pipe discharge, can reduce the vertical flow of water where it becomes excessive and brick saturation is a problem.

The introduction of brick aprons or splash resistant surfaces at the foot of walls can carry off excess water and assist in maintaining a uniform regime at low level. In wet climates mud splashing has the double disadvantage of being both disfiguring and of nurturing plant life while blocking the surface pores which allow evaporation from the lower courses.

The improvements of detailing by the introduction of damp-proof courses behind sensitive areas of masonry can do much to improve the longevity of brickwork and other associated materials such as timber and plaster. Where an existing detail allows a horizontal path through porous brick and mortar which provides too short a length of travel, this length may be increased by the cutting in of a damp-proof course, perhaps concealed behind the jambs, head or cill of a door. Climate, exposure and nature of the brick will give a guide to acceptable distances of travel. In temperate climates 300 mm may typically be a satisfactory minimum under normal conditions of exposure whereas 100 mm will be quite inadequate. The careful introduction of a barrier may well increase the water path sufficiently to allow the historic fabric to continue to function satisfactorily.

The filling of aquifers may be necessary where the persistent influx of acidic rain water has eroded a path through lime mortar and possibly through cracked brickwork. The careful introduction of a suitable grout may be necessary perhaps in conjunction with other detailing or in some cases in conjunction with a 'waterproofing' additive.

It is not normally the conservator's mandate or objective to perpetuate the faults of bad detailing or materials which have failed but there will always arise arguments in favour of historical integrity as against practical suitability. Brickwork coated with limewash will need annual renewal in areas of high rainfall whereas an alkaline lime-compatible paint may produce the same visual effect with very much longer intervals of maintenance. The art of building conservation will always involve careful evaluation of improved detailing, the use of advantageous materials being set against arguments of historic authenticity which incorporate the demerits of poor design or the use of inadequate materials available at particular historical periods. Particular attention will always be given to precipitation arriving on cappings of brickwork and terracotta because the greatest volume of water arrives on top surfaces with the maximum impact and then continues its path of penetration with capillary attraction and gravitational pull acting together. For this reason changes in material and detailing which might be inappropriate elsewhere may be essential at roof, verge, parapet and valley. It is generally held that changes in materials or detailing 'where they cannot be seen' are not exceptionable. This philosophy does not satisfy the purist.

8.4.1 Roof tiling

It is the function of terracotta used as tiling to discharge large amounts of water to be carried clear of a building via eaves and gutters. Where tiles are suspended (i.e. not bedded) it is their function to deflect water and discharge it even when they have become totally saturated by absorption in which case the water running down the underside is carried to the lowest point, surface attraction allowing it to remain as an underskin, the angling of the tile being sufficient to allow water to move downwards at a greater rate than it accumulates so avoiding the formation of drops. Thus it is that pantiles and other forms of roofing succeed when only a single skin of porous material perhaps less than 10 mm thick serves to protect the interior of the building. The consistent strength of the material and its integrity are, therefore, crucial to its success. Other forms of roofing – such as plain tiling – are designed to interpose two or three layers of surface between the outer and inner air spaces. This simple system of deflection of water depends on the integrity of the components which, therefore, are manufactured to a small size within which the normal stresses can be comfortably absorbed. Generally failures occur due to delamination, impact or ferrous expansion in the fixings and except in the rarest cases the remedy for the failure of a component is its replacement with a matching unit.

Many European tiles have high permeability and absorption characteristics, their performance depending upon physical design and methods of interlocking. Many of them derive from high lime clays which typically produce porous terracotta.

Much terracotta used for surface protection, however, is integral with the substrate. Tiles are directly bedded on to terraces and on to roof surfaces so that there is a full continuity of absorbent material between the outside and the interior. Water penetration is sometimes reduced by an external glaze but as this inhibits evaporation the procedure can be as much a disadvantage as an advantage in anything other than very dry conditions. Techniques of this type can only be successful where the rate of evaporation is capable of exceeding the absorption following precipitation. In many parts of the world this is an acceptable condition. Where there are substantial surfaces available for evaporation such techniques can work even in areas of high rainfall. Inward evaporation is acceptable and even desired in climatic conditions such as those of central China and Indonesia, where roofs are left open internally for this reason.

8.4.2 Embedded tiles

Tiles bedded on to the tops of walls and provided with a sufficient overhang can be adequate protection from rainfall even in wet climates and if the upper section of wall has two evaporative surfaces and is not unduly thick this arrangement can be durable, particularly because the zone immediately beneath the overhang provides for continual evaporation avoiding the introduction of further moisture. In such conditions there is no need for the introduction of an additional damp-proof course underneath the coping tiles or bricks. Creaser (or increasing) tiles used under copings are usually of low absorption and permeability.

The condition in which brick or terracotta is bedded directly on to an absorbent porous base such as a flat roof may be controlled by the introduction of a damp-proof layer immediately beneath the top surface. This technique has some historic basis in areas where bitumens have been available. Generally these have been areas of low rainfall where high rates of evaporation have allowed the brick or terracotta to provide initial run-off and a good wearing surface but do not cause them to be saturated. In circumstances where the material above the damp-proof course becomes saturated and is then subjected to frost action, however, failure of the bedding mortar layer and frequently of the brick or terracotta is almost inevitable. Consequently the introduction of a layer or membrane beneath the upper courses in historic buildings should be attempted only where the combination of saturation and frost action is unlikely to occur or alternatively where the surface layer is to be regarded as sacrificial, as might be the case in the capping of an historic structure to protect visibly earlier masonry, brick or earth beneath. Even in such circumstances the longevity of the capping material, can be significantly increased if adequate drainage and porosity allows the absorbed water to escape before the brick or terracotta is saturated. With this objective in mind the porosity of material should increase with depth. Thus a surface of tile bedded on to a layer of clay should be carried on a further layer such as granular earth or sand or a fibrous support from which water can escape.

Figure 8.12 Pressed bricks of the industrial era (probably early twentieth century) with a distinct fireskin, failing due to saturation from defective rainwater goods with consequent frost action and spalling of the surfaces.

In the choice of surface a fired clay product with a skin of high density, such as terracotta with a good fireskin or even a glazed material, offers the advantage of discharging a high proportion of precipitation as run-off. Saturation can, therefore, be avoided. The worst condition commonly experienced is that a genuine historic coping is rebedded on to a waterproof membrane causing the historic material to be saturated at intervals – conditions for which it was never designed which may destroy it quite rapidly. This problem illustrates the care that must be taken in the introduction of any method of water prohibition in a historic structure.

Historic buildings are often made more complex by the changes that have been imposed on them during their history. These may have altered the patterns of drainage and have changed the patterns of loading. Very frequently these changes will have been made with less care and skill than the building justifies and the conservator, anxious to correct faults, overcome inconsistencies and improve the structure generally in the interests of reducing maintenance and longevity will be inclined to introduce amendments by way of additional flashings, damp courses, seals and discharges. First consideration should always be given to a return to the original intentions of the builders as opposed to improvements or amendments. In practical terms the balance between absorption and run-off must always be evaluated to anticipate the long term consequence. If the original system was unsatisfactory in design this will be apparent. Failure due to inadequate maintenance must not be confused with a design fault.

8.4.3 Water repellents and waterproofing

Among the more attractive and apparently simple solutions to the problems of water in masonry is the application of clear liquids to 'waterproof' brick and mortar. The nature and general characteristics of these materials are dealt with below. They can work in two different ways – by reducing pore size and, therefore, inhibiting absorption, and by reducing the hydrophilic properties of pore surfaces to counter the capillary effect even to the point of reversing the attraction so that it becomes a repulsion. They must be evaluated on the basis of their effects upon brickwork by colour change and change in the weathering pattern, as well as by their efficacy. They will also be

Figure 8.13 A capping to a massive wall. The complex bonding and regular pointing required by this weathering detail is crucial to its success.

evaluated on the basis of stability and longevity. Finally their ultimate breakdown must be understood in terms of the residual products – which may become pollutants. They may perform differently on brick and mortar. Their performance on the combined structure will be a crucial part of the assessment.

In one group come the long-established silicate compounds such as potassium silicate which deposit inorganic molecules (colloidal silica) within the pores on drying out, physically reducing pore size, and in the other category come silanes in which organic radicals with silicon in the molecule are the essential components. Both types of material can be effective and have a proven record. The effects of reduction in pore size in the surface layer of brickwork are to reduce the inward flow of liquid water in accordance with Poiseuille's Law in which the resistance offered by the size of an opening reduces to the power of four whereas the reduction in evaporative effect is directly proportionate to the reduction in area. In practice the 'edge effects' distort the simplicity of this picture, increasing the resistance to gaseous (evaporative) movement although not in parallel to resistance to liquids.

Both types of material are available in various proprietary mixtures and are liable to become increasingly ineffective with the passage of time – the first group by solution and the second by decay and evaporation. This latter process occurs more quickly than the former and a crude generalization of an anticipated life of 15–30 years by comparison with about 5 years is perhaps a reasonable rule of thumb. Both materials can allow brickwork to remain porous and, therefore, to 'breathe'. This means that the processes of evaporation can continue although the absorption of liquid may be inhibited.

Dangers associated with the use of such materials are frequently rehearsed and can be summarized as the creation of zones of unequal moisture content and of different density. The distribution of these inhibiting materials in brickwork is achieved by the use of a solvent which acts as a carrier, evaporating to leave the inhibitor in place within the brickwork.

It is now generally accepted that the breakdown of applied chemicals, and the repetition of such treatment is likely to introduce substances which will cause unforseeable reactions and in some cases have already caused damage by the products of their decay. Brickwork is generally sufficiently robust to withstand some treatments of this sort, but the short term advantages must be weighed against unknown long term problems caused by catalysts and their reagents. In important historic brickwork caution is essential.

8.4.4 Evaporation

The phenomenon of movement of liquid in brickwork can be understood quite simply in practical terms by analogy with the evaporation of water from a cloth. It is common knowledge that two different effects are produced; these depend upon a balance between the amount of water involved, its placement and the rate of evaporation. In the one case evaporation proceeds uniformly so that the frontier of wetting moves back from its edge towards the wettest point. Where evaporation proceeds more slowly the liquid moves outward towards the frontier between wet and dry zones and evaporates at that point. The effect of uniform evaporation is to leave a reasonably constant deposit of any solute across the affected area whereas in the case of the liquid moving to the perimeter a high density of material is concentrated along that frontier leaving, in the case of water on cloth, a visible stain due to the other materials which have been carried to the perimeter and then left behind. The phenomenon of evaporation within brickwork is unlikely to be quite as simple but the effect can be to produce a zone which is rich in deposited material which consequently acts as a moisture inhibitor in precisely that zone and much less in other areas. The consequence can be delamination due to the unequal performance of the brick across the boundary. As there is no ready method of determining how the densities of deposition will occur within brickwork the casual applications of this type of material can result in disappointing and damaging subsequent performance. Used with care and understanding, however, such materials may control absorption and, therefore alter the balance between evaporation and input in such a way as to achieve significant results in controlling the internal moisture content of brickwork. Changes achieved in this way can affect the biological regimes of brickwork by reduction of moisture and can, therefore, be conceived as an alternative to the use of biocides in addition to their potential usefulness in the general inhibition of damp. A colour change due to biological decay or loss may result. Clear liquid surface treatment is rarely possible without some change in surface reflectance and hence a perceived change in colour. In brickwork this may be unimportant.

9

Biodeterioration

9.1 Life cycles and ecosystems

Living organisms cannot subsist without moisture. Brickwork so dry that it will not support some form of life is very rare and, therefore, almost all standing brickwork must be regarded as being host to plant life – and usually also to small animal life. Normally life establishes a system in which several species and types of organism are related in function. Any one species in such a system will not thrive alone. Some species are so closely interrelated that they form one living element and are described as symbiotic. For simplicity this present analysis does not differentiate between the plants and other orders of microscopic life forms which biologists now categorize separately.

Characteristics of the environment determine what type of ecosystem can be set up and maintained. Since these systems, in which organisms and micro-organisms flourish, can play an important part in the weathering processes which achieve the decay of the rocks of the earth's surfaces it is evident that similar processes will affect brickwork. Fortunately these generally occur at a very slow rate, but occasionally they work with dramatic effect, as gentle persistent pressures create sudden fractures and larger plants heave and overturn substantial structures. The physical forces, which through the effects of plant growth cause dramatic results of this sort, are generally of less overall significance in historic buildings than the slow processes which transform sediments and soils and break down minerals. They have a beneficial side also producing the pleasant effects of weathering, patination and the overgrowth of mosses, lichens and algae which give much-loved characteristics to brickwork. The mellowness of old brick walls has much to do with the softening of contours, discolorations and mortar decay which accompanies biological activity.

Environmental characteristics and the nature of the base materials determine which ecosystems are set up, together with the processes of seeding. While most microbiological organisms are present environmentally to such an extent that they will take hold wherever a suitable opportunity affords, larger plants are not so ubiquitous and much depends upon the availability of the seed and the circumstances in which it is delivered. Atmospheric and water-borne pollutants may also be critical in creating environments in which ecosystems will be created and these, in turn, may create ranges of secondary pollutants which allow further systems to flourish. Biodeterioration is not an isolated phenomenon and is perhaps the most complex of all the processes of breakdown affecting brickwork. It always occurs in conjunction with other chemical or physical processes of deterioration, taking advantage of them and contributing to them. In terms of conservation, however, their interrelationships must always be understood and taken into account in general and sometimes in very specific terms. The smaller organisms, however, are often difficult to identify and laboratory analysis may be necessary.

All organisms have limits of environmental tolerance – to circumstances of temperature, moisture, light, pH values, solutes, pollutants, aerosols, disturbance and to many more subtle factors such as toxicity, and effects of other life forms. The limits of tolerance create a pattern of delimitations which express the capacity of the organism to flourish. These are

Figure 9.1 Surface erosion of tiles beginning to be apparent on the lower surfaces of tiles after about half a century, by which time, on the same roof 4 lb lead has been eaten through by organic acids.

expressed in Shelford's Law which is a definition of life tolerance. This definition is important to the conservator who can seek advice both on the survival of organisms if parameters are amended and what alternative bio-organisms might flourish under changed conditions. It is not sufficient simply to alter a condition to exclude one organism without considering what other life will also be excluded and what will flourish in the changed circumstance.

The broad groupings of biological agents of decay in brickwork are animals, plants (i.e. higher plants which are generally root forming), mosses, liverworts, lichens, together with fungi, algae and bacteria, the last three often occurring in microscopic forms which may not be visible to the naked eye. The effects of plants on brickwork are physical and chemical. By the pressures generated through osmosis and leverage they can break apart brickwork: by the production of chemical substances, principally organic acids, they can dissolve components of brick and mortar, principally carbonates as well as metals and stone; by secondary reactions of their products and by products generated by their decay they can create further systems of degradation; by amendment of ground conditions they can create heave and subsidence.

Some bacteria are even capable of concentrating sulphates from atmospheric sources, giving rise to the formation of gypsum in the surface layers of brickwork and the production of sulphuric acid. In addition some plant life is capable of fixing atmospheric nitrogen and oxygen and withdrawing water vapour from the air, depositing it in the brickwork. Most plant life withdraws energy from the surroundings as light and ultimately emits it as heat or as a chemical compound in which the energy is stored The interactive processes that go on during this energy transfer can be described diagrammatically.

An ecosystem may survive on brickwork simply as a point of location. Without drawing from the brickwork any physical matter, a plant may produce significant physical substance by collecting water from precipitation and using light energy to synthesize an organic structure, perhaps with the aid of key trace elements obtained from the atmosphere if not from the brick or water. At this point the brickwork can act merely as a host for location and a cycle is set up in which the pH value of moisture associated with the plant life changes: acids are made available, living organisms decay by oxygenation and through the activities of other organisms. The process continues, accelerating in circular systems until a balance is reached.

Environmental factors interact to limit the extent of the pattern. The space available is one limitation: others are the parameters defining conditions to which species are sensitive,

Figure 9.2 Sonargong, Bangladesh. A constantly hot and humid climate encourages in brickwork and stucco-work the vigorous growth of all plant life from algal slimes to very large plants.

for instance the availability of minerals, humidity, insolation and pH values. The presence, absence of and extent of these factors will delimit the suitability of the environment for particular species – a number having very narrow tolerances. These may act as an indicator of the precise nature of a localized environment. Biological growth will depend upon species characteristics, climate and the availability of primary energy (radiation and ambient temperature), and secondary energy in nutritional material.

9.2 Controlling factors

The wide variation in climate consequent on geographical location is varied by microclimatic conditions. Buildings themselves and their urban surroundings amend the microclimate by affecting wind speeds, creating pockets of warmth, producing situations of high rates of runoff and providing unnatural ingredients from air pollution. These localized factors are overridden by general diurnal and seasonal regimes with changes in temperature, rainfall, radiant heat and length of daylight.

Nutritional factors are equally complex. Some microflora may flourish on lime-rich mortars whereas a silicaceous environment (which is slightly acid) on a brick surface will support other types. Bacteria generally favour alkaline conditions while fungi often favour those which are slightly acid. Minerals are always needed for nutrition. These may be dissolved out of brick, particularly where iron is available and where sulphurous compounds can be found. The availability of organic materials is also important and these may come from other plant colonizations, pollutants and materials deriving from other parts of the building structure or surface treatments. The availability of these types of materials can favour or inhibit the growth of heterotrophic bacteria, fungi and microfungi (actinomycetes). In the absorption of light energy, its duration, intensity and wavelength are the key parameters and these can vary over short distances.

An organism suited to living in surface conditions might be incapable of survival a few millimetres away in the depths of a pore. The availability of different photosynthesizing chemicals and photooxidizing enzymes then become controlling factors. Plant life contains different photosynthetic pigments (chlorophylls, carotinoids, xanthophylls and phycobilins) with differing factors of light sensitivity and ability to convert light energy to organic matter.

Rate of growth is generally controlled by ambient temperature and to a lesser extent by available water. High levels of available water and high temperatures usually favour growth and variety of species, some plants being adapted to wide ranges by death and regeneration. Algae, fungi and bacteria can tolerate wide climatic variations by retrenchment or encapsulation during adverse conditions with major resurgence in favourable circumstances. Fluctuations of this type can produce changing stresses which, associated with physical variations, may act powerfully towards the disintegration of the host material, brickwork.

Pollutants, which are not necessarily man-made, may sustain and accentuate various forms of plant growth by the provision of a wide range of particles and substances dissolved in the available water or arriving as gases. Plants may thus absorb nitrogen, sulphur, carbon, fluorine, chlorine, phosphorus, sodium, potassium, magnesium as trace elements in addition to hydrogen and oxygen, all eventually to enter the chain of reactivity in the brickwork. All aerobic organisms produce carbon dioxide dissolved or dissolvable in water to form a weak acid, the emissions taking place within the interstices or the immediate vicinity of the brickwork; other absorbed materials form more complex organic acids many of which are released on or are the products of decay.

9.3 Chemical attack

Consequent chemical attack is of two similar types, acidic and chelating. The principal acids produced are sulphuric, sulphurous, carbonic, nitrous, nitric and oxalic which then move into solution with calcium, magnesium, sodium, etc., breaking down crystalline structures as they do so. The effects of actions of this type are more visible on the mortars in historic buildings than on the bricks themselves. Chelation is a phenomenon associated with complex organic acids such as lecanoric (produced by lichens), citric, ketogluconic and also oxalic, which are usually the product of plant life. In essence the process involves the absorption of a metallic ion, such as calcium from the brickwork and although the process takes place in a slightly acidic environment the chelating group is essentially basic. The term refers to the sharing of a hydrogen or metallic atom within a stable ring compound. The hydrogen or metallic ion has two bonds with the radical thus achieving the ring. The organic compound is the chelating agent and may be an organic acid such as EDTA (ethylene diamine tetra acetic acid) used in the laboratory in testing for calcium for exactly this reason. Chlorophyll and haemoglobin are chelates and as such are analogous in the organic sense to the 'salts' of inorganic chemistry. Chelating agents can be aggressive. Tri-ammonium-citrate is, for instance, used for cleaning in this way, being 'hungry' for iron in particular, and capable of causing the breakdown of grime and also of materials whose retention is desired.

Biological destruction looked at closely exhibits specific characteristics depending upon the agent. Many agents and many processes are interactive, one feeding off the other or consequent upon earlier events.

9.4 Effects of animals

Historic brickwork suffers more from man than from any other agent. Erosion, abrasion and deliberate actions apart, passive effects such as continual exhalations from visitors can change the circumstances of monuments – particularly of plasters and renderings which carry paintwork, stimulating the growth of bacteria and algae by water vapour and carbon dioxide emission, so that even where direct pollution is avoided indirect pollution produces immediate damage.

The excrement of all mammals is rich in urea, providing available nitrates which foster further growth or enter the reaction chain immediately to produce soluble materials from substances previously rigid in mortar structures. Calcium, sodium and potassium are all available as highly reactive ions. Uric, nitric and phosphoric acids particularly react with carbonates and to a much lesser extent with silicates to damage the structure of brickwork. Dependent upon the location, the depositing agent may vary – bats, birds and to a very much lesser extent mammals. Submerged brickwork is at risk from direct boring by molluscs and polycheate worms. Some anchoring and boring creatures are able to achieve penetration of mortars by secreting organic acids either for the purpose of creating voids or locking themselves into the structure. Boring bees, termites and other insects excavate masonry – particularly mortars – and in so doing open up channels for weathering, plant life and further boring creatures. Below water level molluscs can behave similarly. Many such agencies act very slowly and virtually none is independent of the others. The interaction between the activities of plants and animals is continual and interlocked with mutually supportive sequences involving both life forms.

A further complication lies in the laws of protection which vary between countries. A number of species such as bats are protected to the extent that their environments may not be disturbed and this will prohibit the use of fungicides in their roosting places. Such protection may be seasonal. Where permissible remedial and preventive actions vary from toxic additions which will kill termites or bees (see below) to deterrents such as wire to prevent birds nesting under tiles. The essence is careful management and good maintenance. Control of a microclimate in heavily used areas may prevent condensation; cleaning down, repointing, protection against hazards and other aspects of care are generally sufficient to safeguard brickwork from undue damage by animals, other than the extreme exceptions of war and violence. Far greater hazards derive from plant life.

9.5 Large plants

Large plants cover the range from trees to rooted plants on the scale of grasses and succulents. Dramatic destruction can follow the growth of the larger plants in brickwork. Leverage is exerted by the penetration of roots, and there are a number of clasping or grasping plants which will envelop brickwork in dangerous embrace, relying on it as they would rock for stability and support. These can vary from trees to vines of which the ivy (*Hedera helix*) and similar species is widely distributed in temperate zones. In tropical zones large trees, such as species of ficus, can enwrap brickwork with roots which may not penetrate it but reaching to ground level use it as a support. The consequent windage caused by the weight and air obstruction of the plant mass may produce turning moments greater than the adhesion of mortar and the deadload of the combined mass is capable of resisting. Likewise brickwork can carry trees which grow so large that they topple their host structure. Other effects include ground movement caused by falling trees, which will carry with them heavy root plates. Large plants will change the microclimate of brickwork – sometimes with advantage where sheltering effects are caused – but frequently with the disadvantage of the products of decay being delivered to the surfaces to host yet further growth.

However, the prime destructive mechanism of large plants is root penetration. This is effected by the expansion of the growing tip at and by the delivery of available hydrogen promoting chemical reactions with soil and mineral fragments. The effects are not consistent but, broadly, root tips are acidic having a negative charge which the surrounding water balances by positively charged hydrogen ions. The high acidity (2–4 pH) produces an etching effect on minerals containing metallic ions – barium, calcium, magnesium, caesium, potassium, sodium and lithium and also ammonia which behaves as an ion in the same way. The medium of transfer can be colloids in soil conditions or solution; and the roots produce amino acids, carboxylic acids, chelating agents and acids such as tartaric, citric and oxalic acids. Alumino-silicate minerals such as the micas can be deprived of potassium (K) to form vermiculite. There are two classes of mica with 7 and 8 metallic and silicon elements respectively e.g. KAl_2Si_4 and KMg_3AlSi_3. Essentially the destabilizing of the minerals around the root tip allows the root to expand by virtue of the very high pressures generated within the cell structures. Mean pressures may be 15 atmospheres (1500 kilopascals) but can rise to twice this value (3000 kilopascals or 30 000 millibars). This pressure continually maintained allows the root to expand into momentary voids created during chemical activity – the 'foot in the door' effect, resulting in a powerful destructive force in mortar and in brick. Structural 'heave' may result from these forces of expansion, a small root in or under a wall expanding as it grows to lift many tonnes of brickwork, displacing the mass and breaking the mortar bedding or cracking the brickwork. Tree growth can have well-known secondary effects of drawing moisture from a clay soil, causing shrinkage in the localized section of the soil affected by its

Figure 9.3 Spanish Town, Jamaica. On an abandoned nineteenth century barracks plant life becomes established at points of maximum water entry – head of wall and cills. The larger plants create a root spread across the wall surface; the smaller create pockets of vegetable material within the structure. In each case root penetration will break the mortar–brick bond.

rootspread. Usually in periods of reduced rainfall the shrinkage becomes significant, causing the subsoil to shrink differentially with consequent uneven settlement and structural damage. A building which has accommodated the gradual effects of the growth of a substantial tree may also suffer from changing ground conditions on its removal, although the expansive forces of wetting clays are much less likely to be damaging than the converse loss of support settlement through shrinkage.

The straightforward remedies lie in the careful control of large plants in the vicinity of historic buildings. When the damage has been done there may be no other remedy than underpinning and rebuilding the damaged brickwork on amended foundations. This normally is a method of last resort.

9.6 Mosses and liverworts

These lower plants have systems of rootlets which are similar to those of higher plants but less complex. Being capable of producing higher acidity around their tips they are more capable of extracting mineral cations. For this reason they are primary colonizers but with very modest root penetration. However, they break down surface structures in brickwork and open the way to the more deeply destructive effects of true roots rather than rhyzines – primitive rootlets (also rhyzoids). They are generally not tolerant of highly basic conditions, and therefore are secondary colonizers of mortars and other basic surfaces. Mosses tend to occur on clay tiles while lichens are found on concrete for reasons of basicity (high pH values).

Figure 9.4 Overwhelming creeper. The entire external brickwork of this house has been submerged in creeper. Its Dutch gables resemble topiary.

9.7 Lichens

Colonization by lichens precedes those of liverworts because they are tolerant of the basic environments created by mortars, and among the 15 000 species there are a number of endolithic types which live within calcareous structures such as lime mortar (e.g. *Kyrtuthix dalmatica*). These penetrate by boring, on a small or even microscopic scale, into bricks and particularly mortars. As they die or are browsed a superficial layer of mortar is removed and a small depression created before the same process perhaps is repeated. With their decay organic acids penetrate the mortar softening it by the breakdown of crystals and amending the basicity, opening the way to the invasion of species less tolerant of alkaline environments. Lichens are symbiotic structures containing large numbers of algal cells in a matrix of fungal cords and filaments. They tend to reproduce by fragmentation and will tolerate harsh environmental conditions. Species by species they are very sensitive to quite specific conditions and in consequence they can be used as indicators of microclimatic change and of the incidence of pollutants. Specialist knowledge of types and tolerances is a valuable method of identification of atmospheric deposition of pollutants. Their sensitivity dictates their absence from one part of a structure and their growth on another apparently similar. A factor as simple as a shadow-fall may dictate whether a particular type will flourish on one spot or be entirely absent. Lichens can act on the most durable of materials by the depletion of calcium, sodium, magnesium and potassium ions and by destroying the cementicious bond. In addition the fungal component produces ranges of polyphenolic compounds – fatty acids, tetronic acid derivates, and aromatic compounds with a chelating property in addition to more common organic acids such as citric, oxalic, gluconic and lactic, which erode the crystal structures of cementing compounds and consequently soften the mortars. The effects are superficial but continuous. Vitreous bonds are affected and brick will become etched and pitted as a result. The combination of carbonic acid (dissolved carbon dioxide) and oxalic acid for instance can produce calcium/ manganese/magnesium oxalate dihydrates, removing the cations from stable structures to produce granular materials which are lifted from the surface by the thalli (suckers) contracting as the plant dies.

9.8 Algae

Primary colonization is often carried out by algae, of which there are a very large number of species adapted to almost every environment. Of the many that colonize brickwork some live entirely on the surface, some live within superficial pores and others actively move into internal spaces which they seek to enlarge. They are the prime agents of photosynthesis and on brickwork generally are

Figure 9.5 Mixed mosses in competition on plain tile.

categorized by forming a thin superficial layer either dark or light grey on strongly sunlit bases or forming a thicker green mould on damper, less well lit surfaces. They are seen as a patina rather than individual plants, unless examined under a microscope. Generally they cause weathering by chemical action but there are significant effects due to their trapping moisture and pollutants to form a stratum of organic materials on which other organisms such as bacteria, lichens, mosses and ferns can subsist.

Variously they secrete ranges of acids and chelating metabolic compounds including polypeptides, amino acids, uric, oxalic, glycolic, aspartic, glutamic and citric acids which react particularly with carbonates, breaking down some components of brick and all mortars. Algae flourish in both alkaline and acidic environments creating or enhancing their own acidic background.

9.9 Fungi

Having no chlorophyll or other photosynthesizing chemicals fungi are not regarded as plants deriving their food energy input from other organic material. Their essential structure is a series of filaments (hyphae) which can be very extensive and can penetrate the pores in brickwork. When the hyphae swell they can exert pressures within the pores and it is claimed that they can act mechanically on mortar. They may contribute to acceleration of the destruction by freeze–thaw processes as do algae in the pores. Their main damaging effect, however, is also in the production of acids and chelating agents.

Carbonic, nitric, sulphuric, citric, oxalic, gluconic, glucuronic, lactic, fumaric and many other organic acids are recorded as being produced by fungi and the structure of most fungal material is such as to retain these acids within or close to the surface of brickwork. Oxalic acid is particularly responsive to iron and will break down ferruginous salts and clay minerals in brick and earths. Amorphous gels carrying calcium oxylates and iron oxides are deposited. On the outer surfaces of fungal hyphae, hydrated calcium carbonate is built up ultimately modifying the structure of the brickwork and possibly accounting for the accumulation of calcium sulphates identified by recent research in the outer zones of historic brickwork.

9.10 Bacteria

Actinomycetes, originally classed as microfungi, are now regarded as bacteria. However, their structure is based on rigid cells and they form colonies with filaments like branching moulds. They include the penicillin family and those many groups which are capable of inhabiting brickwork and masonry act on it in much the same way as fungi except that they do not produce the deposits of hydrated calcium carbonate (calcium bicarbonate). Bacteria, too small to be visible other than in collective groupings, inhabit the pores and depths of brickwork as well as surface regions. While capable of protein synthesis and in some instances capable of survival by using carbon dioxide from the air, most bacteria require sources of organic nutrients. Therefore they tend to be secondary colonizers but being single celled organisms capable of self-propulsion with rapid rates of reproduction (as short as 15 minutes) they can populate large areas of wetted brickwork very rapidly. In adverse conditions they can remain dormant. Bacteria function very largely in breaking down organic matter and in brickwork this occurs using nitrogen or sulphur drawn from living matter. These give rise to two separate cycles or repetitive chemical processes. In the nitrogen cycle of an ecosystem ammonia is formed from which nitrogen trioxide results. In the sulphur cycle hydrogen sulphide leads to the production of sulphuric acid. Elemental nitrogen and elemental sulphur are also formed in parallel and it has been demonstrated that chelating agents, inorganic and organic acids and alkalis (ammonia and amines) are produced. Metals such as copper, nickel, manganese and iron among others are linked into 2-ketogluconic acid produced by bacteria in addition to the ranges of organic acids typical of higher organisms. In addition, it has been shown that bacteria can dissolve silica and silicates as well as iron, aluminium and calcium phosphates. The processes of decay are subject to investigation and theorizing. One theory ascribes the production of gypsum (calcium sulphate) to bacteria of the thyobacillus group – the postu-

Figure 9.6 Spanish Town, Jamaica. Failure to maintain rainwater goods can provide the opportunity for spectacular plant growth in humid conditions, solidifying the blockage and allowing the transfer of the plants into saturated brickwork.

late being the oxidation of hydrogen sulphide, thiosulphates ($H_2S_2O_3$) and other sulphurous compounds to sulphates which combining with the calcium of calcium carbonate result in the deposition of gypsum found in the outer surfaces of older walls. It has been further hypothesized that the anaerobic bacterium which reduces sulphates to sulphides in the ground is also a source of the sulphur carried into wall structures. The damaging effects are those of acidic compounds, as with the higher forms of plant life and the formation of soluble materials (e.g. gypsum). Bacteria can damage brickwork only by chemical attack.

9.11 Mechanism of chemical decay due to biological activity

In the recitals of acid attack and chelation there is a common pattern which can be summarized thus: chemical decay in well-burned brick is very much slower than in historic mortars, which partly accounts for mortar loss and the need for repointing. Biodeterioration of this type is a complex pattern of several interacting processes of degradation, including physical microdamage. In aerobic conditions organisms will produce aqueous solutions of carbon dioxide – carbonic acid – which are the basis of chemical activity (e.g. CO_2 by several phases to $Mg/Ca/(HCO_3)_2$). Soluble bicarbonate compounds are transported out of the originating structure leaving it granular and friable. This is a straightforward acidic reaction, particularly applicable to mortars. In the more complex processes of chelation one atom of a metal such as sodium or calcium performs a dual function. Many organic acids offer the opportunity for stable ring relationships, drawing metallic ions out of their inorganic structure. The strength of the chelating group is critical and tends to be highest in neutral conditions – pH7, at which level the stability of the chelate with its double linkage tends to localize and fix the ion preferentially. This process is slow and must be regarded as more important in tiles and mortar than in brickwork. There is some evidence that in tiles damage of this type is more significant than the rate of activity alone would suggest because of penetration along planes of lamination where an already weak bond is further weakened.

9.12 The inhibition and prevention of biological attack

Where it has been determined that biological decay is actively harmful or is not tolerable, the possible courses of action are either to counter the life forms by a destructive agent or to change their environment. Many actions of either type, however, have other consequences that require secondary actions and further maintenance. A tree may be cut out of brickwork before it does significant damage but the root structures left in the mortar will be lost to microorganisms as they rot, leaving a void for tunnelling animals and providing food for other life forms. A wood-destroying fungus such as *Serpulla lacrymans* may be killed chemically or by changing its environment, eliminating the water essential to its life. The former introduces foreign and complex chemicals – the latter requires a new regime dependent on altered conditions.

In many instances the first action may be to remove the biological growth physically by cutting out and cleaning down. The follow-up action will then be to counter the growth of other organisms. The widest ranges of direct control are chemical. Most of the chemical products applicable to work of this type are synthetic compounds designed to kill the organism which has to be controlled. It must be recognized that at this stage of progress – and despite very considerable advances in the chemistry of toxicity – the methods employed in the use of biotoxins – the nature of the materials and the extent of the effects vary from crude and poorly understood to barely adequate. Legislation and codes of practice concerning the use of such materials is in place in sophisticated communities and is to some extent effective but much has yet to be learned. As a result the legislation tends to be heavy to the point of being hamfisted, or else wanting. It is also subject to rapid change as the encouragement towards the use of organophosphorous compounds followed by a reversal of policy has shown.

The first question to be answered, therefore, must be whether the use of such materials is justified in brickwork in the circumstances. For all the evidence of the production of organic acids, erosion and boring it is a general truth of brickwork that it will survive attractively for long periods with nothing more than physical maintenance to eliminate tree growth, to repoint eroded mortar bedding and to clean down excessive incrustations of lower plants. The ideal methods are straightforward environmental controls which modify microclimates. The case for going further and using biocides on brickwork is small. Only in special circumstances should it be considered. These special circumstances might include the elimination of fungal transfer in association with failure of timber, the elimination of growth prior to bonding other materials, including paint, to the surface and the importance of contamination which may spread from brickwork to other artefacts or works of art. Occasionally special materials must be introduced to combat specific threats. The tropical termite, of which there are some 2000 species, may threaten other aspects of a historic building's fabric and the denial to the termite of a burrowing path through mortar might require the insertion of a metallic anti-termite course (copper) or a zone of chemical impregnation.

Before deciding on a strategy a number of salient questions must be asked, namely:

1. What is known of the history and performance of the proposed material in similar circumstances?
2. What is its toxicity? Measurements of specific toxicity are based on the amount applied to achieve lethal doses in upwards of 50% of the organisms being treated. It is to be noted that each definition will relate to a particular species in identified circumstances. It may be the case that a specific toxic dose will eliminate 50%, that a further similar dose will take the total elimination to 75% and so on, but it does not necessarily follow. Further uncertainty arises over persistence and therefore over build-up. The key question, however, is regeneration – how, in what circumstance and to what effect?
3. What residual effects are there, including those of degradation of products?
4. To what extent is pollution likely to follow in the local and general environment and what is its permanence once the initial function has been performed?
5. Does the mode of application show how the material has been delivered, whether it has been successfully applied and whether it has strayed from its intended target (e.g. fugitive colour inclusions, etc.)?
6. Will it have any residual effect upon the brickwork itself?
7. What is its persistence as a biocide?
8. Will any reaction have an adverse effect on the brickwork or associated materials, e.g. stone by the formation of surface coatings, crusts, etc.?

Figure 9.7 Sufficient water is retained beneath the paving of this footbridge to provide for vigorous plant growth.

9. What is the best time for application? This is usually at the beginning of vegetative activity.

In choosing the material for application careful regard must be paid to legislation and manufacturers' recommendations as to codes of practice. The choice of biocide must be made on the basis of minimal disadvantage environmentally, certainty in application and minimal toxicity. It is also important to identify the agents within proprietary formulations that produce active effects.

Biocides may be selective, but generally they operate over a broad spectrum against different types of life and the level of toxicity will vary from circumstance to circumstance and species to species. A mixture which would be adequate to deal with one species in one condition may be inadequate in another circumstance. Further, compounds with the ability to attach themselves selectively to tissues or to penetrate growth systems may be added to the commercially available mixtures to maximize their effect and prevent washing away. Many are effective only at specific points in the life cycle and dealing with an infestation in this way is therefore a matter of carefully considered treatment and specialist advice – often from the manufacturer whose response may not be impartial.

9.12.1 Prevention of animal damage

The erection of physical barriers and control over the routine of beasts and vehicles are self-evident. The visual and practical consequences of such structures must be considered. Birds present problems principally associated with the deposition of faeces which changes the pH value of the surface on which it is deposited and provides material for the growth of plant life. Physical devices which prevent or inhibit landing and roosting are usually the sensible response.

Other than for use against boring insects and termites no toxins are recommended for use in brickwork against animals. Deildrin, Chlordane and Durspan are proprietary formulations of high general toxicity and increasing disrepute, which can be introduced into brickwork and earths to eliminate the burrowing attacks of small insects such as masonry bees and termites (Isoptera). Termites feed on plant material including cellulose and may obtain sustenance from earth structures which generally contain plant material and from organic binders in mortars. Some members of colonies are flighted and therefore can bypass brickwork which has been impregnated. They are particularly likely to burrow beneath plaster surfaces. They also penetrate deep into masonry liking nothing better than living in voids presented by loosely built cores in moderately humid brick walls. They shun light and where forced to traverse an open surface will construct conspicuous runs or tunnels formed of grains of sand cemented together with a salival secretion.

The application of suitable chemicals is usually carried by out drenching or a drenching spray with aqueous or aromatic hydrocarbon carriers. These toxic chemicals are generally persistent and are dangerous to man. Great care must be taken in their use and proprietary formulations only should be used, always in accordance with manufacturer's instructions and all local legislation. Care must be taken to guard against accidental and unwanted dispersal, unsuitable disposal of excess, ingestion, the presence of susceptible animals, entry into the food chain and all secondary effects, including the misplacement of containers. The least toxic but effective product will be the preferred choice.

Lesser pests, such as masonry bees (*Osmia rula*) which burrow into mortars and earths, may be dissuaded by the use of fine mesh – metallic if necessary – covering a local infestation. If unduly persistent they may be eradicated with a repeated dosage of Permethrin or a similar insecticide, usually applied as a drenching spray.

9.12.2 Herbicides

Agricultural research has produced wide ranges of herbicides, some of which are selective and specific to taxonomic groups. They are generally short-lived and only moderately invasive of the general ecosystems. The only likely application in brickwork generally will be the destruction of root growth which has been cut off and which may otherwise persist. There are situations where the destruction of brickwork to eliminate roots physically would be unnecessarily damaging to a historic building and an appropriate tested biocide specific to woody species may be used.

There is no case for controlling small plant growth in brickwork with herbicides, other than weed growth in brick paving and aprons where proprietary materials may be used subject to such washing procedures as ensure that no build-up of soluble salts is introduced into walls. Among the commonest herbicides are nitro-organic compounds, but many are type-specific, and must be applied in a suitable manner at the right stage of growth. The advice of specialists or manufacturers is essential, particularly in those which act by altering or affecting the enzymes controlling plant growth.

9.12.3 Algicides

Most mosses and lichens can be tolerated, even welcomed in moderation on historic brick and tile. Where their removal is necessary they are susceptible to compounds which interfere with the photosynthetic processes and these are often painted or sprayed on to colonized areas. If this treatment is needed on brickwork the most commonly used are Quaternary Ammonium compounds which are provided in proprietary mixtures. Proprietary mixtures containing phenols (powerful antiseptics) are used, sometimes in great dilution. Their residual products, including ammonia, can result in the formation of nitric and nitrous acids.

Applications may be by poultice, powder or in solution. Recent research suggests great caution in the use of these materials. It must always be specific to an identified problem.

The processes of application will vary but should always include the removal of loose material by gentle brushing or pressure spraying. Iron wire brushes should not be used. Spalling brickwork and lifting renders should not be subject to high pressure washing. The adhesions of some mosses and liverworts are sufficient to justify their individual removal by hand. Proprietary mixtures are supplied usually ready for dilution (typically 1 in 20) and application should be by fine spray working downward, saturating the surface briefly. The process must be repeated several times. A biocide may be applied initially followed by retreatment after a week following further cleaning down. The wide range of materials used include carbonic acid compounds and quaternary ammonium compounds including hyamines, alkylbenzil trimethyl ammonium and benzyl, alkyl, dimethyl ammonium chlorides and bromides. These have the advantage that they are generally relatively harmless in relation to higher plants and animals, but full protective precautions should always be taken and specialist advice on the actual products must be sought. Secondary and by-products should be washed into safe disposal on completion of the treatment.

9.12.4 Fungicides

Other compounds regarded as efficient in controlling fungi and bacteria are derivatives of benzothriazoles, benzalkonium chlorides, isotiazolinone chlorides and formaldehyde. Some of these, particularly the last, are not easy of application. Formaldehyde is a gaseous aldehyde derivative of methylalcohol, in aqueous solution.

Lichens are susceptible to attack either on the fungal or algal component of the symbiotic material. Among other recommended materials are zinc and aluminium silicofluorides (which suffer from the disadvantage that they can cause crusting on mortar surfaces), ammonia and formaldehyde, pentrachlorophenate and orthophenylphenate and lithium hypochlorite. There is a risk of secondary reactions with chlorides and chlorites in the presence of mortars with the consequent production of highly soluble calcium chloride. Formulation and proportioning are important. Coupled with the need for testing and long term experience the conservator must take every advantage of the experience of manufacturers as embodied in proprietary combinations and the relevant instructions. However, for relatively simple purposes of initial cleaning some straightforward methods are available. A brief change in regime may be justified. This may be achievable by the application of sodium chloride solution subsequently washed away or of a highly alkaline solution (e.g. ammonia) which causes the thalli of lichen to swell and be removable. Subsequent washing is essential.

9.12.5 Sampling

For microbiological analysis laboratory techniques and expertise are necessary. It is usually necessary to apply not only qualitative analysis, giving an understanding of the species involved, but to apply quantitative techniques in order to interpret results adequately. Laboratory analysis employing counting techniques is the only reliable method. The extent of the presence of microorganisms and their relationships may determine the role they play in deterioration.

Where samples are collected their origins should be noted and pinpointed and they should be taken with sterilized tools (flame or solution sterilization is usually sufficient) and put in sterile containers (petri dishes or tubes if possible). Unless the material is going direct to the laboratory it should be maintained at 4°C by refrigeration and in conditions of normal humidity. If the sample has been taken at a high ambient temperature and high humidity the reduction in temperature to 4°C may induce saturation by condensate unless adequate ventilation is allowed as a precaution. Extended retention in polythene bags is likely to result in the decomposition of the sample. For quantitative purposes a sample of not less than a gram and preferably several grams is needed in each case.

In the laboratory stereo-microscopy may be applied and the microorganisms may be cultured on media specifically adapted to provide the nutrients required. Investigations of plants, mosses, lichens etc. require similar care in collection and the sampling should be environmentally homogenous. Samples from different microconditions should not be mixed.

Identification is carried out by comparative sampling and by microscopic observation. Specialists are likely to make field visits as the interpretation of interconnection is critical to understanding the processes involved and deciding whether the observed effects are progressive or regressive. Elaborate investigations are justified only if the scale and potential of the problem is large. Within the context of a modest single historic building the activity will be much more at the level of the intelligent gardener utilizing the best commercially available products.

9.12.6 Passive and natural techniques of control

Modification of an environment by the introduction of a different regime can be an effective method of controlling the circumstances of biodeterioration. An archaeological ruin field of brickwork might be kept clear of tree growth, grass and large weeds by mowing and the application of appropriate sprays at specific times. Equally effective, much less costly and damaging would be the presence of a suitably managed flock of sheep coupled with capping and repointing of walls, physical protection of vulnerable structures and like operations.

Interiors which might be sealed for security purposes can, as a result, become damp and smeared with algal slimes causing immense damage to plaster, paint or brick. The avoidance of hermetic sealing and the provision of adequate ventilation may be the alternative. There is an important interaction of management, maintenance and conservation. Some results can be fortuitous and surprising. Red clay tiles on roofs in southern England will weather brown or even green as a consequence of the microscopic algae or lichen which inhabit them. This is an effect which is entirely acceptable and might be thought to have been historically accurate. In the recreation of an open hall house by the repair and re-erection of a timber frame, a roof of plain tiles, weathered brown and green would appear to be entirely satisfactory. However, after maintaining a small open fire in the hall for about two years the browns and grey greens of the roof give way to orange and flame reds inherent in the original tiles revealing an historical effect which would have been typical of medieval houses before the introduction of chimneys but had never been suspected. The effect of the smoke passing upward through the tiles is to kill the lichens and algae which change the tile colours. Ambient conditions can be used as a deliberate biological control with less drastic side effects and greater permanence than many chemical methods whose application can be hazardous, labour intensive and short term.

Changes in regimes of humidity, insulation, ventilation and food input may be quite simply achievable with the result that the adverse effects of growth are eliminated. Cleaning may be involved initially. Steam lances can be used to good effect on growth on brick and tilework but, once cleaned, it is essential to change the microclimate or other conditions to prevent a recurrence of the situation.

In some cases the active introduction of inorganic and organic inhibitors or accelerators may be effective. In a rainy climate a telephone wire stretched across a roof will leave a lichen-free zone where water from the wire has dripped on to tiles, the copper having been mildly attacked by the weak carbonic acid in conditions of condensation, forming copper carbonate. This mild, quasi-natural biocide can be deliberately introduced, perhaps with a secondary purpose by the introduction of copper drips or soakers in salient positions. The organic acids produced by algae and lichens may damage other materials as they are discharged from roof surfaces. The corrosive effects of such acids on lead soakers and flashings is a ready instance of the problem.

The inhibition of plant life by the effects of lead or zinc may be adequate justification for the introduction of passive inhibitors, care being taken to avoid other side effects, such as staining and corrosion induced by electrical polarities. Distilled hydrocarbons, such as fuel oil, can effectively inhibit plant growth, ultimately evaporating totally, and at a more sophisticated level the introduction of enzymes capable of accelerating the growth of species to the point of self destruction has been advocated and experimentally tested with success. Proteolithic enzymes have been satisfactorily used for the removal of lichens but the techniques at the moment are confined to laboratory and research conditions. Likewise some micro-organisms may be used to counter other micro-organisms by producing antibiotics. Antibiotic preparations have been used in laboratory circumstances for cleaning and the destruction of unwanted micro-organisms. These techniques are likely to be applicable only for use on special artefacts which become the subject of conservation technology on a museum scale. Long term evaluation is an essential part of the preliminary examination of materials of this type.

10

Natural and man-made pollution

10.1 Atmospheric pollution

A pollutant is a material deposited at a point where it can be damaging. Turbulent air movement of the lower atmosphere, extending well above the height of all parts of the earth's surface, ensures that its composition is uniform as to the prime elements although it is subject to variation at points where other materials are introduced and dispersed. The composition of air at sea level, excluding water vapour, is 20.9% oxygen, 78.09% nitrogen, 0.93% argon, 0.032% carbon dioxide and other gases totalling approximately 0.004%. Of these neon, helium, crypton and xenon are, like argon, inert and can be disregarded. Other pollutant gases such as methane, nitrous oxide and ozone (O_3) may be derived from the primary constituents.

Natural sources account for the injection of substantial volumes of material into the air, much of it on a localized basis. Sea spray produces minute solid particles of salts, the residue of evaporated droplets. Volcanoes produce large volumes of pollutant gases and ranges of solid particles, some of which are so fine that they persist in the atmosphere over very long distances and for long periods of time, ultimately being deposited as soil components or being absorbed into the oceans. Likewise smokes from natural events are minute particles of condensate, some liquid, some solid, moving as aerosols and the gaseous organic productions of plant and animal life such as methane are delivered into the air as pollutants. Sulphur in substantial volume is derived from volcanic sources. Ozone and nitrogen oxides (NO_x), important in acid rain chemistry, are created in the upper atmosphere as the result of solar radiation and reach the earth's surface, as does hydrogen, by processes of diffusion.

Two methods of measurement are available for the description of atmospheric pollutants depending whether they are solid or gaseous. Solids are measured by weight of substance over volume of air (grams/m^3 $\times$ 10^6) whereas gases are considered volumetrically in parts/million. Solids may also be defined by size, in microns. Particles of 1000 microns will be carried only by high speed air, but particles of less than 10 microns may remain airborne in quite light airs.

The most variable component of the lower atmosphere is water. It is the medium for the bulk of chemical reaction in the atmosphere and its physical presence is important in the removal of trace materials. It is always present as vapour and may occur as an aerosol ultimately falling out as drops.

Reactions between materials account for the modification of some primary pollutants. The availability of carbon dioxide and oxygen in solution is generally beneficial and in terms of the setting of lime mortars absolutely necessary. Nitrogen oxides and sulphur dioxide produce ammonia, nitric and nitrous acids and sulphuric acid which even in the extreme dilutions in which they are carried in precipitation can have an erosive effect, particularly upon mortars. The formation of calcium nitrate is not readily detectable because it is a highly soluble product readily washed away whereas calcium sulphate builds up demonstrably in calcitic material subject to acid rain. The currency of the term 'acid rain' due to its effect upon large vegetation is indicative of a swing in the balance of pollutants. The effect of fuel-burning emissions – particularly coal from power stations – is to augment the sulphuric

Figure 10.1 An emulsified bitumen used as a water barrier on the inner face of a solid brick wall. The barrier has broken down due to the passage of vapour through micro pores, leaving behind it accumulating salts. The consequent powdering of the brick surface breaks the bond with the rubberized bitumen which then contracts, breaking the skin, predominantly on the line of the mortar joints.

and carbonic pollutants and the oxides of nitrogen. The oxidation of sulphur dioxide is the most important process in the formation of sulphuric acid in atmospheric droplets. In this process ozone (O_3) and hydrogen peroxide (H_2O_2) are critical as intermediate reagents. Without them the formation of sulphuric acid from sulphur dioxide (SO_2) is a very slow process.

Trace materials acting as catalysts (mainly manganese and iron) energized by sunlight may be a further stimulant to the formation of pollutants in urban atmospheres and particularly in inversion layer conditions where smoke, haze and fog provide conditions for accelerated production. The displacement of calcium and aluminium ions in mortar may be balanced by the build up of gypsum in the outer layers of stone and brickwork. Processes of this type appear to be one of the principal sources of the breakdown of mortar. The formation of hydrated calcium sulphate from calcium carbonate, for instance, is expansile and is therefore disruptive.

It is frequently thought that atmospheric sulphur arises principally from the activities of man through the burning of fuels. However, the estimates for 1976 suggest that the biogenic emissions of sulphur (usually as the result of action by bacteria in breaking down organic matter) substantially exceeded the products of man-made emissions. To this total the sulphur derived from sea spray and volcanoes must be added, producing a powerful preponderance of natural production at that time. Since then the balance has shifted to a preponderance of artificial production. Man's production of atmospheric sulphur heightens the problem in particular localities of high industrialization, to provide the atmospheric conditions in which the formation of acid is favoured and to provide that additional contribution which, in addition to natural production, significantly lowers the pH value of precipitation to the detriment of plant life, limestones and mortars. Brick, although it can be damaged by subsurface calcium sulphate, is less at risk.

'Sulphate attack' in brickwork is normally visible first in crumbling mortars and the bursting off of renders. It will not occur with normal lime mortars but can be evident with powerfully hydraulic mortars and much more so with Portland cement. The problem derives from the expansion consequent on the reaction between soluble sulphates (sodium, potassium, magnesium and to a much lesser extent calcium) and the tricalcium aluminate (Ca_3Al) formed in the setting of cement. It can only occur in wet conditions.

The sources of the sulphates may be gypsum, sulphate-bearing water or even the unoxidized black centres of industrial brick. Fletton bricks are also a common source of sulphates. The remedies are to eliminate wet conditions, to rake out affected mortars and to repair mortars and renders using sulphate resisting cement. The problem is often manifest in a close-knit reticulate pattern of cracking in rendering, reflecting the expansion of the mortar joints and in blistering effects. The blistering of applied damp-proof membranes on the inner face of walls will arise similarly if wet conditions have built up behind its impermeable skin.

It is typically a problem of historic industrial structures and buildings, made with coal-fired brick and rendered.

Figure 10.2 Industrial brickwork suffering the effects of sustained pollutants entering from behind. The structure is a century-old retaining wall to a railway viaduct. Rainwater enters at high level, eventually reaching ground level and maintaining high internal humidity, causing continual polluted evaporation. The effect is due to crystallization behind a crusted surface.

The activities of man can be much more dramatically evident in acids deriving from halogens, particularly fluorine, which are naturally present in minimal concentrations but are measurable at much higher concentrations where industrial processes emit them. Fluorines emanate from the burning of coal, from phosphate fertilizer for plants, from brick and tile works, glass manufacturing and the smelting of steel and aluminium. As hydrogen fluoride the pollutant is gaseous and water-soluble, ultimately reacting with calcitic material to form calcium fluoride (CaF_2) which is insoluble, forming a hard crust on the surface of brickwork and masonry. Although this fluoride is stable and insoluble the breakdown of both the silica, silicate and calcium crystals to form new crystals degrades the intrinsic material. It is believed that hydrogen fluoride reacts with silicaceous compounds to form gaseous silicon tetrafluoride which then moves into further reactions with aluminium, sodium and iron to form additional crystalline compounds, inducing in the process microcracks and etching vitreous structures.

Other unusual concentrations of gases, destructive of brickwork and plaster, can arise from excessive human activity. Rooms subject to intensive use by visitors can accumulate carbon dioxide concentrations many times greater than normal atmospheric conditions in conjunction with humidity approaching saturation. Condensate forming on cold surfaces can then convert calcium carbonate into calcium bicarbonate, a more soluble material, which will recrystallize as humidity falls to form much larger crystals causing internal pressures within the render or mortar, which can cause quite rapid breakdown. Where significant artworks are carried on plaster this process represents a real danger, and it must be remembered that condensation will occur within micropores when the general relative humidity is well below 100%.

Unusual concentrations of saline materials occur within coastal zones but where there are high prevailing winds significant pollution can occur well inland. Sea spray becomes airborne in tiny droplets which evaporate leaving airborne crystals of the salts of the sea so small as to be aerosols suspended to become in turn the condensation points of rain or mist. Arrested in their path by wetted brickwork salts are absorbed and on drying out the sea salt crystals form as columnar structures within the mortar and brick. While the outer surface of the brickwork may be kept clear of salt formation by high levels of precipitation, evaporation towards a dry interior will leave both brick and mortar surfaces heavily polluted with growing salt crystals. Brickwork which may appear sound, well pointed and unharmed externally may show internally continual loss of mortar from the joints and spalling from the faces of less well burned bricks to the point where the slenderness ratio of the brickwork is severely compromised.

Condensation represents probably the most dangerous condition (although still of only modest effect) for erosion of masonry and brickwork because acid concentrations can be very much higher than in conditions of precipitation or as a result of dry deposition of particles. Temperature and humidity gradients (Stefan flow effects) accelerate condensation towards the coolest available surfaces in warmed air conditions. Beads of moisture on the surface of mortar or brickwork can attract pollutants forming near perfect sinks, building up unusually high concentrations which remain in contact with the host material for a considerable time, allowing acid and saline solutions to penetrate. Further, when the bead dries it leaves behind it a deposit of salts which attract the next phase of condensation immediately providing a concentrated solution in which further reactions will take place. Erosion is thus concentrated, with the result that pitting takes place, concentrating the damage and carrying it below any protective outer surface.

The formation of salts following the reactions within the pores brings saline material near to the surface, possibly building up a silicaceous crust or carrying gypsum indicating the prevalence of sulphur compounds. The crust may be porous and slightly soluble with minimal integration with the substrate. It is subject to different thermal movements to the main body of brickwork, generating pressures of its own. Ultimately it will force the outer surfaces of pointing and mortar away from the wall in a way similar to other forms of sulphate attack.

10.2 Porosity, condensation and hygroscopic salts

In some circumstances condensation from the atmosphere may occur at temperatures above dew point and at less than 100% humidity. Within very small pores condensation will occur within parameters of temperature and relative humidity which would not lead to condensation in normal atmospheric conditions. The moisture will first appear in the finest pores and the range of values within which condensation can be anticipated is calculable. When relative humidities are above 90% and particularly above 95% the volume of take-up is likely to be significant. Further and significantly, some salts associated with the pollution of brickwork are known to be hygroscopic. The presence of hygroscopic salts within masonry and particularly within the fine pores will accentuate condensation effects so that brickwork apparently well above dew point may well be absorbing moisture on account of both phenomena. As condensation occurs the concentration of salt will be at its highest and the solution will, therefore, be in its most reactive possible condition. Mixtures of salts tend to be more aggressive than their individual components. While there is no direct evidence of the breakdown in brickwork for such reasons, apparently inexplicable failure of some types of brick structures may arise from causes of this type.

10.3 Ground water

While the consequences of atmospheric pollution are principally associated with breakdown due to the creation of an acid environment and are mainly chemical in nature, those caused by ground water drawn up into brickwork are related to the availability of salts in solution, their crystallization and consequent physical pressures.

When this phenomenon takes place at the surface of brickwork it may be entirely harmless. It is often evident as efflorescence (sulphates of magnesium, potassium or sodium). The growth of crystals in the air progresses outwardly from the crystal tips causing no pressures in the brick itself. Such crystals may not be restricted to salts but may be simply ice crystals. Water always plays a part in the process and the action of water in the pores of the brickwork is therefore critical. Where similar growth occurs beneath the surface of brickwork the situation is potentially more damaging. Even if the chemistry of the saline solution passing through the pores has no effect upon the material itself, the growth of crystals within the pores has a continually erosive and destructuring effect, which, perhaps combined with other events such as the freeze–thaw cycle and the presence of microorganisms or plant life, causes breakdown. This breakdown is more rapid in underburned bricks simply because they are less strong and perhaps also because there are more freely available salts providing iron, aluminium and manganese ions which can enter combination to form secondary crystals.

Capillary attraction draws moisture out of the ground and up through brickwork. Historic buildings have commonly lacked damp-proof courses, relying necessarily on evaporation from the lowest courses of walls, which over the years have accumulated richer and richer concentrations of available salts. An equilibrium is reached in which the higher levels of salinity local to the areas of evaporation diffuse salts back towards the lower levels. This diffusion can take place through any continuous channel of liquid so that the effect, once equilibrium is obtained, is that each absorbent wall acts as an evaporation zone with highest salinity at the surfaces. Walls can, therefore, be seen as sites of the emergence of ground water and the active detrimental effects on brickwork and earth structures are visible within the height to which these solutions can be lifted by capillary action countered by gravity.

Over these surfaces further dynamic balances are set up. In particular the conservator is concerned with the level within the structure at which evaporation takes place. This is directly related to the evaporative conditions at any one time which reflect the relative humidity and temperature and the resupply of liquid to the evaporation. The rate of this resupply is determined by height, viscosity of liquid, pore sizes and surface tension, which is itself a product of the nature of the solution and the hydrophilic qualities of the surface. Any electrostatic field in the wall will also affect the flow but this effect is rare, occurring only in association with embedded metals – pipework typically.

Within the pores of the structure there is a continual movement of water upwards and outwards to resupply the surfaces from which water has escaped into the atmosphere. The currents in the pores carry with them ions representing the disassociated salts in solution while, at the same time, other ions are attempting to move from the zones of high salinity at the surface (where saturation is achieved) towards zones of lower salinity. This interpretation may be compared to people attempting to get near a spectacle while others are attempting to escape from the event. The analogy is unduly crude because there is no conscious effort involved and the phenomenon relates simply to the higher pressures of molecular activity where there are a greater number of molecules of larger size in motion. The effect is osmotic, there being a greater number of impacts from large molecules at the surface than in lower levels. Therefore there is a return movement towards the areas of lower salinity and this motion itself impedes the current of replacement solution moving towards the surface, carrying with it further saline material.

10.4 Crystal formation

The internal balances thus set up determine where evaporation will take place and, therefore, the nature of the ultimate decay. At lower levels where water can move more freely upward, sufficient solution emerges at surface level for evaporation to take place only outside the host structure. Crystals, therefore, form at surface level and grow outwards relatively harmlessly. At successively higher levels the rate of available flow is diminished

Figure 10.3 Soft (rubber) brick used for visual effect with stone (alternating red and white voussoirs). High levels of industrial pollutants have caused rapid erosion of the brick, ill-chosen for this situation.

to the point where evaporation takes place below the surface. In such a zone there is some restriction upon evaporation and, therefore, a band of equality is visible. Above this band evaporation is taking place well below the structural surface. In this zone crystal formation takes place within the pores and even higher in the wall evaporation takes place at greater depth, although due to the protection of the structure it occurs at a much reduced rate. Within the pores themselves pressures are created by the formation of crystals. These pressures are limited by the physical processes involved and where this takes place within the depth of the material superficial loadings can effectively counter any pressures set up. At the surface, however, these pressures of expansion are countered only by the tensile strength of the (vitreous or semi-vitreous) structure between the pores. Where this is insufficient the material can break away. Consequently the material fails as flakes, laminae or spalling of similarly thin surfaces which decrepitate. Experiments carried out with sodium chloride illustrate the process, although sodium chloride is not necessarily the most typical salt involved in masonry decay other than in coastal structures.

The mechanisms by which the pressures are created within brickwork have not been fully demonstrated. The general principles, however, are that the regular structure of a crystal derives from the stability of three-dimensional relationships between atoms in a molecule which in turn depends upon the angles at which the bonding between atoms takes place. This is a process of physical arrangement dependent upon electrostatic forces on the molecular surfaces. It is not a chemical bonding. The geometry of the molecule presents a specific external surface in which the regularity of the crystal structure is reflected, producing the smooth surfaces and sharp angles associated with all crystals. The surfaces may, however, be complex. Some have a geometry with salient and re-entrant points. The spaces between the salients are, therefore, available for other molecules to join the assembly. If the available space can be filled by lodgement the assembly becomes larger. The enlargement by the lodging into suitable spaces of water molecules produces a structure in which the embryo crystal is capable of interlocking with other similarly shaped molecules in a regular lattice. This assemblage grows as water molecules in continual movement take up such spaces within larger molecules and the electrostatic forces hold them there. They can be dislodged by an increased input of energy (i.e. heating) which causes them to break away. These energies are very much less than those required to break up molecules in which the chemical linkages have been formed. Thus it is that calcium sulphate will part with water of crystallization simply by being heated to 400°C and at lower temperatures will recover its crystalline form on the availability of water; hence its setting as Plaster of Paris. To break the calcium–sulphur–oxygen bonds requires much higher temperatures. When it is created chemically within the pores of brickwork by the processes of atmospheric pollution by sulphuric or sulphurous compounds, the initial chemical bond forming calcium sulphate ($CaSO_4$) is succeeded by the accumulation of water molecules embraced within a rigid lattice which completes a regular structure. Multiples of these regular structures then assemble as crystals. The incorporation of water of crystallization produces significant enlargement. Many molecular structures offer various opportunities for the formation of lattice solids with the consequence that at different conditions of temperature and humidity the base molecule will embrace different numbers of water molecules. As these crystals grow they exert and sustain disruptive pressure on adjacent material.

A phenomenon which is not clearly understood in the creation of these pressures is the interaction between the crystal and structural surfaces. However, it is evident that this process can take place only within the dynamic conditions of molecular vibration. At the point at which the force fields of the molecules constituting the pore structure (probably themselves crystals or regular molecular arrangements) and the terminal molecules of the growing crystal interact there should be no further addition to the structure since no space is available. However, if it is allowed that every component of this arrangement is in constant movement there will be instants when the terminal gap widens to permit entry of a further molecule of water, which by statistical chance has been projected to that very point. As the gap closes the interaction of the extended crystal and the structural surface creates a mutual pressure transmitted down the length of the crystal reacting on the wall of the pore. The pattern then changes and the greater pressure at the point of growth reduces the pressure at some other point on the growing crystal allowing the phenomenon to be repeated in an adjacent position. Thus the growing crystal forces its tip against the internal surface of the pore in the brick.

If this surmise has substance it can be readily understood that the concentration of liquids by evaporation within the pores of brickwork allows and inevitably causes expansive pressures. The crystals may be growing randomly: their numbers may increase. It would appear that the strength of the pore is able to curtail the growth of the crystal simply because the necessary voids cannot be opened, and this represents a point of maximum pressure between pore, wall and growing crystal. In some circumstances the cumulative pressures are so great that the structural strength of the pore is insufficient to contain the pressure and its structural cohesion is broken. This rupture usually occurs within the superficial layers of the brickwork and not where crystallization occurs deep within the brick where the structure is more strongly retained.

Some disruptive effects of deposition within pores has been extensively studied, particularly in terracotta. Deposition may not be aqueous; for instance, carbon deposited out of carbon monoxide in furnace flues at 400–600°C may nucleate on iron-spots, and similarly vaporized sodium sulphate from the combustion gases of poor coals will be deposited in the pores of terracotta flues.

However, the effects of deposition vary with variations in external conditions. The rate of availability of ground water or salt-laden depositions, the rate of evaporation and the growing obstruction of pores together with microorganisms and many other factors affect the zone in which the damage occurs. Daily and monthly tidal cycles, surface precipitation and conservation activities will all interfere with and vary the evaporation frontier within the brickwork and as this frontier moves so the zones of damage move. It is generally clear in brick structures and even more positively clear in earths that the most pronounced effects occur just above ground level or water level. The phenomenon does not affect saturated brickwork and is relatively insignificant in the upper zones although there may be localized circumstances such as brickwork above a cornice on which moss and water collects or brickwork against a bank which remains damp, where the same type of damage can occur. The failure of bricks in a wall saturated at low level therefore tends to occur within 2 metres or so of the ground or water level, the weaker bricks failing first to leave pockets in the wall. These then induce other forms of decay through plant growth or wind erosion.

Crystal growth is generally detrimental to brickwork but there is an exception. Epitaxial growth in crystals describes the addition of further material layer by layer on the axis of the crystal, and this process does not necessarily cease on construction of the brickwork. The crystal structure of cement involving the complex interrelationships of aluminium, iron, silicon etc. does not provide ready conditions for the extension of the growth of crystals but calcite can take up carbon dioxide from rain water in sufficient quantity to create significant epitaxial growth, i.e. growth on the tip of a crystal, or longitudinally at a fracture point. Thus it is that fractured lime mortar can over the years redevelop some strength – perhaps 10% of the original value – by the re-establishment of crystals across a fracture line. It is this process which gives lime mortar some self-repairing property. Epitaxial growth within mortar structures can be encouraged by the introduction of solutions containing calcium hydroxide – lime water – and this technique has now been developed for the conservation of limestone. It can be further enhanced by the introduction of artificially enriched atmospheres of carbon dioxide but such techniques have been used so far only on a laboratory or experimental scale. Artificial enhancement of crystal growth in this way is unlikely to be useful in the consolidation of brickwork on any significant scale due to the deposition of calcite on the face which will have a significant staining effect. However, it will be important in the consolidation of plasters and renders and with the experience gained in stone repair is an incipient tool available to the conservator of specialized monuments. The growth of calcium carbonate crystals is also a pollutant – a positive disadvantage in some circumstances. Mixed construction of brickwork frequently incorporates limestones and sandstones bound by calcitic material. Acid rainwater produces solutions containing calcium ions, sulphate ions and carbonate ions which precipitate as the solution becomes richer. Thus water running down off limestone across brickwork and evaporating as it goes leaves behind it deposits of calcite and gypsum together with lesser deposits of (soluble) calcium nitrate and possibly calcium salts of organic acids.

These are soluble in water to a considerable extent. The gypsum precipitate is not readily carried away by rainwater and the calcium carbonate will cohere within the pores of the brickwork causing staining. Evaporation normally takes place on the surface and the brickwork is disfigured rather than damaged. Such stains can be removed by prewetting followed by the application of dilute hydrochloric acid and the soluble calcium chloride formed must then be washed off.

'Lime stain', a white and usually streaky deposit, derives from the transport of calcium hydroxide in solution. This may originate in limestone, lime or as a by-product of the setting of Portland cement mortars. As a surface deposit this is usually seen on historic brickwork of low porosity. High porosity brick absorbs an excess more readily and the deposit will occur within the brick affecting its weathering qualities.

Except in laboratory and museum conditions it is unlikely to be of assistance to the conservator to be able to predict the precise depth and volume of decay in brickwork due to the crystallization of solutes. It is, however, possible. The calculation may be made on the basis of the formulae of T. L. Poiseuille, known for some 150 years, which determine the amount of water capable of being delivered to the surface. By identifying the rate of evaporation based on porosity, aqueous tension and diffusion layer thickness and by applying parameters relating to the position of the site and characteristics of the solution, the rate of evaporation and deposition can be determined. Hence it can be determined that evaporation will take place at a particular distance beneath the exposed surface. By proportioning the crystal growth to available space and relating these to the nature of the solid, the calculation will yield values for the depth, extent and rate of structural breakdown. Calculations of such complexity demand extremely careful measurement of conditions and solid structures as well as analyses of pollutants and presumptions about ambient conditions. Consequently their practical value in conservation is limited to theory where they serve to promote and verify material behaviour. Any useful prediction which might be achievable by such theoretical means in relation to brickwork is likely to require monitoring and cross referencing in such a way that the original calculation is overtaken by empirical observations.

When salts are deposited out of solution they may constitute upwards of 20% by weight of the water. Deposited within the pores they

Figure 10.4 The soffit and face of voussoirs in a window arch of a church in a marine environment after 150 years' exposure. The decrepitation occurs at its maximum closest to the junction of evaporative surfaces.

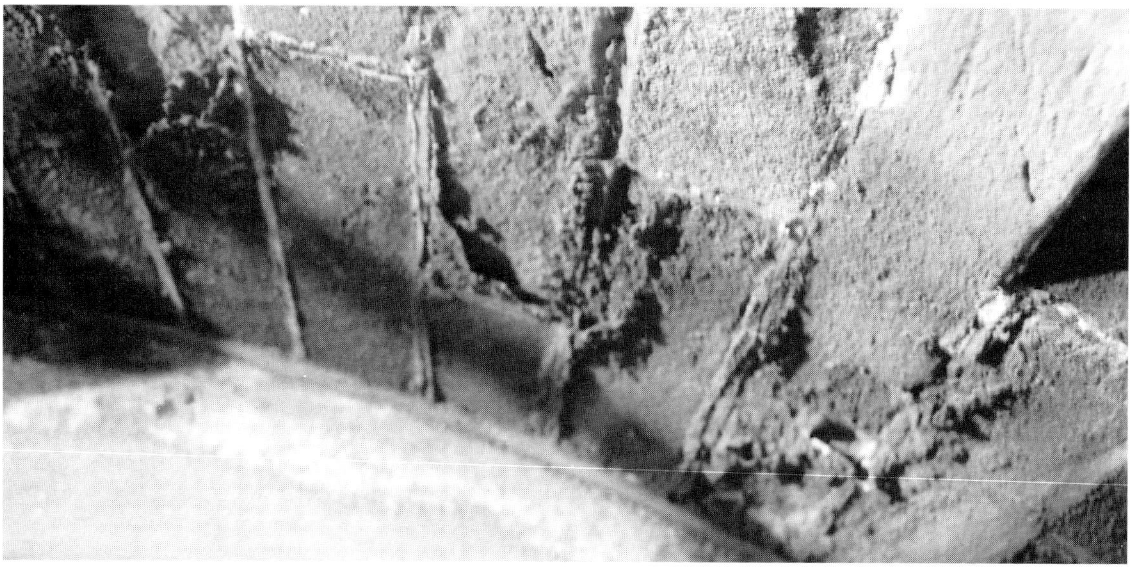

become an inhibitor to movement of moisture in its own right if the pores remain intact. This means that a strong brick structure capable of resisting the internal pressures of the crystallization will lose its porosity, not so much by lining of the pore structure but by the construction of new materials lying across the pores and since the ability of liquids to move through the pores under the driving forces of capillary attraction and, perhaps, gravitation is limited by the inverse of the pore radius to the power of four, the reduction in pore spaces becomes an acute inhibitor of the movement of liquid. Thus strong brick can become increasingly 'waterproof', i.e. resistant to the movement of water through its pores as a result of long exposure but this does not necessarily apply to the wall. The reverse may occur as the calcitic content of lime mortar is leached out forming aquifers while the calcite removed finds its way, in part at least, into the actual brick. The total permeability of the brickwork will be very high as the mortar itself fails and ultimately its strength disappears as well.

As the ratio of pore sizes varies the response of the brick to the cyclical effects of frost action may be affected and an historic brick originally frost resistant may become more susceptible to failure. The failure of pressed and extruded brick under frost action occurs under the surface laminae as the result of the process of manufacture, possibly accentuated by changes in porosity.

10.5 Brick surfaces

Bricks and terracotta are manufactured with a surface which differs from the underbody to some, and in many instances very considerable, degree. It may even be glazed and it may incorporate materials aimed at giving it a different visual quality. The 'fireskin' of terracotta and brick is measured in microns (μm) or thousandths of an inch, and is not to be confused with surface lamination which, in general failures, is thicker.

In use bricks and terracotta also develop surface characteristics which differ from those of the underbody due to the deposition of atmospheric pollutants, the accumulation of salts and the generation of materials such as organic residues and gypsum within the outer faces. These materials amend the outer skin. Weather conditions – insolation, air temperature, humidity and precipitation – alter the ambient circumstances and for combinations of all these reasons it follows that the outer skin of brick and the outer face of mortar or any superficial skin, rendering or plaster may well behave differently to the substrate in terms of dilatation or thermal movement. If sufficient internal stresses arise the outer face will spall away. This may result in an unsatisfactory visual appearance and open the way for the continuing erosion of the body.

Common disfigurements include sooting and streaking due to run-off from other parts of a structure, particularly windows, stone courses and cills or facing materials. Smoke and oily deposits from hydrocarbon vapours can disfigure brickwork in industrial and urban areas and the growth of bio-organisms can exceed their attractions as a weathering and become obtrusive or damaging. Staining may arise from carbonate and silica which form white deposits removable with hydrochloric and hydrofluoric acids, and from metals such as copper and iron.

Where it is necessary to restore a wall surface an individual brick or other component may be cut out and replaced. This is a normal but occasional activity in dealing with bricks but a regular process in mortars which generally decay at faster rates and therefore need repointing. The control or restraint of organic growth will, in some circumstances, be achieved by modifying environmental conditions.

10.6 Cleaning

Neither the brickwork nor any other sections of building should be cleaned without good reason. The objective of cleaning a building surface is usually to return it to something apparently close to its original condition. In the process, however, brickwork and glazed or highly finished terracotta may lose their original surfaces and, being materials which are not consistent throughout, may suffer irreparable damage, particularly if abrasive techniques are used. To restore the surfaces of brickwork to a cleaned condition will normally involve the removal of patination and weathering built up over a long period of time, possibly attractive in itself and in normal circumstances a defence against erosion.

Occasionally it is necessary to clean brickwork to remove a pollutant which otherwise will wash off to cause damage or danger. Sometimes other offensive material, such as graffiti or paint splash, must be removed. Cleaning is sometimes carried out for reasons of vanity or even to absorb an available element of budget! While in principle there can be no objection to buildings looking well-tended nor to having useful investment in them, both these reasons may on occasion be poor justification for the operation and in some circumstances may be no justification at all. It is better not to clean a building than to cause more damage than would otherwise be the case. Where the right skills are not available and where there is doubt over the efficacy of treatment it is better to await an improved opportunity provided only that progressive loss and damage is not being caused or condoned.

Cleaning should be regarded as a separate conservation activity, subject to initial evaluation and separate decision-making. Comparative circumstances may, of course, be taken into account. The availability of scaffolding which would otherwise be separately required, the presence of site facilities and the integration of cleaning with other conservation programmes must all be relevant. The place of cleaning in the conservation programme is not necessarily at the end. Brickwork should be cleaned before rather than after painting or repointing. For budgetary purposes, however, cleaning is sometimes regarded as optional and is, therefore, treated as the final operation. Buildings requiring re-rendering or repointing may be in need of thorough initial cleaning, and the process of cleaning brickwork will cause the whole structure to be examined very carefully off scaffolding in a way that might not otherwise be possible.

A cleaning survey, therefore, must be undertaken as a separate operation, but should always be seen in conjunction with the overall conservation of brickwork. It is particularly important to understand the extent to which cleaning is to be associated with other techniques of conservation such as the elimination of salts, repointing and decorating. A preliminary survey should establish the nature of any soiling or painting of brickwork and the occurrence of associated failures and should envisage the results of a cleaning operation. There may be merit in carrying out detailed surveys, including laboratory tests and chemical analyses of deposits. This will be particularly useful where there is evidence of salts that may require removal. The operation of poulticing for such purposes may be entirely separate from the more superficial aspects of cleaning. Sample cleaning may fall into one of two categories – typical operations carried out by operatives or available contractors to specific requirements or under monitored conditions or alternatively limited subcontracts in which specific areas are cleaned so that the results can be evaluated.

At this point a number of crucial issues are relevant:

1. The nature of over-painting, disfigurement, soiling, encrustation or dirt. There may well be a number of contributory causes including forms of previous treatment such as inappropriate or unwanted painting.
2. The extent of the operation which would be necessary, having careful regard to the protection of vulnerable surfaces.
3. The extent to which any historic surface will be amended. The type and softness of brick must be evaluated; any cleaning being directly related to the ability of the brick to withstand treatment.
4. The practical implications of methods to be employed and the protective measures necessary.
5. Subsequent remedial or surface treatments that may be necessary immediately or after cleaning.
6. Other forms of remedial work necessary or desirable to be included in the operation. Pointing and localized repair, the removal of ferrous fixings etc. are often appropriate to a contract for cleaning, which may also extend to repainting of joinery.
7. Necessary approvals, protection, access and notices. No cleaning operation should be carried out unless the proper equipment, skills, protection and approvals are in place. The conservator must be aware of subsequent hazards, in particular forms of staining, that can appear as a consequence of the excessive use of water or the failure to remove from brickwork the materials applied for cleaning purposes.

Methods available for cleaning brickwork include dry rubbing with absorbent materials, air and water pressure jet and vacuum cleaning, steam cleaning, mechanical brushing (dry or wet), washing using detergents or other solvents, chemical solvents and specialist techniques such as laser and ultrasonic removal of dirt. This list does not include mechanical or physical abrasion by sand shot or grit blasting. These latter are methods to be used only in special circumstances or as a last resort since they can remove the fireskins of well-baked brick and terracotta leaving a surface containing minute fractures. The final surfaces are not those intended historically.

The techniques of cleaning brickwork and masonry have advanced very considerably and expert guidance in specific circumstances is available from technical advisory services of governments and research institutions as well as consultants and contractors. Care, sensitivity and skill in application or the reverse can radically affect the life of a building. Methods which are satisfactory and reasonably applicable in some circumstances can be disastrous in many others and, as with the use and application of synthetic consolidants, sensitivity to each specific problem coupled with experience and prudence determine a zone between generalities of the textbook and the precise circumstances in the field which no conservator should ignore. There are, however, many principles which should be understood in terms of technique and practice and some essential generalities. It is, for instance, frequently not understood that the reason for multiple applications of washing is the simple mathematics of dilution.

If a brick surface retains one unit of a cleaning fluid and 1000 units of water are available to remove it a single application will leave a solution of 1:1000 dilution as a residue. If the washing water is divided into two equal parts, a solution at a dilution of 1:500 will be left after the first washing, and if one unit is left the residue will be at a dilution of 1:250 000 after the second part has been used. Further subdivisions greatly increase the dilution of the final residue.

The length of time that the water remains on a surface controls the extent of its absorption. This is known as the 'dwell time'. The effect of subdividing washing operations and repeating them is to enhance enormously their cleansing effect. Small amounts of washing water, and moderate dwell times minimize absorption: multiple applications ensure that the ultimate dilutions are as great as possible. The consequences of these simple mathematics apply not only to the volumes of water used in washing down but to the volumes of the expensive solvents used in special applications and cleansing of excesses.

A further simple mathematical principle relates to the energies carried by moving particles. Water lances, particle blasting – even air jet cleaning – all rely on the energy carried by fast-moving molecules. While the actual nature of the particles is significant in terms of the nature of the impact the prime consideration will be the energy involved. Energy is the product of mass and speed. Each moving particle decelerates suddenly on impact with the brick or mortar, its energy being transferred into the structure with the intention of abrasion – that is, of the breaking away or fracturing of the surface particles of dirt and brick. The abrasive effect, therefore, of any given material can be directly related to the energy input required to accelerate it to its impact velocity. This energy requirement is in relation to the square of the velocity and it follows that an increase of speed carries very much more energy in the upper ranges.

Very low energy techniques, such as the rolling on of silicone rubber, are always to be preferred.

10.6.1 Air and water jets

Depending on the nature of the soiling, the effectiveness of soft abrasives may be enhanced by the use of swirl head nozzles which deliver the particles obliquely, increasing their power of detachment and reducing their direct impact on the surface.

The application of a high pressure air jet is non-pollutant, straightforward and generally not attended by danger and it would, therefore, seem an ideal method of cleaning brickwork. It can be satisfactory for removing relatively dry and light material, particularly in dusty climates, and if satisfactory can be used repeatedly. High pressures, however, can build up even higher pressure in superficial pores and result in abrasion. The alternative of vacuum cleaning should always be tried first, partly because it removes to security the unwanted material and particularly because it makes use of a small differential in air pressures – 100–200 mbars. It is usually aided by physical disturbance by brush. Straightforward brushing techniques are always applicable with a preference for non-ferrous wire brushes where these are needed. Mechanically powered brushes can be quite strongly abrasive but brick and mortar which cannot withstand these techniques must in any case be due for significant repair.

The simplest abrasive technique involves nothing more than a knife or brush operated by hand. A knife or similar tool is used to clean off substantial plant growth from the size of lichen upwards. The brush is an effective loosener of dirt which is not ingrained and may include mechanical spinning. This technique is dangerous if discs or metal brushes of substance are used, but it can be remarkably effective with plastic (nylon) bristle or similar filaments and in conjunction with water sprays can be effective and non-damaging for the removal of the soiled material. Guards must be provided. Wet or dry sand blasting, shot blasting or grit blasting have been used disastrously on brickwork and façades of mixed brick and masonry construction, but also with success, provided that care is taken in the choice of the abrading material and pressures used. The techniques are essentially methods of surface disturbance and generally do little to remove deeply ingrained dirt unless applied persistently. Undue persistence should not be allowed because the effect is to erode the brickwork in depth. The cleaning system should be designed to do no more than disturb lightly embedded and superficially weak material. Grain size is critical, as also is grain shape. While abrasives softer than sand or metal are available there is less point in using them to reduce the impact than there is in reducing the pressure of delivery. However, on delicate features in terracotta and stucco softer abrasives should be specified in order to avoid erosion of fireskin or delicate surfaces. Granular chalk is commonly and successfully employed on brickwork in air-water streams. Crushed marble, crushed olive stones and crushed grape seeds have been used in air streams satisfactorily. Even rubber crumbs can be used. The methods of delivery of air vary in volume and pressure. Up to 65 lb per square inch are commonly offered, although figures at half this value are sufficient for almost all purposes. High speeds may provide a faster operation but at the much increased risk of damage. Water jets may be added to air streams to minimize dust or the entire stream may be water-borne. Silica sand is a common abrasive but it is too powerful and sharp to be used on brickwork. A non-quartzitic abrasive less on a scale of hardness than the brick itself is still powerful enough to dislodge embedded dirt, gypsum and salt-bound accretions. In addition quartz is regarded as hazardous to operatives in this usage and there are some limitations on its application. On brickwork it should be used only in circumstances where there is an overcoating of a material which will not otherwise be touched – a paint or epoxy-bound materials. When used, scaffolds should be enclosed with sheeting and the entire operation kept within a single air space which is mechanically vented with fresh input air being directed at the operator, preferably via the helmet.

Figure 10.5 A brick surface cleaned by dry grit blasting with evident surface damage.

Air abrasive techniques have the advantages of minimal wetting. They are fast, relatively safe and are not dependent on weather conditions. However, they can be noisy, produce considerable amounts of dust if care is not taken and present risks of silicosis. On completion of the work the brickwork should be washed working from top to bottom using a low volume, high pressure water spray. Techniques have been evolved using vacuum heads in conjunction with blasting so that cleaning can proceed without the disadvantages of screens and seals. The abrasive and residual products are effectively removed and a brush shield masks the operation. Consequently the device can be used from a mobile platform.

Despite the apparent advantages, the presumption in cleaning brickwork should always be against the use of abrasives. In the exceptional circumstance of there being an overcoating for which these abrasives are necessary, great care should be taken and the work should be stopped prior to the full

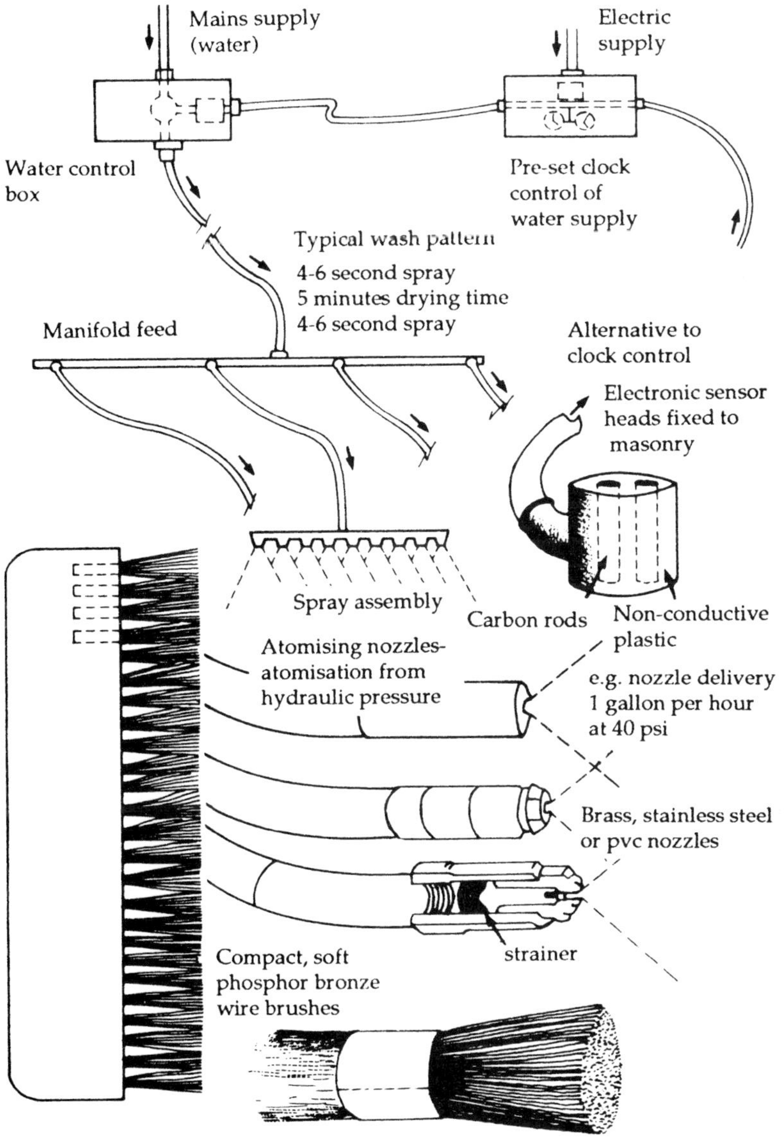

Figure 10.6 A description in a public advisory note of a system of spray control designed to deliver and maintain pre-determined conditions of surface wetness to achieve softening and washing.

removal of the overcoating, which will then normally be cleaned off by solvents or similar methods.

One specific specialist technique overcomes many of the objections to air jets with abrasives. It derives from the aircraft industry and is available only on a limited basis. The 'abradant' used is 'dry ice' – solid carbon dioxide in small particles. These are introduced into the air-stream at the gun nozzle and are instantaneously vaporized, either on the surface creating a cavity pulse effect or even behind the dirt particles which are thereby loosened and carried away. This treatment may have future importance.

10.6.2 Steam cleaning

Although currently out of fashion, steam cleaning can be effective and the conjunction of high temperature with a high pressure vapour jet producing a low water volume can soften and shift materials which a water spray will not touch. Bricks impregnated with plastics, hydrocarbons, grease, bituminous stains and a variety of synthetic chemicals may well be responsive to this type of cleaning which seems to have no adverse effect on sound brickwork. Skilled application limits the time during which heat is applied. Thermal stress can affect brickwork in poor condition if application is careless or extended. No special precautions are necessary although it may be desirable to provide additional water jets to remove the products once they have been detached. Even if they are not water-soluble the small particles formed must be removed.

10.6.3 Washing

Water is a remarkably wide-ranging solvent. Although many of the materials which have to be cleaned from brickwork are water-soluble some of those commonly found, such as gypsum, require prolonged exposure before they have softened and dissolved sufficiently to be moved by water jets. Prolonged water exposure causes saturation which is undesirable and may be unacceptable. The problem with brickwork therefore, is to provide a sufficiently moist environment at the surface and to maintain it for a sufficient period without saturating the structure. This is generally a less significant problem than with stone masonry but the techniques developed for stone are of value in cleaning brickwork.

Prolonged or heavy saturation can draw salts to the surface unnecessarily. Wetting without saturation is the objective. 'Atomizing' systems are available in which extremely fine sprays produce an atmosphere in which the water is virtually an aerosol. To be effective very high humidity levels are maintained and, therefore, a sheeted enclosure must be constructed against the face of the building. Spraying is usually devised on a moving head or rotational basis and is designed not to impact upon the brickwork but to create a fog-like condition which can be maintained over a long period. In alternative systems very fine sprays are directed at the building surface, being switched on for short periods at regular intervals or alternatively switched on automatically by sensors actuated by the effects of drying. Experiment determines the saturation period required. The dirt is removed by brushing with bristle, plastic, brass or phosphor bronze filaments. Alternatively, high pressure water jets may be employed using very low volumes with pressures up to a maximum of 1000 psi. Lower pressures than this are advisable with brickwork unless it is particularly well-fired and durable. Water and detritus should be removed from the building face by immediate collection. The washing operation must work downwards on the brickwork, sheets below the operation being brush-edged to intercept moisture and direct it away from the wall for disposal.

It is vital that the strength of the brick is assessed properly. 'Rubber' brick, for instance, in a wet condition, will be abraded by bristle. Preliminary testing is vital.

10.6.4 Chemical cleaning

Acids and alkalis are used to clean brickwork by chemical reaction and proprietary mixtures are available which should be used strictly in accordance with instructions and with particular care in washing after cleaning. Saturation of brickwork is to be avoided by short washes and efficient removal. To flood brickwork is to invite surface saturation. Small volume, repeated application and efficient removal are the keys to success.

Hydrochloric acid is the principal constituent of most widely used and effective cleaners of brickwork but alkalis are increasingly used. Acids and alkalis are used as cleaners only where there is a specific and recognizable application for them which is not readily met by alternative means. Hydrochloric acid is the most commonly applied and it should be used specifically where mortar stains, calcium carbonate deposits and other rigid, inorganic salts must be removed. Brickwork should be wetted before application. Dwell times of about five minutes are acceptable at normally specified dilutions. Some proprietary products contain hydrofluoric acid in addition to hydrochloric. Both materials are corrosive but hydrofluoric acid is immediately injurious to glass and dangerous to the operative. It may be delivered in less dangerous form as ammonium bifluoride solution, hydrofluoric acid being released on the surface by decomposition of the ammonium radical. Paintwork is damaged by both substances. Use of chemicals of this order

Figure 10.7 Brick used for alternating voussoirs in conjunction with stone cleaned and uncleaned.

should be restricted to the areas which are affected, for instance stains on brickwork due to limestone or to cement which has become ingrained on brick or floor tiling and other forms of mortar staining. These acids will etch mortar and therefore damage the finish. All safety precautions should be thoroughly observed and on completion of the work the salts and surplus acid must be washed off effectively. This is best done by repeated applications of small volumes of water with efficient collection. Neither the acids nor the cleaning water should be allowed to persist on the brickwork and they must never dry off before the final wash. Washes should be brushed while on the surface to agitate the mixtures. A proprietary acid cleaner may also contain a buffering agent (EDTA) and a detergent.

Alkalis – sodium, ammonium and potassium hydroxides – are used on brickwork for degreasing and the removal of organically bound paints. For the latter purpose they are used in paste form. Their persistence on the surface may be increased by the use of fuller's earth or bentonite clays, or by cellulose binders in some instances. Alternatively, proprietary paint removers may be used. Caustic alkalis may be combined with detergents and are effective in degreasing and the removal of grease-bound materials. There is a danger that they will, however, leave residues providing unacceptably high concentrations of sodium or calcium ions. As a counter to this effect acetic acid or solutions of oxalic or citric acid may be allowed to reside on the surface for up to half an hour before being washed out. As with all washing techniques the work should be carried out from top to bottom. However the toxicity of oxalic acid removes it from recommendation for general use.

Other reagents may be applied for specific circumstances. Ammonium bicarbonate may for instance be used to react with salts producing more soluble products which can be absorbed into a poultice or washed out. Surfactants (wetting agents) should always be considered at an early stage for localized and specific forms of cleaning. Careful washing or poulticing is desirable if such materials are used regularly because residues can build up to provide concentrations capable of reacting with other pollutants, which subsequently ease the flow of moisture into the pore structure. Some varieties are particularly rich in sodium ions which can leave a bloom if not thoroughly removed by washing. For occasional cleaning and specific application they can be an important aid to mechanical methods, or a complete system in themselves.

In all applications care must be taken to determine the maximum appropriate dwell time (e.g. 5 minutes for dilute hydrofluoric acid). The times must not be exceeded.

10.6.5 Special cleaning methods

Dry cleaning at its simplest involves no more than hand brushing the brickwork with bristle, nylon or in stubborn cases brass wire brushes, followed by intensive vacuum cleaning. Silicone rubber may also be applied to the surface to bring off the dirt by attachment to the crumbs which must be collected and can be reprocessed.

In some rare circumstances materials can effectively be removed by the application of heat. Organic compounds may be burned off and the splashing of molten lead where it has penetrated brickwork (particularly in fires) may require the application of short term heat blasting. This may be achieved using propane flame or preferably an electrically heated air jet. Application of heat should be sufficiently gradual to minimize thermal shock.

In museum conditions tile and other terracotta artefacts may be cleaned slowly but very effectively by the use of lasers focused on small areas of a surface and used to vaporize deposited material. At a maximum, areas the size of postage stamps can be cleaned with each 'shot', and the surface must be relocated with great accuracy between shots. The laser vaporizes the dirt layer without affecting adjoining material. Its application is very rare and applicable only to valuable objects. In industry old and dirty terracotta is sometimes cleaned by the relatively simple and cheap method of refiring.

A further museum technique, also applicable only to relatively small objects, is the use of ultrasonic cleaning. The object may be suspended in an ammonia bath or detergent, sometimes with superfine corundum (aluminium oxide) in suspension, and subjected to ultrasonic pulses for a short period. Glass and glazed materials may be cleaned in this way with care after initial trials. Alternatively an ultrasonic waterhead may be used. In this device a stream carries the pulses on to the object, removing the loosened material in its flow. These techniques have no significant application to brickwork. The purpose of cleaning baked clay objects by such techniques may be to facilitate their repair. If they are cleaned in an ammonia bath as opposed to a water stream the removal of all ammonia is essential. This may be accomplished by heating gently in a closed oven with suitable exhaust apparatus. No article which is not fully coherent should be trusted to this process. The effect of the ultrasonic high energy impulses is to produce cavitation within the liquid – that is to say, the formation of gas bubbles which collapse immediately. This phenomenon was first discovered in association with high speed marine propellers where it caused the rapid erosion of bronze, the surface material being sucked away in the collapse of the bubble. On a minute scale this process detaches surface particles. The danger is that these can include laminae of glaze or fireskin in which invisible and microscopic cracking is developing. Careful testing is essential as zones of pigmentation can be removed irretrievably. Treatments of between one and three minutes may be acceptable under laboratory conditions.

10.6.6 Poulticing

Two different functions may be served by a poultice – the sustained delivery of liquid to a surface, and the withdrawal of liquid from the surface. At its simplest the process involves placing two absorbent masses together to allow a moisture balance to be achieved between them. On the removal of one mass, part of the liquid content has been withdrawn from it or delivered to it. In brickwork poulticing may be required to withdraw saline material but it is rarely required to deliver liquids, although occasionally the technique is useful in maintaining a reagent such as ammonium carbonate on the surface to maintain a reaction with unwanted salts over a period.

To be effective in withdrawing salts the salt-impregnated mass of brickwork will be wetted over a sustained period. It may be saturated. An absorbent mass is then placed against it, usually by trowelling on. Cellulose fibre (pulped paper), sepeolite or attapulgite clays (fuller's earth, bentonite) are the most convenient effective media. They may be applied layer by layer (damp pulped newsprint) or by trowelling on a stiff mix of clay perhaps against chicken wire reinforcement. The poultice can be up to 40 mm thick but half this thickness may be ideal. While the poultice is in situ evaporation takes place from its outer surface and in compensation liquid containing dissolved salts migrates into it from the brickwork. The poultice will remain in position until it dries, at which point no more salt-laden liquid can move across. The poultice is then removed, carrying with it deposited salts, and is disposed of by washing or by destruction.

The process may be repeated up to four times to achieve significant reduction in salt content. The process is effective because bentonite, attapulgite and fuller's earth have a molecular structure which has a very strong affinity for water and oils. Low temperature calcination improves their response.

In practice it might be thought that maintaining a body of water against the brickwork would have a similar and suitable effect, but this is not the case because the capillary attractions of the brickwork are such as to cause the water body to move into the brick and to saturate it while relatively little salt migrates into the washing water. Effectively the salts are driven back only to reappear on the next washing cycle. The only way in which water can effectively be used to reduce salt content is to provide short frequent washings which carry away salts already deposited on the surface without providing sustained surface wetting, allowing sufficient intervals between the washings for further salts to migrate outwards and appear on the surface. This practice, which corresponds to the removal of efflorescing salts by regular rainfall, can be effective if maintained but where a substantial body of brickwork in a hot climate delivers to the surface considerable volumes of salts the only way in which the content can be effectively reduced is to apply a poultice so that the liquid in the surface pores of the brickwork is pulled outward by the absorbency of the poultice. The driving force in the salt-bearing liquid is the capillary attraction of the poultice transmitted through the continuous water channels which run between the pores of the brickwork and those of the poultice. It is along these channels that the solute moves from zones of higher concentration to lower. The process of equalization is slow because movement through the pores is itself slow. Attapulgite clays are particularly fluffy – and therefore absorbent: fibres are even more so. Bulked waste paper is perhaps the best cellulose tissue. In all cases a heavy paste should be achieved using clean water. A number of proprietary pastes and gels have been prepared with thixotropic properties which allow them to cohere to the brick or masonry body. It is possible also to introduce other products which may facilitate cleaning or it may be thought desirable to allow them to migrate into the brickwork while outward diffusion of salts is taking place. Other than in the simplest conditions and circumstances poulticing will be carried out on the basis of initial experiment and experience and, where proprietary poultices are involved, with the manufacturer's guidance.

10.6.7 Stain removal

Poulticing offers the further opportunity of sustained chemical reaction over a long period and this can be valuable where chemical reagents are required to reduce or eliminate the effects of metallic staining. Iron and copper are the principal offenders. Over the years the wash-off from bronze, brass, copper and rusting iron can produce stains on brickwork and stone which can be most effectively eliminated by long-sustained chemical reaction. In such a case the poultice material will be left in place for some considerable time, its damp condition being maintained until removal. For this purpose an impermeable outer sheet will be necessary. The removal of green copper staining – verdigris – usually a complex pattern of copper salts encrusted among calcium carbonate, may be achieved by use of ammonium chloride activated by ammonia solution. Mixed with the poultice material it is normally applied several times, being allowed to dry as a means of drawing the salts into the poultice. The operation can be repeated on a daily basis.

Ferrous stains may be removed by the use of sodium citrate, possibly in the presence of sodium dithyonite which has an intermediate or catalytic effect accelerating the reaction. The more soluble compound formed moves progressively into the poultice and any residue is washed out. The inclusion of a buffering agent such as EDTA often helps in the removal of complex metallic stains. Some other chelating agents, such as tri-ammonium-citrate, which are advocated in cleaning may prove very difficult to remove unless taken out by repeated poulticing.

11

Protection, consolidation and repair

11.1 Physical protection

Brickwork may require physical protection against accident, protection from vandalism and protection from the erosive forces of nature. Against vehicles, machinery and pedestrians defence is often by the erection of barriers, usually metallic. Historic brickwork should never be required to carry iron fixings directly. They should be bedded in a flexible waterproof material. Iron embedded in any rigid cement concrete must not be used or allowed to survive. Historic fixings, if necessarily retained, may be protected against rusting and bedded in a flexible material – lead wool, bitumen or a semi-solid permanently flexible filler. A thiokol-based sealant may be used to bar water entry to a packing such as lead wool. Wherever iron is embedded in historic brickwork it is likely to cause problems of expansion and should be replaced by phosphor bronze, stainless steel of appropriate formulation or a non-corrosive alternative material. This applies to ferramenta in windows, railings and all other physical devices attaching to brickwork. Historical authenticity may require the replacement of the original material – cast iron for cast iron – in which case non-setting waterproof mastics may be used in conjunction with traditional poured lead or lead wedges to secure the fixings. Phosphor-bronze tips may be applied to historic ironwork where it is still serviceable and is therefore to be retained.

Brickwork bedded in lime mortar is not sufficiently strong to restrain the forces generated by the movement of lead in a socket under the strain of rust development. The historical success of such fixings (where they were successful) depended largely upon the use of high carbon content cast iron which resisted rusting to a considerable degree. Where wrought iron was used expansion was permitted by lead wool packing. Any steel and much wrought iron will rust sufficiently to constitute a danger if solidly embedded. For this reason historic ironwork is often 'tipped'. Stainless steel may be used to provide a solid non-rusting tip (tipping) as an alternative to phosphor–bronze.

11.2 Management

Passive methods of protection against human behaviour must include control of access and nature of use. The exhalations of visitors include carbon dioxide and water. Visitors will contribute moisture to the atmosphere at the rate of some 50 g each per hour if they are modestly active and at double this rate if they are vigorous. Therefore an enclosed space with minimal ventilation will gain water at significant rates and typically may then be allowed to cool over night, the water being deposited as condensation and gathering first in the smaller pores of brickwork and plaster. The cumulative effect of persistent human use can therefore be significant as the amounts of moisture will build up. Condensation will also occur on cold surfaces such as glass and brickwork and the effects upon wood, plaster, canvas and artefacts will be damaging. Water running down internal surfaces will carry soluble salts with consequent stains and bio-organisms will appear. Management is, therefore, a vital part of the protection of historic brickwork and historic buildings generally, ensuring that adequate ventilation maintains suitable atmospheric conditions. The important

Figure 11.1 Massive cast iron protectors designed to defend vulnerable masonry against the impact of cart wheel hubs, fitted to the corners of piers on a nineteenth century commercial building.

criterion is that the level of humidity is never allowed to build up sufficiently to produce condensation. Two methods of control are possible: regulation of visitor numbers and adjustment of air conditioning/heating/ventilation under careful management.

Management is the critical safeguard against wear. The human foot is a destructive agent on earthenware, particularly if damp, and where the material is precious – a Roman pavement, medieval encaustic tiles and glazed tile flooring – the wear can be reduced by the use of soft coverings (felt slippers) and cleanliness to minimize the presence of grit. An over-floor may be necessary. Management, perhaps at a draconian level, may ensure that the fabric is not touched at all. High wear follows the concentrated and near point-loads of small-heeled shoes (however light the wearer) and abrasion is much increased by the turning movement of the foot at sharp corners.

Cleaning materials and washing processes can be a danger. Milled quartz is often used as an abrasive in cleaners. Abrasive cleaners constantly applied are the cause of a high rate of surface wear on material as soft as most tile, particularly if the abradant is not conscientiously removed. Removal however often involves washing, usually with water containing a pollutant. Detergents and soaps with concentrations of free sodium ions or other reagents are often carried into cracks, crevices and fissures from which they are not to be retrieved and their effect may be reinforced daily. Bonds between tiles and mortar may be broken and deterioration is carried unnecessarily deeper into the structure.

Management is also the principal defence against vandalism, but no defence is complete. Where the risk is high a sacrificial rendering of lime mortar or a limewash might be considered a reversible protection. Where the continued appearance of the brickwork is important impregnation with microcrystalline wax may provide a base which will diminish the effect of painting and allow removal to be effected. An alternative pore lining technique used with success has been a spray coat of a polyurethane film on an isocyanate prepolymer. Chemical cleaning can remove paints from such surfaces. Protective compounds of this sort should be used only on sound substrates. A friable surface likely to be attacked may be treated with a suitable consolidant, such as an aryl-alky polysiloxane polymer. In some circumstances surfaces may be cleaned by the use of pencil jet air abrasives using a soft abradant such as chemically precipitated aluminium oxide (fused aluminium oxide–corundum – is very hard). It should be a general rule that paint solvents are not applied because of the danger of enhancing penetration. The paint should always be broken down chemically before being removed. Surface abrasion may assist.

11.3 Paint systems

While paints and paint systems (combinations of different but compatible paint materials) are a subject in themselves, the painting of brick-

Figure 11.2 A simple technique of physical protection is to anticipate impact and remove the vulnerable section. Here the brickwork has been cut to form a rounded corner.

Figure 11.3 High cart loads have, over a long period of use, damaged this arch and capital by impact.

work has been so long an established practice that general principles must be considered. The smoothest brick and terracotta surfaces are, like vitrified surfaces, poor paint receptors. An etching primer, which bites into the surface to provide a microscopic mechanical key, is therefore advocated and some paints can contain substances which will achieve this effect. Most brickwork, however, has a texture ideally suited to taking up the least adhesive of paints, although porosity and suction may be a problem in initial coating.

The simplest, most basic traditional and primitive paint is limewash, self-setting and self-colouring (white) but capable of carrying some moderate pigmentation. The achievable colour ranges are narrow and are all light, drying to much lighter tones than the mix. As a coating it has the advantages of porosity, allowing moisture vapour permeation and compatibility with mortars. It is not durable and will not adhere well to a poor base. Its colour density is low and it is removable. For external use it is sometimes made more waterproof and durable by the addition of linseed oil (which oxidizes to form the leathery sodium lineate) or tallow, an animal fat which must be dispersed in the hot mixture (usually at slaking) and decays organically, usually within the year. Its effect is largely hydrophobic.

Virtually all other paints are compositions of binders, pigments and protective materials, designed to meet specific conditions. The surface of new brickwork and renders is alkaline, tending to become acid with time. Paints suitable for application to brickwork include silicate based paints (with potassium silicate binders) of which early formulations were the Keim range (Continental) distempers and the multiple ranges of masonry paints designed to resist abrasion and atmospheric conditions while allowing sufficient porosity for the escape of moisture vapour. It will normally be the function of a paint to discharge liquid water arriving on the surface but to allow moisture vapour balances to be attained between internal and external surfaces. A paint may be given hydrophobic qualities for this reason but this is not necessarily an advantage and the simple reduction in pore size is usually sufficient to ensure that the inward movement of liquid is minimal.

The most widely used modern paints are based on resin emulsions, usually polyvinyl acetate (PVA) or acrylic-based perhaps in combination and with modifiers (styrene, styrene-butadiene-polystyrenes and vinyl copolymers are also used in this way). The loss of water on application causes the resins to cohere, binding the material into a powerful flexible coating. The loss of water has the effect also of reducing the volume by as much as 70% and this causes the coat to acquire tensions during the later stages of drying. These smooth out the surface but any break in this flexible skin will retract the surface, breaking it away from any insecure substrate. On such underlayers cleaning down and/or the initial use of very weak applications of the material as a consolidant may overcome the problem by providing a greater depth of penetration and bonding to the base material.

A wide variety of masonry paints are based on such binders, some also including alkyd resins and oils which oxidize as the water is removed. Primers may be required. These paints are given extra body by the addition of particles of silicaceous or similar materials – ground up slate, gneiss, feldspars, mica, bentonite and others. They tend to be 'high build' paints, i.e. to develop a thick coat and for stability in application often are formulated thixotropically. Their effectiveness depends considerably on this thick coating which offers vapour permeability coupled with the water resistance of non-hydrophilic small pore structures that allow only very slow penetration by liquids.

Of the binders used, vinyl copolymers are the least stable in wet conditions.

Potassium silicate-based paints penetrate into the brick, lime or earth substrate of historic surfaces, forming an inorganic bond with the base. They fall therefore into a different category of material, and should be applied direct to the base material to allow the continuous silica bond to form. Their formulation derives from a silicate glass, autoclaved in aqueous solution. When dried it has a high resistance to solution under atmospheric conditions; particularly after carbonation.

In choosing any paint system, longevity, appropriate porosity, water resistance, bio-inhibition, compatibility and historical appropriateness must be considered in conjunction with visual effect. The comparative cost of material is a minor factor. Paints may be regarded generally as reversible treatments although removal is not always easy. Before considering a paint application the mechanisms by which it may be broken down should be understood so that its removability can be determined. Provided that it can be removed and the original brickwork can be restored the painting or repainting of historic brickwork can be considered. Paints should not be used for their consolidant effect as the depth of penetration is minimal. It may be convenient that a friable surface may be made coherent by paintwork but this is not in itself a justification for the use of a paint which visually alters the building. The application of an effective masonry paint to historic brickwork will change the visual quality of the structure but may at the same time greatly improve its weather resistance, the thermal properties of the wall by the exclusion of damp and the longevity of the structure at large. These considerations may be seen in the larger context of the visual change and structural permanence and no absolute rule can be laid down. An arguable proposal might be a panel of brickwork built into a timber framed building in place of a lost wattle and daub panel which would have been limewashed. White painting of the brickwork might be seen as a return to the visual character of the original building without destroying the historical evidence of change, therefore being justifiable in terms of historical accuracy and weather resistance.

Removal of paints is often carried out with solvents many of which are alkali-based, being caustic solutions in water. They soften the binders, often after an extended dwell-time, and a typical application may not penetrate more than one or two coats so that several applications may be necessary. They work more effectively in higher ambient temperatures but under heat give off fumes. Direct heating should be avoided and no method of paint removal employing high temperature alone should be used on brickwork with the possible exception of glazed and highly vitrified smooth brick, where the bulk of the material may be removed after heat softening, preferably with a hot air jet. A solvent type stripper must then be used for the final cleaning down. For final cleaning a high-pressure low-volume water application is essential, and the principle of multiple washing without saturation should be observed.

11.4 Water-inhibiting compounds

Considerations which relate to the retention of the building with minimal visual loss and the effective exclusion of water have caused the extensive advocacy of transparent solutions which are capable of performing the same role as a paint in excluding moisture. Early attempts to achieve these advantages have been at best modestly successful in the long term. Applications of materials such as potassium silicate and a wide variety of oils and water repellents have a history of failure where clumsily used, interspersed with some successes. Of the proprietary materials advocated by their manufacturers for waterproofing brickwork and mortar, formulations based on the deposition of silicates carried in spirit-based solvents have been successful in the relatively short term. The evaporation of a hydrocarbon carrier leaves highly concentrated silicaceous materials dissolved in water. These then crystallize, some forming less soluble compounds, and by pore reduction the rates of flow through porous brickwork is reduced.

There may also be consolidant effects. However, deposits can have some water solubility and eventually the leaching out of parts of the deposit breaks down the water resistance. Proprietary formulations frequently include silanes (silicones) in order to enhance water repellency. These too break down ultimately. The effects on brickwork are generally colourless or nearly so, with minimal to modest yellowing with age, but there is often a change in reflectance – a varnishing effect. Such materials may be used appropriately to minimize water entry in situations where brickwork is particularly vulnerable; on the tops of string courses, ledges, drip moulds and hoods. The reduced water absorption will change the biological regime and reduce internal erosion. In cases where water penetration has led to the internal decay of lime mortar and the formation of minor internal aquifers the life of the structure can be extended, perhaps indefinitely, by the careful surface repair of the brickwork coupled with giving it increased water resistance by the application of water repellents and water repellent grouts. The application of unsightly gutters or lead cappings may then be avoided. Other repellents may rely on heavier organic coatings, some historic in themselves. Of these the principal is of paraffin wax. Repeatedly applied to the granite of Cleopatra's Needle on the Thames embankment, this material has preserved a major artefact of great historic importance against the effects of a corrosive industrial atmosphere for the largest part of a century. Paraffin wax, is, however, a material hardly suited to brickwork and its natural successor, microcrystalline wax has not so far been prominent in the defence of brickwork or terracotta against weather conditions. Microcrystalline wax differs from normal paraffin wax only in its physical chemistry, the size of each molecule, being a multiple of that of the normal paraffin wax molecule although their chemical composition is the same. The principle on which these waxes are effective is simply that they provide the pores of the material with a hydrophobic lining reversing the effects of capillary action so that the normal attraction due to the angle of adhesion is converted to a repulsion. Waxes do suffer however from the fundamental disadvantage of attracting and holding dirt. The material also tends to fill the pores. Like many organic substances they are evanescent, although to a very minor degree – that is to say their molecules can pass into atmosphere without moving through the liquid phase. This loss is coupled with and may be associated with oxidation, the other fundamental destructive reaction. Assiduous householders have been known to maintain terracotta with wax polishes without ill effect for long periods, however – the key being regularity in maintenance.

The new ranges of materials offered by modern chemistry exhibit powerful advantages as protective substances over their predecessors. Foremost among them are the ranges of materials known as silicones. While initial long term tests have shown that a number of these materials are ineffective, the relatively short period of experience allows no clear conclusion beyond the comfort that successful results have been obtained, that new products are coming forward and that the techniques and practices of use of the available materials demonstrates that improved performances can be anticipated. Conservation has moved beyond the phase in which recent generations of sophisticated organic water repellents are regarded as 'cure-alls' to the stage of careful analysis of longer term problems and wider understanding of the potential of the materials. In this the use of water repellents, consolidants and materials available for organic repair are following a pattern to be seen in other periods of industrial progress and the general advance in human skills and knowledge. A pattern which begins with a modest success and evident potential can expand to high levels of effectiveness.

Silicones are siliconesters (organic salts of silicon). Another material of similar water repellency is a range of fluoro-polyethers, polymers with very large molecular structures which have been tested in application dissolved in trichloro-trifluoro-ethane. Complex organic molecules tend to break down by oxygenation, under bacterial attack and under exposure to ultra-violet light. The deposition of such materials in the pores of brickwork after repeated treatments and continual breakdown poses serious questions of longevity. As a matter of principle it follows that the use of water repellents of these types will be confined to those areas of structure which are not conservable by conventional means and only where the structures are not of the highest levels of historical importance. The further caution that conservators will take into account is the need to confine their use to a minimum for effectiveness and a careful judgement has, therefore, to be entered based on the extent to which water entry should be inhibited. It is rarely necessary or desirable to totally preclude water entry. The objective will normally be to achieve the ideal balance between absorption, diffusion and evaporation. In making such assessments other sources of water entry, such as internal deposition on wall surfaces, may justify the application of a different type of inhibitor or of alternative management techniques. It must be remembered also that a surface rendered impermeable or less permeable may cause a build-up of water in the brick and/or mortar with the consequent deposition of salts and spalling away as the brick breaks down.

The silicone molecule is a long chain structure in which a continuous hydrogen bond is available for linkage to the molecular surface of the material. Consequently silicones adhere to that surface with a continuity reminiscent of large numbers of caterpillars on a leaf, forming a very thin film of molecules each perhaps upwards of fifty atomic components in length. Known collectively as silanes, they vary in constructional detail and in methods of polymerization. It seems certain that silanes will become more widely used in the general field of building as consolidants and as water inhibitors. Their use in historic buildings will lag behind more general applications and prudence dictates that they will be used in situations of least risk to important surfaces.

In the complex chemistry of silanes the use of different radicals (carrying members such as phenyl, ethyl and methyl roots) provides for different ranges of the initial compound to be carried into the brickwork. To these radicals are linked the polysiloxanes which after hydrolysis and polymerization, perhaps in the presence of a catalyst, are to remain in the pores of the brick or mortar to inhibit water penetration or act as consolidants. As these processes (hydrolysis and polymerization) proceed the products of reaction such as ethyl or methyl alcohol and the catalyst must be able to escape. Some of these may be complex and persistent. Phenyl and other ring radicals (generally described as aryl, e.g. aryl-alkyl-polysiloxanes) may by their inherent stability seem attractive, but the potential for the creation of a chain of long term chemical reactions invites caution.

While the short term effects of such materials can be beneficial, the long term activity of replacing them periodically to maintain their performance suggests that there may at some point arise acute problems of pressures created by the swelling of materials remaining from previous treatments and secondary compounds being generated within the surfaces of brickwork. Such pressures may lead to losses of surface due to spalling. The science of water repellency is still young, as is the interrelated study of consolidation.

Silicones may have little consolidant effect and although they may cause minimal visual change initially they can produce brown discolorations as they decompose – an effect relatively unimportant on much brickwork, but damaging on mortars. The presence of iron-based compounds in the brick is a possible danger and the removal of such decomposed material is not yet an available technique. There is a substantial body of opinion which regards organic polymers as ineffective and a significant number of conservators go so far as to regard them as a potential danger. Brickwork has for long been the poor cousin of masonry in the use of consolidants and advances of knowledge in handling stone monuments must be a guide in the conservation of historic brickwork.

At a day-to-day level a number of well-established proprietary aqueous solutions maintain their place in public affections. Their usefulness derives variously from potassium and sodium silicates and binders. They can be a useful adjunct to more substantial works of maintenance if used after careful trials. Situations in which they may be of particular value occur where the original builders have left a section of structure – for instance a brick mullion – of insufficient thickness to resist weather penetration and no structural amendment is possible.

11.5 Consolidants

The functions of consolidation and water repellence go together. Materials recommended have included acrylic resins, alkyl-aryl-polysiloxanes, alkyl-trialkoxy-silanes, methyl and ethyl silicates (the most successful to date) and other materials such as tetramethoxy and tetraethoxy silane which break down to deposit silica initially as an amorphous gel, but ultimately as a glassy or crystalline deposit. Mixtures of these materials and others, the availability of catalysts and the variety of methods of placing have given chemists and experimenters very substantial ranges of and choices of materials. From these experiments have arisen a number of proprietary products, of which a formulation of the Building Research Establishments, Brethane, has been claimed to have significant advantages. 'In this system water sufficient for hydrolysis is added to alkyl alkoxy-silane and the liquids are rendered miscible by the addition of a very small quantity of acid. An organo-metallic catalyst is also added'. The system is covered by a British patent. The viscosity of the system remains low until a certain induction period has elapsed. After that it begins to rise rapidly as hydrolysis and polymerization take place quickly under the influence of the catalyst. This sequence keeps evaporation loss to a minimum and freezes the advancing liquid so that it does not spread too far. The system is restricted to application by specialists. It can result in the creation of a layer of discontinuity and is in a stage of long term trial and experiment. It is applied only under licence but the material has been withdrawn from the market. This description is useful as an indication of the present state of knowledge and the techniques involved in further advance rather than as an immediately available process.

Alkoxy-silanes of the Brethane type (not now recommended for other reasons) have been reported as penetrating up to 50 mm in porous stone and it would appear that a significant depth of penetration could be obtained in porous brick although penetration depths inevitably must be less in bricks of higher density. A further problem is that application to brickwork leads to varying depths of penetration where there is, as is commonly the case, a significant difference in porosity and absorption characteristics between the bricks and the mortar beds.

Two groups of organic polymers are available as consolidants. In the first group are those which are entrained in their solvent and are already synthesized as polymers prior to placement, and second those which are introduced as monomers to be polymerized within the pores of the host material. The advantage of this second form of treatment is that the monomers are more fluid when appropriately diluted in a solvent and therefore deeper penetration is possible. Polymerization – the joining together of the chemical into long chain compounds – then takes place within the pores, the larger molecules being formed in intimate association with the structure. There are a very large number of available materials and a wide variety, therefore, of techniques of application of mixture and of calibre of product. The timing and method of polymerization of a monomer can be critical to its success and if, for instance, it is drawn back towards the surface by an evaporating solvent before polymerization has taken place there will be an undue concentration at the surface and the advantages of the use of a monomer in penetration will have been lost.

Thermoplastic materials used include polyvinyl chloride, polyethylene, polyamide, polystyrene and polymethylmethacrylate: and these substances are reversible to some degree. Thermosetting polymers include polyurethanes, polyesters and the wide range of epoxy resins. Some thermoplastic polymers can be cross-linked to become thermosetting. These are generally not reversible in the conservation sense. Once in place they are removable only by high temperature methods.

Acrylic polymers available include methylmethacrylate and butyl methacrylate. For work, on individual objects such as wall tiles where a flaking or peeling body carries important art work, laboratory conservation with acrylic copolymers is the favoured and probably the most promising method. Detailed studies of rates of polymerization, depth of penetration and the conditions of application are available from specialist suppliers. The building conservator faced with a consolidation problem of this type will be faced with choices of acrylic based polymers, acrylic copolymers, fluorocarbons and silicon esters or alternatively vinyl polymers such as polyvinyl chloride, polyvinyl acetate and chlorinated polyvinyl chloride.

A brick, tile or terracotta object requiring treatment at this level should be referred to specialists and removed to the laboratory for conservation. The techniques of polymerization may include irradiation or heat treatment and the controlled conditions of a laboratory will be the key to successful work where a consistent and high level of impregnation is required. Even after laboratory treatment there is a danger that a conserved object might contain zones of varying density of impregnation with consequent risk of varying behaviour under changing conditions of temperature and humidity. It is also possible that any such layer would form a barrier or inhibitor to the movement of salts and therefore would be likely to increase the internal stresses on boundaries within the brick or terracotta. Epoxy resins which are by definition irreversible are polymerized to form thermosetting cross-linked polymers of extraordinary toughness. Common epoxy resins are derived from diphenyl-olpropane and epichlorohydrin – viscous liquids diluted with organic solvents such as acetone. They react with a hydroxyl radical supplied in an aliphatic compound and cured by amine or anhydride curing agents. The epoxy chain may be aliphatic or aromatic and it may vary widely in its behaviour in critical aspects such as water vapour transmission and gas permeation, solubility (up to 5%), softening point, thermal coefficients (always high by comparison with brickwork) and other important characteristics. To generalize can be misleading. The material must therefore be known by its established performance or by a chemical design from the hands of acknowledged specialists. To resort safely to a proprietary formulation is to leave out of consideration a possible tailor-made material at probably higher cost.

The foregoing remarks apply equally to the alternative very successful range of consolidant resins which are grouped as polyesters. The five types available have been grouped (Dibenedetto 312) as (i) general purpose, (ii) isophthalic, (iii) bisphenol-A, (iv) chlorine bearing, (v) vinyl. Their mutually typical properties include adhesion, high strength, permanence, water resistance and rigidity with lesser thermal coefficients than their epoxy equivalents, but their variety in all the permanent characteristics related to solubility, transmission etc. is wide and the same problems of careful design and formulation apply. There is a range of good products in both resins commercially available and aimed more at use with timber and stone than with brick.

The many epoxy and polyester resins with widely varying characteristics of performance and a variety of secondary products can cause difficulties, such as the formation of white, chalky excrescences or surfaces by secondary reaction. The principal practical difficulty in any attempt to use them as consolidants will be the problems of entrainment due to their viscosity, control of curing and subsequently the compatibility of the consolidated material with any more friable adjacent brick or terracotta.

Consolidation of terracotta by such synthetic materials is likely to be a laboratory technique and used only rarely, but this is by no means true of repair in the field. For making good defects, re-adhering flakes or fractured pieces, bonding fractured elements and restoring missing elements of glazed material specialists continually rely on proprietary formulations of these ranges of polymers and copolymers. For materials which are to be housed or retained in controlled conditions or internal environments there is now so substantial a record of success that the laboratory conservator will have little hesitation in adopting the appropriate formula confident of the permanence of his work. Repairs which are to be exposed to the hazards of external conditions present greater problems. In tiles, terracotta and elaborate brickwork where fractures, flaking, spalling and similar defects are apparent it is often possible to introduce organic consolidants as bonding agents, perhaps by injection with syringes, by straightforward application on to broken surfaces or by capillary attraction into cracks. Frequently the bonding agent will be stronger than the adjoining materials. Inevitably its coefficient of performance will be different, making its bond susceptible to stress as conditions change. In its life on the building, however, the organic polymer is likely to be subject to ultraviolet light and cycles of water intrusion, thermal expansion and contraction, freezing and thawing, chemical attack by mild acids and alkalis and oxygenation often due to the presence of bioorganisms. No organic polymer is known to withstand these often combined sources of attack with the same longevity as fired clay. While there may be no better method of restructuring or repairing the object in question such potential vulnerability must be considered at the point of specifying the repair.

11.6 Patching

Polymers and their composites are particularly attractive for the formulation of filling and patching materials. Both polyesters and epoxy resins with crushed brick and appropriate pigmentation can be compatible with a brick substrate and form a powerful filling material. The greater the volume of crushed brick or similar material the closer the physical behaviour of the filler will be to that of the host material. A darkening in colour is unavoidable but careful handling of the surface finish, texture and materials can provide suitable colour matching. The bonding to the substrate of filling compounds of this type may be enhanced by the inclusion of a bonding agent such as an organo-silane. Such materials may bond with silicate surfaces to provide a linkage between the inserted filler and the substrate, the coupling reaction being through the hydroxyl groups on the silica surface. The advising specialist will consider not only the chemical bonding but the physical action during bonding where the wetting, viscosity, the surface tension and the time over which pressure is maintained all determine the degree to which the pores are penetrated. Where patching and substantial filling is involved the surface bonding is particularly important because there will always be a mismatch of physical properties between the inserted material and the host brickwork. The wide variety of epoxy, acrylic and polyester materials available makes the chemical design process itself one of great complication and in the present state of knowledge, there is considerable uncertainty particularly in relation to longevity.

The very small volume of experience in the use of organic consolidants, bonding and filling agents in brick and terracotta by comparison with the very wide range of conditions, considerable range in materials and the extensive ranges of available chemicals demands careful methodology from the conservator, whose analysis must seek answers to the following questions.

1. Nature of failures and reasons for their occurrence and whether further similar problems will arise. Spalling is a typical instance of a recurrent event.

2. Nature of base material and susceptibility to similar decay.
3. Rate and characteristics of degradation.
4. Critical performance characteristics required of material to be applied.
5. Justification for application of organic binder.
6. Performance characteristics required in the substrate, in penetration of adjacent surfaces as well as in the mass.
7. Range of materials meeting required performance characteristics wholly or partially and their compatibility.
8. Design chemistry advice available: whether it is independent or commercially driven and consequently specific to a manufacturer's product.
9. Extent to which performance criteria are satisfied.
10. Possible areas of risk and failure.
11. Knowledge of accelerated weathering tests, field tests and evaluation of comparative performance.
12. Degree of reversibility, colour change, curability anticipated in material (meeting performance criteria).
13. **Decision** as to whether to proceed with the use of organic polymers.
14. Special techniques of application.
15. Suitable availability of laboratory and site skill and the capacity for the control of site operations and monitoring of performance.

The conservator should also be prepared to ensure ultimate reporting of performance to trade and professional circles and to data banks.

A patching material should be as near in character as possible to the base material and it may be necessary to achieve a colour and texture close to that of brick and terracotta. Since polyesters and epoxy resins, with their widely varying performance characteristics, are almost certainly the synthetic binders to be used the problem can be contained within their characteristics. Generally, by comparison with brick or concrete the compressive flexural and tensile strengths of quartz aggregates also bonded with resin will be between three and six times greater. In consequence there is a capacity to weaken the polymer composite by the addition of lighter and softer aggregates in larger proportion. Earth and other inorganic pigments may also be introduced provided that highly reactive metallic ions (such as sodium, calcium, potassium and magnesium) are not made available in subsequent exposure to water. Judicious blending and careful experiment will lead to selection of an optimum mixture. Porosity in the substrate may be a further hurdle even after a mixture of the right strength, colour, workability etc. has been achieved.

A patching material is likely to be a paste with poor penetration. Mechanical methods such as keying, undercutting and pin linkage are possible. More sympathetic however is the design of a bonding coat which has good pore penetration and is compatible with the patching material itself. Penetration is not simply a matter of viscosity. A thin liquid may not penetrate porous brick any more readily than one which is thicker. Penetration by a fluid of pores of specific size depends upon viscosity, surface tension, molecular sizes, the contact angle, reflecting wetting capability and the time in which the fluid is present prior to setting. Penetration may be minimal, or at the best, slow. This difficulty is worsened by the large volume of fillers such as sand, brick dust, pigments, PFA, vermiculite, etc. which may be incorporated in the patching material, making it intractable as a bonding agent.

The possible response to this problem is a preparatory resin, compatible with the main mix, with good wetting properties and adequate setting times which can be applied as a primer.

11.6.1 Proprietary polymers

Proprietary polymers may be used as binders. In some instances, and bearing general principles firmly in mind, commercial emulsions of acrylic and acetate polymers may be effective bonding agents. Some have been designed specifically for this purpose and have achieved a high measure of success in the right hands. Their behaviour in exposed conditions is sometimes disappointing due to fungal activity but emulsions have the profound advantage of providing a porous structure.

Other commercial formulations with a proven record of success derive from concrete hardeners and synthetic screeds.

These may simply contain sands and epoxy resins. Careful grading and selection have made available tough strong materials typically with tensile, flexural and compressive strengths more powerful than a high grade concrete by factors up to six times. While these are likely to be excessive for normal patching and repair purposes they may provide a basis for a more sympathetic material (despite negating any manufacturer's guarantee) and they may provide a material suitable for infilling where strength is important and attachment less so. Pads under tiles or tesserae might in some circumstances be a case in point if the original construction could not be repaired or reproduced. Other cases might relate to mixes required for rapid strength in historic brick subject to intermittent stress or loading.

Strengths obtainable can be:

compressive	55 MN/m^2 compared with	40 MN/m^2 (concrete)
tensile	15 MN/m^2 compared with	3.5 MN/m^2 (concrete)

These catalyst-cured two-part compounds are not tolerant of high temperatures but are very resistant to dilute acids, alkalis, solutions of salts, sugars and solvents such as alcohols, oils and aromatic hydrocarbons. They are not reversible but may be destroyed by high temperatures. They may depend upon oxygen for curing and must therefore be used in layers not exceeding 20 mm thick.

11.7 Thermal stress and fire damage

Brick and terracotta are fired at temperatures of between 800°C and 1200°C generally speaking, and although these temperatures may be exceeded in burning buildings it is unusual for masonry to reach temperatures in excess of 1000°C except in extreme conditions. Fire damage to brick and tile is, therefore, not so much a consequence of temperature as of thermal stress due to rapid heating, structural movement, collapse, removal of support and inequalities of temperature caused in many instances by rapid cooling in fire-fighting. Bricks do not change profoundly by reheating up to 1400°C but above 1600°C they can become semi-molten. Fuel oil burned efficiently reaches 2200°C, at which temperature brickwork may become a viscous liquid. This condition is exceedingly rare. The general level of failure relates to mortars and physical displacement.

Bricks may contain some calcite and gypsum which will be chemically affected by the heat of building fires. Unless there is some ready availability of fluxes however the silica in the bricks will not be affected. While this is true for the sand in mortars, their calcite will readily be oxidized above 800°C forming quicklime, possibly to be suddenly hydrated by water from the fireman's hose. This can cause a violent and expansive reaction in the joints breaking brickwork apart although, if quicklime has been formed, the brickwork is in any case condemned. Although in theory the mortar will rehydrate, in practical terms it is essential to return through the process of the formation of lime putty and relaying. Atmospheric hydration which will occur after heated lime mortar has been allowed to cool and remain exposed to air does not produce the compacted structure which was characteristic of the original mortar nor does it bond to the bricks.

The very substantial temperature rises which occur in fires cause substantial thermal movement, often breaking the horizontal bonding along lower courses where buried sections of wall may not experience significant temperature rise and, in both expansion and contraction phases, stressing the walls across perforations, commonly producing vertical fractures. Within the bricks themselves intense heating of faces can cause surfaces to spall as the result of excessive thermal gradients. Under-burned bricks which will have been used for internal walls will be exposed to the highest temperatures in many building fires and, particularly if exposed to sudden cooling, may fail by decrepitation rather than spalling. Under-burned brick may withstand rapid change in temperature in the lower ranges more readily than a hard-fired vitreous brick. The greater porosity and lesser glassy bond give it a useful degree of resilience.

Loss of support brings down tile roofs. Timber failure will also affect the stability of brickwork by loss of support, variation in loading, loss of tie strength and primarily by leverage. A timber, burned through, is likely to exert leverage on brickwork and in historic buildings the linkages provided by structural timbers within the walls and between them are the most vulnerable structural elements. While timber stringing courses and lintels may burn out in walls, perhaps dangerously affecting the slenderness ratio in some instances, the greatest source of danger lies in the collapse of timber floors with the consequent leverage of girders and of wall plates providing dynamic forces which unbalance the structure and cause it to topple. Brickwork is, however, astonishingly durable in fire with remarkable instances of structures surviving in precarious isolation, sometimes in apparent defiance of all structural principles. Expertise in assessment of the stability in brickwork which has survived is increasing. This task is not to be undertaken lightly and involves the methodical assessment of the condition of mortar bedding joints and the brick itself. Slenderness ratios may become critical and initial support may be given by pads of epoxy resin mortar introduced at salient points without stressing the structure by mechanical devices.

The drilling out of cores, testing the calibre of mortar and the evaluation of the integrity of brickwork by impact are the tasks for structural engineers experienced in the field, usually in association with loss adjusters. Their decisions will be a basis for the work of the conservator who must then decide on methods of cleaning down, removal of stains of molten metal, scorch and soot, stabilization and repair.

Lime and gypsum plasters are vulnerable to fire and with their loss the exposure of wall surfaces previously concealed may sometimes provide significant further information about the history of a wall. All such information should be carefully recorded before remedial work obscures the evidence. The extent of the remedial work and its nature will be the consequence of policy decisions made by building owners and advisers, including conservators.

In the important stage of recovery after a fire careful removal of material, retrieval of artefacts, recording of architectural details and equipment should extend also to the recovery of all usable bricks which should be carefully cleaned and stacked in groups that identify their original place in the structure and, where possible, identify their specific function. In the instance of some significant or crucial items it may be possible to be precise as to their earlier position in the building and there is then a prima facie case for their reinstatement. A major fire may provide historical opportunities which can be some compensation for the disaster itself.

Recording of artefacts, original condition and condition of recovered material is one of the most painstaking aspects of post-fire and post-destruction activity.

11.8 Electrolysis and cathodic protection

While electrolysis describes the movement of electrons through a solute and therefore embraces the failures due to salt deposition, this section is concerned only with the effects upon iron in brickwork.

Because it is impossible to eliminate moisture completely from masonry, electrolytic action leading to the oxidation of iron can be avoided only by using an iron alloy such as stainless steel and some forms of cast iron, or by coating the iron impermeably, or, further, by maintaining around it an environment which is alkali rich (pH upwards of 9). This is the condition in which steel embedded in concrete or mortar first finds itself. The persistent presence of acidic water, however, gradually overcomes the alkalinity allowing the iron to oxidize through the medium of acid ion exchange. The consequent expansion can be disastrous, its most damaging general characteristic being failures of the fixings of terracotta. Tile pegs fracturing their tiles and terracotta fixings forcing apart the terracotta units are the largest source of loss and failure in burned clays due to rusting iron.

There is, however, one active method of protection. While the iron normally acts as an anode (electrically positive), attracting electrons, this effect can be reversed by using artificial anodes. In masonry this may be achieved by introducing metallic rods into the masonry and connecting them electrically. The solute completes the circuit. While it is possible to achieve a static reversal of potential by exploiting the appropriate metallic relationships (e.g. zinc/iron), steel in buildings is efficiently protected by using a direct current applied to the steel frame which thereby becomes a cathode – hence cathodic protection.

The principal application of this technique is in the protection of steel in reinforced concrete or steel frames, and it can be an important method where such frames are sheathed with terracotta. Provided that the polarity is maintained permanently the effect can be indefinite.

11.9 Structural reinforcement externally

Obvious methods of providing additional strength to brickwork are usually temporary but occasionally permanent additions are made by fixing stanchions to the face of retaining walls bulging under pressures or by banding chimneys. In the latter case the objective is to supply a tensile component to ensure the recovery of the rings of brickwork following expansion.

Engineering advice must be complemented by the care of the conservator in controlling visual effects and ensuring protection of steel from rusting.

Figure 11.5 External reinforcement. Steel banding of brickwork in which there is possibly inadequate tensile strength may be employed, as in this mill chimney, to provide extra security. A continual regime of maintenance is required.

12

Mortars, renderings and plasters

As mortars, earths set by frictional cohesion, to which traces of crystalline interlocking will usually give some further strength. Other forms of cohesion are dependent upon the formation of an interlocking matrix of inorganic crystals. The simplest derives from the enlargement of a crystalline structure with water of crystallization, a physical change, while the formation of crystalline carbonates and the more complex aluminates and silicates involves the formation of new compounds.

These setting materials, which take up a permanent shape after being placed, may also be used to form articles of building technology – bricks, tiles and blocks.

12.1 Mortars: an introduction

Gypsum has been used as a mortar throughout history because it is easy to make and has the specific property of rapid hardening due to the coherent setting action of the water of crystallization.

Gypsum is calcium sulphate ($CaSO_4$) which occurs naturally in a hydrated condition, each molecule carrying two water molecules as part of its crystalline structure. At the relatively low temperatures of 120–150°C the water is partially driven off, leaving only one water molecule attaching to two $CaSo_4$ molecules ($2 \times CaSO_4 + 2H_2O = 2CaSO_4 + H_2O + 3H_2O$). This hemi-hydrate can exist as a white powder (Plaster of Paris) and on the addition of water it recrystallizes, slightly expanding, to the original hydrate which is a rigid solid. As the process occurs at virtually the same speed throughout the material a very 'sharp' set occurs providing a rigid adhesion within a few minutes of placing the mortar. This quality, together with the ease of firing and low temperature of manufacture has caused builders to use gypsum as a mortar and as a plaster despite two significant disadvantages – its low strength and moderate long term solubility in rainwater. Its usefulness in situations of regular wetting is, therefore, limited. However, in relatively dry climates the material (often known as juss) has been much used for vaulting and complex brickwork. It has the further advantage that it expands on setting, thereby avoiding the problem of shrinkage cracks. Variants of the material derive their different properties from impurities, fillers and the methods of preparation.

Calcium carbonate – calcite ($CaCO_3$) – has been until modern times the material upon which the most permanent mortars have depended. Calcite in the form of chalk, limestone or marble is converted to quicklime by heating. The quicklime can then be hydrated, reduced to a putty and mixed with some three times its volume of sand to form mortar. It will slowly harden by the absorption of carbon dioxide from the air or more accurately by the reaction of calcium hydroxide with dissolved carbon dioxide: in other words, water must be present. Essentially, therefore, the process is similar to the use of gypsum. The basic rock is heated, transformed into a powder, wetted, placed and it then solidifies in the shape required in its new position, reverting to the original chemical. Since, however, it is not water of crystallization which is driven off, but carbon dioxide in chemical combination, far higher temperatures are required. To change the common forms of calcium carbonate to quicklime or calcium oxide (CaO), temperatures of between 750°C and 900°C are necessary. At these temperatures both limestone and chalk give off their carbon dioxide ($CaCO_3 = CaO + CO_2$). These temperatures happen also to be those at which the various forms of silicate begin to melt in the presence of suitable fluxes. The raw materials for making quicklime are limestone and chalk, both of which were laid down in primeval seas by the action of marine life in extracting carbon dioxide from the air, in order to build up their own skeletal and protective structures (shells, carapaces etc.).

The denser metamorphosed form of limestone – marble – is generally less suitable for the relatively simple processes of burning into quicklime. Some very pure limes are burned from the shells of contemporary marine life.

Pure calcium oxide – quicklime (CaO) – forms a rigid mortar firstly by dewatering (with shrinkage) then by slow setting through carbonation. To achieve this it is first combined with water in a process known as slaking and it is then allowed to set in an air-dried condition which means an inherent water content of up to 15% by weight.

In its simplest form this reaction is expressed as $CaO + H_2O = CaOH + (–OH)$, or $Ca\ (–OH)_2$. On exposure to air the available carbon dioxide produces the reaction $Ca\ (–OH)_2 + CO_2 = CaCO_3 + H_2O$. In reality, the reaction is more complex than these equations suggest. The process is slow and the initial set would be little more than the physical hardening due to loss of lubricant, were it not for the fact that the raw materials always contain impurities. A substantial lump of chemically pure hydrated lime, if left in normal atmospheric conditions will not be converted fully to calcium carbonate for many months. Major Byzantine buildings have yielded samples analysed to the effect that after many centuries the lime in the cores of their enormous piers has never been fully converted. In mortars, however, impurities burned with the original stone, or introduced materials of a pozzuolanic nature such as brick dust, react chemically to produce an 'hydraulic set' (i.e. a set which can occur under water), by which crystalline structures are formed in the material in advance of, and in preference to, the setting which occurs by the absorption of carbon dioxide and the formation of crystals of calcite. This process has been used deliberately by builders in order to obtain a rigid structure at the earliest possible date. These hydraulic sets are in essence the chemistry of cement mortars. Setting is also achieved in pure limes by the addition of pozzuolanic materials which are always the pulverized products of vitrification on firing at temperatures in excess of 800°C giving rise to the production of soluble silica and their action is a consequence of the formation of complex calcium aluminate and calcium silicate crystals.

Varieties of cement mortars are discussed in later sections.

12.2 Concrete blocks, bricks and tiles

Bricks and blocks are now commonly made with cement-based materials and for a considerable period have been formed using lime. Lightweight blockwork has been made using aggregates of expanded clays, clinker, coke, vermiculite, PFA, wood dust and polystyrene. Although found in some historic structures these and similar materials may be regarded as being sufficiently modern to fall outside the scope of this work. They are all manufactured mechanically in controlled conditions, usually in steam autoclaves. They are normally laid in a cement mortar, which desirably is gauged with hydrated lime to give it some softness and a lesser rigidity than the blocks themselves. In parallel industry now produces immense ranges of highly specialized precast components of concrete and similar materials reinforced to meet the circumstances of their use. These are relevant to this work only in the context of the conservation of brickwork where they are used in intimate association with the objective of insulation or particular strength. Historic concrete is becoming an increasingly important area of conservation activity in its own right.

The manufacture of roof tiling in cementitious materials is likewise now at a stage of high technical achievement. In various forms both bricks and tiles have been made of pressed lime, sand and cement/lime/sand mixtures for over a century, some products proving as durable as their fired clay equivalents and having become historic in their own right. Mechanically pressed and produced in a steam autoclave, bricks of this type are known generically as calcium silicate products. Their cohesion is not due to a cementitious crystallization but arises by a distinct silicate reaction in moist conditions under heat. Roof tiles, however, are bonded with Portland cement and are formed with minimal moisture content by a semi-extrusion method on to a profiled pallet cured in warm conditions. This passing reference must note that American, French, German and British patents have all led to the production of mechanically accurate sand–lime bricks and tiles, the significant and effective processes always involving a high degree of mechanical compaction achievable with a low water content. The mechanical compaction overcomes the lack of lubrication by excess water. Curing at temperatures of between 70°C and 80°C induces rapid crystallization and a set achievable in hours rather than weeks or months. Concrete blocks and blocks containing lightweight aggregate or with entrained voids are often simply air-cured or sometimes put in hot chambers with steam although they are not autoclaved, i.e. cured at high temperature under steam pressure. Although the finished products appear similar and behave similarly it is necessary in conservation to differentiate between calcium silicate bricks, which are the result of the interaction of silica and calcium carbonate at high temperature in the sustained presence of water, and concrete products which contain a matrix of aluminosilicates, curing in the presence of moisture at normal ambient temperatures. In both materials the density of the initial mix has a significant bearing on the final product.

Careful manufacture involving selection and refinement of sands, the use of broken down limestone, chalk or rock or volcanic aggregates, air entrainment, efficient mixing, pressing and curing offer products with special qualities of density and insulation value at prices that often command a substantial market share. Many are deliberately made to compete visually with the natural or traditional product and even where they are regarded as visually inferior they may well be incorporated in concealed sections of structure. With the vagaries of the commercial marketplace many such components which have been efficient and incorporated in historic structures may no longer be available but it is usually possible to replace them, where necessary, by special manufacture at some expense. Many small works are prepared to experiment with sideline products – patterned, coloured floor tiles and complex roof tiles of which a typical example is a patented German tile laid diagonally and manufactured on the continent in the late nineteenth century. It seems to have been the first concrete roof tile brought to Britain, where it proved the foundation of the 'concrete' tile industry. Large numbers of this particular tile were manufactured before the industry moved to designs more closely imitating traditional tiles. Replacement can be achieved only by special manufacture.

Precast and mechanically pressed cement-bound materials have been used to perform every function in the building industry for which fired clay materials have been used. While their detailed consideration is beyond the immediate scope of this work they are so intimately involved in construction that they must be considered as close alternatives. Their weathering is very characteristic and in wet climates some types build up substantial growths of mosses and lichens, particularly after the amendment of initial high alkalinity (pH 11).

A wide variety of materials have been used as mortars in which to bed bricks and blocks. With the exception of pure earths and bitumen they all depend on the formation of interlocking crystalline structures for their stability. They are all, including earths and bitumens, intermiscible with various degrees of advantage and disadvantage in the finished product so that the total range is very wide. Burned brick in various crushed sizes has been used extensively in concrete and a specific expanded sintered clay aggregate has been available for reducing the weight and improving the thermal insulation qualities. This Leca concrete has also been used to form blocks.

12.3 Earth mortars

Earth mortars have been used very widely in historic buildings, particularly in the more arid tropical zones where excessive wetting is rare.

The principal advantage of earth as a mortar is that it provides a uniformly tolerant matrix throughout an entire wall structure, the maximum load-bearing capacity being determined by the compressive strength of the earth in a confined space. The largest historic brick structures are laid in earth or modified earth mortars. Mud mortars will normally be laid in a wet condition, the drying shrinkage being taken up by movement under the superimposed loads of the upper layers being placed, cracking therefore, being minimal. Substantial strengths can be achieved, structures several hundred feet high being both theoretically possible and having been attained in practice. Muds will normally be alluvial material containing at most 30% clays, the remaining being a silt/sand content. It is very common for mud mortars to be partially stabilized with calcitic material, sometimes augmented or replaced by gypsum-based additions. These all give sufficient additional rigidity and hence structural strength improving performance noticeably. Historically there is no clear line of differentiation between a mud mortar containing a setting material and the same setting material containing a sand and silt rich earth to make the mix. In some parts of the world where the materials are available crushed or powdered chalk or limestone may be used to provide this calcitic component. Even crushed sea-shells have been used. The 'burning' process by which carbon dioxide is driven off only to be regained from the atmosphere is thus bypassed. The calcium carbonate delivered into the mix in that form can be partially dissolved in water which is a weak carbonic acid, and in the wetting and drying processes a redeposit allows recrystallization and crystal growth to provide a weak set. Much depends upon the intimacy of the mixing process, the length of time it takes and the circumstances of setting. The clay structure is also modified by saturation of the laminar spaces with calcium cations, causing a restructuring of clay particles.

When an earth-type mortar with a substantial calcium carbonate content is found to have been used it may well be that part or all of the calcium carbonate was originally included as crushed marble, limestone or chalk, rather than its burned equivalent. Raw material with a high content of small calcium carbonate particles is still readily obtained. In the chalkland areas of Britain it is dug commercially from deposits formed by recent natural erosion and although not specifically used as mortar it is supplied as a hard fill and used under floors, drives and pathways where a solid bed of moderate strength is achieved after a period of stabilization and drying. It is also the basis of cob walling. These same materials have been used as additional aggregates in many forms of mix in which limes and cements have been the prime setting component. Pure crushed chalk, which is upwards of 98% calcium carbonate, is known commercially as whiting. This is supplied as a very fine powder for applications outside the building industry but has also been extensively employed in the formation of finishing materials particularly in

Figure 12.1 Earth mortar stabilized with gypsum in a massive brick structure. Turfan, China.

the surfacing of plasters and in paint applications. Whiting is produced by water washing and screening or sedimentation to remove larger particles followed by drying. In simulating a calcitic earth mortar it is a useful specifiable material. A content of between 1% and 3% is usually sufficient. The earth must have a low clay content – not exceeding 10% – with organic materials not in excess of 1%. The remainder should be a balance of silts and sands.

Sedimentation is the simple method of determining proportions of materials. Shaken in water with a trace of a surfactant the sample will settle out in layers which can be assessed from their depths in a glass tube, the larger sands being lowest, smaller sands above them, silts above again and clays as the topmost layer. Waterlogged organic materials will lie on the clays and others will float. The entire settlement process may take three days. Multiple samples should be taken and a hand glass is an important aid in field tests. Proportions will be estimated by depth of deposit. An alternative method is to pour off the vigorously shaken liquid after three seconds. The residue is aggregate. A further shaking and pouring off after three minutes gives the sand content. Pouring off after three hours gives the silt content and a further repeat after three days yields the clays. The samples may be dried and weighed to give a more accurate assessment of the fractions. Laboratory techniques give evidence more precise than is necessary in the field. Sieving using standard screens will provide precise schedules of content by size, and hence an indication of material. The large bulk of very fine residue will be clay, the majority of sand particles being quartz, i.e. silica.

12.4 Soft mortars

The prime function of mortar in structural brickwork and masonry walling is to provide an even transfer of superimposed load for which purpose two qualities are required: an adequate fill of the joint to provide uniform bedding and sufficient compressive stability in a constrained circumstance. At one extreme, and not examined here, is the use of vegetable fibres, grass, straw and moss. They do not have the permanence and stability required in a thick wall with earth jointing even where the outer pointing is adequate. Where a durable water-resistant pointing has been employed the combination is permanently stable while the compressive loadings are maintained. This is the general condition of the majority of historic brickwork and, therefore, much earth or stabilized earth mortar survives often unsuspected within the core. Earth, however, does have cohesive properties where it contains a clay binder and more so where other binders have been employed also. Its use as a mortar has been extensive and, in historic buildings, remains an important sector. Failure may occur by leaching out or by the removal of dry particles by insects – a process which can lead to the loss of a large part of the earth mortar in a wall which externally appears sound, particularly if it has been maintained in a well-pointed condition. In tropical circumstances termites may remove large volumes of sand and silt carrying particles considerable distances for their nests and tunnels, leaving the structure in parlous condition. Grouting with a stabilized earth, impregnated with an insecticide may be an alternative to complete rebuilding. If an earth mortar is not to be replaced (as ideally it will be) by a matching earth mortar a soft lime-rich mortar is likely to be the most appropriate answer.

12.5 Setting mortars

The function of most mortars is generally held to include, in part at least, adhesion. To perform such a function the mortar must undergo a physical change even if only by taking up water of crystallization. Its adhesion to the brick or block is then capable of being used to transfer some tensile stress to adjoining material. Thereby the whole structure becomes cemented into one integral mass. Although this effective cementation may be weak, sometimes desirably so, it does allow the masonry to behave as a coherent physical unit.

Modern science has developed a number of materials which provide cementing effects and mixed with inert material the variety of mortars obtainable has been widened. Historically the materials available in quantity

Figure 12.2 The use of gypsum mortar to construct arches without centring. A craftsman working on the Medrese of Mir-i-Arab, Samarkand is constructing an arch using the set of the mortar to hold the voussoirs.

and with reasonable ease were limited to gypsum, lime and, in the last two centuries, masonry cements. Excepting the use of gypsums in exposed cements all are varyingly intermiscible, all are gauged with sands and/or silts and clays in their general use and all have been used as mortars, renders and plasters. In terms of conservation they present a wide variety of problems, falling within generally similar categories.

12.6 Gypsum mortars

Gypsum is found in a number of forms which originally were laid down by marine deposition. Therefore, while being predominantly calcium sulphate, the deposits contain significant amounts of other materials. The beds are sometimes thick and usually startlingly white. Their structure is distinct and they often contain beds of long fibrous needle-like material. Occasionally gypsum occurs in a more compressed condition where the material is translucent – alabaster – a readily worked material which, when used for sculpture, is capable of taking on a high finish. As a powder or in small grains the material is capable of adding substantial strength to mixes of silts, sands and clays and has historically been used to stabilize earth blocks for construction, an early and important example being one of the first tombs built in Islam, the Sulaibiye at Samarra. This modest, octagonal structure reflecting the form of the Dome of the Rock in Jerusalem is built apparently of earth blocks stabilized with an aggregate from the nearby gypsum beds.

Gypsum has been the basis of a wide variety of internal finishes being easily transformed into a mortar-forming material simply by driving off the water of crystallization which is then regained in the mix, with a sudden set. Its further advantage is that on taking up the water of crystallization gypsum expands, thus avoiding shrinkage cracks. The setting is dramatically rapid if the material is used in a pure form. Its partial solubility in weakly acid water has inhibited its use in external conditions but as an internal plaster it has had a wide application throughout history. Used in conjunction with limes the advantages of both materials can be brought into balance and highly effective non-shrinking plasters can be obtained.

The rapid set of gypsum-based mortars has been used to advantage in the construction of vaults without centring. In the construction of barrel vaulting or the low profiled shallow vaults running between parallel girders (known as jack-arching) vaults may be constructed rapidly by the delivery to the mason of each successive brick tile together with the required quantity of gypsum on the point of setting. Placed immediately in position and held there momentarily the brick is secured by the setting of the mortar. By the simple process of adhesion each successive ring can be constructed and an efficient two man team can achieve a rapid rate of work. Mortars for this type of construction may contain up to 60% fine sands, significantly increasing the compressive strength. Pure gypsum may, however, be used. Traditionally the mortar is known as 'juss'. The crushing strengths of such mortars tend to be low but are generally workable in low rise structures. There is also an advantageous tolerance and slight flexibility in the material. It is sufficiently soft, however, to be penetrable by termites. It may be gauged with sand to proportions of up to 1:6. These techniques are essentially non-European.

12.7 Gypsum plasters

Gypsum is hydrous calcium sulphate – $CaSO_42(H_2O)$ – which can also occur in an anhydrous form. A coherent mass attains standard II on Mohs scale of hardness – it is, in other words, a very soft rock, occurring in localized deposits but being found widely having been laid down in marine conditions. It is the most easily prepared cementitious material. At a temperature of 150°C one water molecule is driven off from each hydrated gypsum molecule to form the hemi-hydrate $COSO_4H_2O$. This in its pure form is known as Plaster of Paris, having been produced extensively in that city, where, in Montmartre, there are substantial high quality deposits. Recombination with water is quick and it has become the prime material for low temperature casting. Its expansion on setting provides faithful retention of profiles. If the hemi-hydrate is heated a further 20° to 170°C the remaining water molecule is driven off and anhydrous gypsum is formed. This hemi-hydrate is found naturally and recombines with water only slowly, so that where it has been formed geologically it will only have produced the hydrated form where water consistently penetrates the crust. These different forms of hydration lie behind the classification of gypsum-based plasters as classes A, B, C and D. A class A plaster contains a high proportion of neat Plaster of Paris – the hemi-hydrate which provides expansion and a temperature rise on hydration. It achieves a rapid set but this can be reduced by modification with lime or whiting (powdered chalk) and it is used to make fibrous plaster mouldings where the finished work is a casting. Class B plasters are also the hemi-hydrate form which are produced with a retardant included in the mixture. The retardant traditionally used was an animal-derived protein extracted from skin, hooves, hair and similar body structures, such as horn. Class C plasters are the anhydrous type whose setting is slow, ultimately producing the hydrated form. Class D plasters are based on the anhydrous material but contain accelerators. A traditional accelerator used with class C plasters is alum and this can be a prime content of a class D plaster. Class D plasters are sometimes considered for external use in protected areas.

As a general rule the conservator is likely to turn to a class A plaster modified with lime as the nearest practical approximation to its historical equivalent. Ideally the new mix will follow the original after analysis to determine lime content, gypsum content, nature of sands, numbers of coats and nature of finishes.

A number of special gypsum-based materials have come into use – almost always for the purposes of obtaining a high degree of finish. In the Mediterranean a common further ingredient was marble dust. Products of this type, sometimes containing oils, waxes or forms of animal protein, were favoured in historic times because they proved capable of taking a high polish. In Britain similar practices were rationalized by nineteenth century patents of which the two principal were known as cements – Parian and Keen's (also Keene's) cements. The prime advantage of these materials was that they achieved a balance between the swelling properties of setting Plaster of Paris and hardening. Modifiers curbed the rapid set so that the material would remain workable for an adequate period. Traditionally hard-burned anhydrous plasters such as Keen's cement are chiefly used to provide a skimming coat over a hard rendering or backing material such as Portland cement, the two materials having closely compatible characteristics of thermal movement. Anhydrous plasters are acidic and, therefore, corrosive to metals. Consequently a combination including lime and other modifiers to produce a neutral material became popular on that ground alone. A further advantage was the balance between expansion and contraction on setting.

Since the nineteenth century Portland cement has been commonly incorporated sometimes as the principal constituent of a rendering and provided that sufficient time of drying out has been allowed before the application of the finishing coats this has proved a successful technique, popular because of the ready availability of materials. Since non-hydraulic lime plasters and renders have no initial set they are extremely sensitive during their first weeks to vibration and structural movements generally, which will damage the slowly-forming keys. These will be enhanced where they are on any form of flexible background such as timber or metal lathing. For this reason gypsum may have been used as an alternative to a hydraulic lime in historic structures. Simple chemical analysis will readily differentiate plasters containing gypsum due to the sulphate content.

Keen's cement contains alum – a generic term for double sulphates of sodium, potassium and ammonium and metallic ions such as aluminium or iron. Commonly it is $K_2Al_2(SO_4).4H_2O$ – a white crystalline powder readily soluble in water. The nearest modern plaster equivalent is a class D plaster which is compatible with historic Keen's cement.

The modifier used in Parian cement was sodium borate (borax) with potassium tartrate to produce a readily-worked finishing material with controlled set and good adhesion to hard materials such as Portland cement/sand mixes. A current compatible equivalent is a retarded hemi-hydrate group of plasters known in the building trade as 'hard wall' plasters.

12.8 Limes in mortars, renders and plasters

Hydraulic limes contain burned clays and sands in addition to calcium oxide (quicklime). These in consequence set chemically whereas calcium hydroxide (hydrated quicklime) in its pure state sets by carbonation alone.

Lime falls into two groups relating to its history following hydration (slaking) – material used as a wet putty and material which has been air-dried after hydration. The former provides the maximum possible carbonation and, where hydraulic, a set within hours. The latter will set following hydration, even under water, hence its somewhat confusing description in these days of hydraulic engineering. If a pozzuolanic additive, e.g. cement, PFA, brick dust etc., is included with the pure well-slaked lime putty it also will produce a full reaction and set by chemical means.

Setting by atmospheric reaction – carbonation – occurs best in damp and warm atmospheres and there is some evidence that variation in the hydroscopic condition of the atmosphere is an advantage – in other words the movement of moisture vapour through the masonry mass carries dissolved carbon dioxide

Figure 12.3 In Sinkiang in Western China quicklime may still be bought as 'lump-lime', here deposited by the cart load at the roadside.

more effectively. This in turn suggests that the action of setting derives from the action of dissolved carbon dioxide – a conclusion that will surprise no chemist. Mortar is wet when placed but will ultimately dry out to a level in which the pores contain between 2% and 6% of moisture by weight in normal atmospheric conditions. This amount will increase dramatically as relative humidity approaches dewpoint. Available water lodged in the microscopic interparticular spaces provides a condensation focus for airborne water vapour which will build up within a wall as well as on the surface, and condensation occurs in small voids when relative humidity is well below 100%. Expressed conversely – dewpoint is amended by reduction in pore size, being reduced by up to 20% below free space values in the microscopic voids of brick, stone, mortar and plaster. Setting will depend on access to the full depth of mortar. It will always be a matter of weeks. In large structures it may be months or even years.

Pure lime putty contracts on setting and in anything other than small volume will show superficial or microscopic cracking which reduces its moderate tensile strength further. It will, therefore, be used only in thin sections, i.e. for joining materials where a fine joint is required or where a thin coating produces the finished effect. As a much higher degree of finish is obtained internally using gypsum plasters a final superficial coating of lime putty is not always employed historically in interiors, although it may have been used externally, usually overlain by a limewash. Commercially available hydrated lime will have been prepared from calcium oxide (quicklime) ground down to a powder, slaked under a water spray, dried and powdered again, producing a slaked (hydrated) lime – an active powdered mixture of calcium hydroxide and calcium carbonate – which differs from powdered chalk and other forms of calcium carbonate in its ability to reform its crystalline structures and thereby acquire a coherence on its own.

However, mixed with cements, it can enter more complex reactions in which the combined crystalline structures modify the hard set of Portland cement in proportion to the mixture. Hydrated lime is therefore commonly used as a modifier of cements to produce a less rigid – softer – material than a cement/sand mix alone. The addition of cement to hydrated lime should not be thought of in the same terms as adding a pozzuolanic material to lime putty where the objective is to induce an initial set after which carbonation can occur. Hydrated lime should be seen as a modifier to the chemical set of the Portland cement, and that set having been achieved no further significant increase in strength is to be expected from carbonation.

It is general practice in building work to use hydrated lime as a dry powder. However an improvement in workability is obtainable by allowing hydrated lime to mature for 24 or 48 hours in a wet condition prior to being incorporated in the mix and in historic buildings workability can be important.

Being directly produced from natural materials, limes vary in content. Because even small percentages of silica, alumina and other clay minerals such as magnesia can change their setting qualities, these variations become important. Only the purest forms of chalk and limestone produce a sufficiently pure lime for the setting to be dependent purely upon carbonation by water vapour from the air. While calcium carbonate can be deposited by precipitation from solution most deposits have been derived biologically from marine life in the seas in the process of the formation of skeletons and the shells of marine creatures and the deposition, therefore, contains other marine-derived minerals to increasing extents moving down the range from a very pure chalk to a calcareous sandstone where the predominant deposit is silica. A source of the purest limes has been the burning of sea-shells obtained as a by-product of contemporary fishing industries. The purer strata of chalk have yielded limes which from the commercial point of view are pure, as does some selected limestone. Marble, which is a limestone modified (metamorphosed) under heat and pressure, is also used as a raw material. Some of the sadder losses of architectural history have occurred when worked marbles have been smashed to feed lime kilns.

Some chalks, many limestones and a number of marbles contain sufficient impurities to produce limes containing substantial percentages of silica, alumina or magnesia which cause chemical reactions of cementitious materials to take place on the addition of water. This ability to set in a wet condition defines a cement and where occurring in a lime is termed hydraulic. Despite their ability to set under water, which determined the appellation hydraulic, naturally formed cements never approach the tensile strengths of the cements manufactured by design. Where a lime is strongly hydraulic, having been burned with silicaceous impurities, it is known as eminently hydraulic. Effectively it is a weak cement with a substantial component of available lime.

The term pozzuolan describes materials which cause calcium oxide to combine chemically with silicon oxides principally. This produces a hardening or 'set' and is primarily caused by the presence of active silica which is produced as a glassy material in firing at 800–1000°C. Lava, brick, ceramics and other similar materials crushed and mixed with lime putty are effective in this way.

Slaked lime will react with silicates before carbonation. The crystalline formations achieved in this way solidify the mortar within hours of the addition of water, rather than the days or weeks otherwise taken for hydrated calcium oxide to form crystalline carbonates, which reaction will also occur with residual free lime. Ground ceramics and ground high temperature insulation (HTI) and pulverized fuel ash (PFA) are mild pozzuolans. Crushed lava and brick dust are stronger. All may be used also to modify Portland cement mortars, the active silicates in these materials reacting with the free lime formed during the setting of the cement. The advantage of these mixtures can include improved frost resistance, and in this sense they may be seen as a substitute for air entraining agents. They will also have an effect on the visual qualities of the mortar and their judicious use will be an advantage in some conservation circumstances.

Historic lime mortars were commonly mixed with brick dust and, although brick dust is now rarely obtainable from mechanized manufacture, finely crushed brick fresh from the kiln is a very good alternative. Good mixing is important but no maturing can be allowed. The process of mixing the very fine particles of lime with the larger particles of sand at an early stage allows the limes, under molecular agitation, to take up positions in the surfaces of the sand particles to a greater degree than would happen in the relatively shorter time between mixing and using in the normal process of preparing and placing mortar. Using a pure lime the resultant mix, known as 'coarse stuff', can be prepared in large batches and stored over a long period. Only pure lime can be used in this way. Any hydraulic set which takes place will be destroyed in the remixing prior to application and therefore eminently hydraulic limes should not be treated as a putty. 'Coarse stuff' is greatly improved if worked (i.e. beaten and chopped) and reworked many times. It will derive a plasticity, if the water content is kept low, akin to that of a clay-rich earth and in this condition will form an excellent mortar. Italian tradition takes this practice further. The very stiff paste achieved after slaked material has settled is worked and reworked, losing water in the process until it has the malleable characteristics of glazing putty. Used in this condition it provides the finest possible neat lime jointing and bedding with minimal shrinkage (J. Fidler, personal comment).

The admixture of pozzuolans to lime has come under increasing scrutiny, which leads to a number of conclusions, for example:

- Among the following additives to non-hydraulic lime putty and sand (1:3) – 'coarse stuff' – the most effective pozzuolan is finely pulverized brick dust from material fired at 800–900°C.
- Refractory brick powder and powder from bricks fired at high temperatures are milder but satisfactory and all forms of pulverized brick are most effective if used in greater volume than the lime putty.
- Due to its high temperature firing, Portland cement is not a good pozzuolan unless used in substantial quantities – i.e. above half the volume of putty when the cementing effect becomes dominant. In trace quantities or small volume, cements have a weakening effect on the final set.
- Brick dust in nodular or gritty form has an air entraining effect, possibly improving carbonation of putty.

While methods of kilning have varied over the centuries the principal bearing on the finished product is in the nature of the raw

Figure 12.4 The methods of making lime can be quite basic. Here at Kashan (eastern Iran) tribesmen are firing a pile of limestone using kerosene piped from a drum into the base of the heap.

material rather than the firing. Early limes were produced with no little difficulty using wood fuel, and the skill lay in providing a sufficiently large kiln, reducing or minimizing heat loss from its external surface and providing adequate flue ways so that the fire would burn intensely. A kiln built into a hillside would lose heat less rapidly than one exposed to the open air on all sides. Increasing the size of the kiln improved the ratio of internal volume to external profile so that heat generated in the centre was less easily transmitted to the outer skin. The core temperature would, therefore, be higher. This can be expressed in terms of a thermal gradient across the kiln. The greater the size of the kiln the longer the thermal transfer path and, therefore, the higher the point reached by the gradient at the centre. Effective flue ways were the spaces between the lumps of rock. The larger the lump the larger the flue way, but conversely the longer the thermal gradient between the outer surface of the rock and its core. An optimum size tended to be a lump with a cross dimension of between a quarter and half a metre. A further problem with wood as fuel was the emission of as much as 20% by weight as water vapour which would then condense initially in the rock.

To provide thorough burning of larger lumps the temperature had to be maintained at the peak temperature for some hours. Limestones which contained silica impurities would form pozzuolanic compounds as the temperature moved above 900°C and at temperatures in excess of 1000°C there was a rapid production glassy silicates and aluminates in the presence of fluxes from wood ash. The resulting nodules would be sintered. On grinding to a powder a cementitious material was produced. To produce a setting mortar this powder might be mixed with dry sand to produce a mix, which would heat up on the addition of water. Into this an animal fat such as tallow could be added. A further favoured ingredient was ash, which contained substantial quantities of potassium carbonate, K_2CO_3. Ash offers a reactive form of potassium compounds in combination with other oxides, i.e. felspars, micas, etc.

With the common introduction of coal as an article of trade the burning of lime became easier due to the higher calorific value, the reduction of water vapour from burning and the consequent ease with which high temperatures could be obtained. Mixing coal with the rock prior to burning produced an ash containing sintered clay shales as a residue. Unless the coal was very pure, which was unlikely because only the cheaper coals were used for lime burning, the resulting lime was inevitably hydraulic. During the nineteenth century lime was commonly produced in shaft or 'flare' kilns allowing continuous production.

These 'flare' kilns were now constructed on a much larger scale being typically shaped as an inverted bell, six to eight metres in diameter at the top and half these diameters below being up to eight metres deep. They would be fed continuously from above with chalk, limestone or even marble on to which fuel would be added layer by layer. Skilfully maintained, an immense hot cauldron was created, its lower surface temperatures partially insulating the hot core and allowing volatile material to be driven off at the early stages as the mass sank with the consumption of the fuel. Air entering from below produced some cooling effect and the lump lime would be drawn off at the bottom free of fuel but impregnated with ash. This residue, often containing fused shale of the coal measures, provided significant impurity of hydraulic character to determine the nature of the lime produced. Where the raw material was pure chalk it would become weakly hydraulic, where limestones with a silicaceous content had been used it might be eminently hydraulic and welcome for that reason.

Modern limeburning in rotary or shaft kilns generally relies on petroleum products for firing. These are generally free from impurities other than a minor sulphate content so that a very pure lime can be produced. Methods of production vary: generally rotary gas fired kilns working on a continuous process can produce a continual flow of fine material to be fed direct to the roller mill. At the other end of the scale in the most isolated regions tribesmen may be seen standing guard over a pile of rocks into which a thin stream of kerosene is piped for some eight to ten hours. Such primitive methods are effective although inefficient. Only the core stones will be sufficiently calcined to produce a usable lime.

Experiments based on additions of clay and other minerals to this basic process led to the production of artificial cements rapidly introduced into buildings now historic by the mid-nineteenth century. Whether the setting process is predominantly carbonation or cementitious may not greatly affect the performance of a historic structure while it is stable and weatherproof. Apart from the initial set the differences between the materials relate to the partial solubility of some compounds and the rigidity of the mortar. While water does not enter brickwork solubility is not likely to be a problem. Where it does however, aquifers will develop by dissolving out the calcium carbonate, which can reappear as stalactites in vaults, the whole material of the stalactite having been removed from the lime mortar to leave continuous voids.

In the search for purer products, methods of burning were introduced in which the combustion chamber was separate from the ore. In kilns built under Belgian, French and English patents taken out by Monsieur de Wit, the raw material would be loaded into chambers beneath which passed flues leading from a furnace to a tall chimney stack. These were known as draw kilns. To be successful this process had to be carried out in a large-scale installation on the size of a major permanent brick kiln. The temperatures necessary were not always achieved reliably and in at least one instance the structure was converted to the flare kiln system.

Lump-lime was easily transported to site to be slaked and involved no crushing. The availability of crushers/pulverizers allowed the convenience of bagged material or tanker delivery to the industry. The best lime mortars are made by the use of a lime putty for the simple reason that a period of maturation provides a very much more workable material due to the coherence in arrangement between particles which develops under water. All mortars benefit from very thorough mixing to achieve an intimate relationship between the granules but the workability or 'fattiness' of the mortar depends primarily on achieving a material in which the particles can slide readily across adjoining surfaces with a reduced mutual interference. Hence a surface bonding between lime and sand is desirable and this may be achieved by the maturation of a mix in wet conditions. It is this mix that is commonly known as 'coarse stuff'.

The particles take up arrangements achieved slowly by allowing the lime to lie for a considerable time under water. This, of course, is possible only with a pure lime: hence the effort put into its production. Two weeks' maturation is perhaps a minimum and there is no maximum. Many traditional techniques have incorporated sand at an early stage allowing the lime/sand mix to mature under

Figure 12.5 Saxon render. St James the Less, Lancing, Sussex. This original wall surface of the ninth (?) century was exposed during conservation works and remains partly accessible.

water or in a sealed condition for a long period. This ensures that slaking is totally complete and allows the plasticity of the material to improve so that its workability is at an optimum when it is placed in its final position. This is not simply a matter of the convenience of the workmen. The placing of mortar, plaster or rendering with a given amount of force will achieve much better distribution and adhesion if the lime is free flowing.

Lime putty is now readily available in 'tubs'. Its quality can be relied upon and it should be specified as tub lime from a specific manufacturer.

12.8.1 Lime renders and plasters

In addition to its use as a mortar, lime–sand mixes have been used as plasters and renders since the Roman period and were conventionally and traditionally used to improve the waterproofing of walls from the medieval period onwards. The purpose of external rendering would have been twofold: to improve the water runoff from stone and brick surfaces and to provide a ready basis for lime washing. The application of renders to early brickwork is uncertain and historically the early applications of renderings to brickwork in Britain were probably more concerned with aesthetics than weatherproofing. Where improved weatherproofing was necessary continual applications of limewash with tallow were the common answer. It was with the self-conscious architecture of the eighteenth century that there arose a wish in the owners of brick houses to emulate the greater wealth of those with stone. This was an age when the quality of brickmaking declined due to the uncontrolled addition of ash and cinders to the clays. For reasons of protection and to permit the use of low grade bricks, rendering became popular. Stucco was often struck or coursed to simulate stone.

It was an age of social pretension. Stucco was applied to brick façades to meet social aspirations as much as to protect. Pure lime was unsuitable, lime plaster was readily eroded and so there was a sustained market and therefore a search for improved cementitious combinations. Where lime stucco was used the original material will generally be found to have been 25–30% lime putty to sand very well worked with minimal water content and a similar mix will be appropriate in the repair with a mild pozzuolanic addition such as a 50% (volume of putty) white or ordinary Portland cement. Portland cement in lesser volumes than this tends to produce a weaker mix than lime alone.

12.8.2 Colour matching

In historic buildings the original material will always be retained as far as possible and in

Figure 12.6 Brunswick Terrace, Hove: stucco on brickwork. Preformed components removed for repair and refixing (above). The reassembled capital before finishing and painting (right).

consequence colour matching will be important. If brick dust was originally used, recently fired, crushed brick should be used as an additive. The ultimate objective must be that the new material blends increasingly with the existing as weathering takes place. Natural colouring materials of the original type should be used in preference to proprietary materials and careful analysis of the original components is usually sufficient to suggest the right balance and source of materials. This subject is considered further below but in the context of limes it must be remembered that as the material dries its whiteness intensifies.

Modern pigmentation offers carefully evaluated ranges of light fast materials in convenient form. Dyes of any type must be avoided – synthetics particularly – but mineral (earth) colours appropriate to the original materials may well be derived from the stocks of colour manufacturers. There is always a tendency to fading and to the development of a 'bloom' in mortars coloured with added pigments. Rates of drying can be critical in this regard and may require control by protection with wet hessian or similar safeguard against rapid drying.

Internal renderings and plasters were traditionally carried out with lime, mildly hydraulic limes being preferred because of their greater certainty of setting on vertical surfaces and on ceilings where the material was in suspension. For this reason gypsum was often added to improve the setting properties. Such surfaces were usually coloured subsequently and colour matching is, therefore, less likely to be important.

12.8.3 Rendering techniques

Where lime mortars have been used as external renders they have not generally been reinforced with fibrous material. For rendering on brickwork reliance has been placed on adequate keying provided by raking out joints and this physical integration provides the best assurance of adhesion. The surface area of each brick is not sufficient to demand enhanced cohesion provided that coats do not exceed the traditional thicknesses of 10 mm with a maximum of 15 mm.

Base renderings are commonly scored while still wet. This action provides an additional mechanical key to subsequent coats and also allows any movement which may occur to take place prior to application of the finish. Finishing coats in lime have traditionally employed richer mixes and finer grades of sand than base coats which often relied on coarse, sharp sands balanced by finer material. While many external coatings have contained organic additives aimed at improving cohesion and weatherproofing, none of these has proved significantly successful in the long term. Oils – usually linseed – waxes such as beeswax and animal fats such as tallow have been used historically, the latter having the most consistent record of success.

Animal glues have also been used and the mixing normally employs the heat given off during the exothermic slaking reaction to allow their dispersion in the medium. Oils and animal fats, being subject to bacterial and fungal attack, cannot be regarded as permanent constituents of historic mortars in the longer term. Their life is extended considerably where the render remains dry, as happens where surfaces are painted and the paintwork is maintained. Finishes with historic additives of this type have often proved to be a very good base for the reception of traditional paints, usually based on linseed oil and white lead, and where the formula has proved successful in this way the material and method of application should be retained in conservation, even where excessive thicknesses of application demand the removal of earlier coats. (The use of white lead is controlled in many countries due to its toxicity.)

Lime mortars used internally for plastering have been applied extensively to wooden laths. Split hardwood was the commonest timber for this purpose until the eighteenth century when, in Britain, the availability of oak diminished considerably and softwood became more freely available. Initially the timber was cleft into broad slabs, using wedges, and then converted to thin strips from the edge of each slab by the use of a cleaver. The availability of power sawing (initially water power) produced the straight edged cleft lath, common in the eighteenth century superseded in the nineteenth century by laths sawn in both directions. The advantage of the cleft lath is the continuity of its grain and the disadvantage the irregularity on edges and surface which affect the thickness of the applied base coat and the key obtained.

Much eighteenth and nineteenth century wall plastering in high quality work was applied to brickwork as a separate skin on lath and batten. Plaster on lath depends for its integrity upon the nails which secure the laths, on the condition of the laths themselves and on the extent to which an effective key has been obtained. The key arises out of the squeezing through a gap of the paste-like mortar. Having been extruded through the gap by the force of the workman's tool, weight and surface tension will tend to cohere the extruded section into a more bulbous shape which when set resists being pulled back. Internal plasters were given additional cohesive or tensile strength by the inclusion of ox or horse hair which had been degreased. Wool was not used. Hydraulic limes or limes gauged with gypsum were traditionally preferred because of their set. Vibration at an early stage, frost action, oily or greasy laths or laths which were too dry could all cause initial failure in bonding although ultimate collapse might be delayed for some long time. A particularly susceptible zone was the neck of each key. If the laths have been too closely spaced the extruded material will have been made into so thin a ribbon as it passes through the gap (if it passes at all) that it is liable to fracture before carbonation has been achieved partly due to premature drying of the over-thin plaster and also due to vibration damage to setting crystals. Examination of failed plasters may reveal weaknesses of this sort which were part of the original construction quite apart from subsequent deterioration of nails and laths. Early vibration can include wind damage to which ceilings under roof spaces are particularly vulnerable, the joists often being light and the interconnection of timbers transmitting windshake into them. A ceiling thus weakened might survive some time before failing.

In the eighteenth and nineteenth centuries lathing and plastering used for lining out walls in major buildings – an early form of dry lining – had the advantage of conserving heat and minimizing condensation superficially although often inducing condensation within the unventilated cavity with consequent problems of dry rot. Warm internal air produces a high vapour pressure in the cavity with consequent occasional condensation on the timber and the cold brickwork of external walls. If a vapour barrier cannot be incorporated in reconstruction of this type of work, vacuum-impregnated timber on an insulating lining to the brickwork is a reliable alternative.

12.9 Modern plasters

Extensive development has taken place nationally and internationally in the design of plaster mixes with the introduction of setting agents, modifiers and lightweight aggregates such as vermiculite and perlite. While these materials are generally not applicable to historic buildings there may be a case for using them in repairs, particularly where there are special conditions relating to weight or adhesion. Manufacturer's technical information will be crucial to their satisfactory use and it should not be assumed that traditional materials can be added with impunity. In particular it should be noted that (some) anhydrous plasters as mixed and sold in proprietary form react adversely with lime and in all cases where an anhydrous plaster is to be used in combination its suitability should be verified and samples should be run with an adequate time interval to allow their maturation.

In general terms there is no reason to abandon traditional materials in repairs to historic buildings unless special qualities are required; in particular anti-condensation, fibrous strength, lightness and acoustic properties may be important. Further it may be a conscious decision that the replacement material should be distinctive and for that reason a sympathetic modern plaster may be chosen.

12.10 Pointing

12.10.1 The heights of accomplishment

Lime putty has served one particular special function in brickwork – the demarcation of bonding by emphasizing the mortar joint. This is not, however, a peculiarly European technique. In medieval Persia brick vaults were plastered and the plaster itself was then incised in complex patterns to simulate brickwork more elaborate than the vaulting. Decorative brick plugs were inserted and simulated joints were struck out in pure lime mortar. This work achieved at the height of Persian brick technology illustrates the importance of brick patterning in surface design. A feature evolved of necessity had become a design concept in its own right, used to express both pattern and directional emphasis. A lime/gypsum/sand mix would provide the render coats and a pure lime or gypsum the contrasting jointing. The 'brick' surface would then be coloured. 'In the Gunbad-i-Surkh (Seljuk, AD 1146, Azerbaijan) we have perhaps the most beautiful example of brickwork known. The angle columns are composed of ten shapes of cut brick involving at least eight moulds perfectly adapted to the curving profile of the colonette' (A. Upham Pope 144). 'Despite its great beauty, however, exposed brick was thought to be a bit harsh. Thus imitation brick bonding was created in order to enhance true brick, even while retaining its genuine decorative quality. This was accomplished by covering a raw brick surface with smooth hard plaster, which was then marked out in interesting brick patterns. Brick end plugs of great beauty and variety were then inserted' (ibid, p. 145). Examples of this technique survive in the Friday Mosque, Ardistan (*c.*1180) and the mausoleum of Oljeitu, at Sultaniyeh (1305–13).

'Early in the eleventh century, in addition to the varieties of flush bonding various sizes of brick, new bondings and joints were devised. The joints were treated in several ways: a deep raked line between the ends created a dark vertical shadow which was in sharp contrast to the thin light horizontal mark between the outer edges of the brick ... Other schemes were developed and the surface was further enriched by bricks both sunk and set forward, whole courses being built out by projecting overlapping units which formed richly shadowed zones ...' (ibid. 146). Not only was this level of achievement far ahead of European; it has hardly been equalled since. The description does justice to this extraordinary medieval craftsmanship.

Pointing has always been a determinant of the visual quality of the brickwork and in Europe, particularly since the sixteenth century, the texture and colour of the finished wall surface has been important to building owners. The natural colour of mortar is determined by the varying yellows and browns of the sands used, the basic white of lime, or the buffs and greys of cements, reds derived from brick dust or blacks derived from ash in urban situations or mining areas. The most painstaking European techniques have involved extremely thin beds of lime putty or the insertion of thin ribbons of putty into a dark pointing.

12.10.2 Finishing and pointing techniques

Where mortar is brought to the surface of the brick as part of the process of laying, the treatment of the surface is known as finishing. Where the mortar is raked or held back and a final bedding of mortar is introduced the work is known as pointing. The terms are frequently confused or used loosely. The earliest technique for which there is good evidence was simply to bring the mortar to the surface of the brick. All early bricks were irregular and the mortar joint was irregular in consequence. Flush surfaces were, therefore, consolidated by pressing inwards with a tool. From such variations in finish evolved the techniques of pointing. By using a trowel to provide a slightly angled surface a clean finish could be obtained and this, when it leaves an overhang on the lower side of the upper brick, is known as weather-struck.

The term 'struck' describes a raked surface set back below and rising to meet the outer edge of the brick. This finish has been less used historically but it has been common to strike symmetrically leaving an arris on the centre line and giving an even-handed finish

Figure 12.7 Early nineteenth century gauged brickwork. The very fine joints are achieved with lime putty between bricks sanded, rubbed or sawn to shape. The original marks of abrasion can be seen crossing the joints, demonstrating that the facing down was carried out 'in situ'.

to the perpend at top and bottom. This is known as double-struck. It was popular in the early centuries of English bricklaying and common in 'Elizabethan' England.

The differentiation between the upper and lower surfaces in sunlight and to a lesser extent in normal lighting conditions provided a regular alignment which did much to overcome the irregularity of hand-made brick – a quality not appreciated in the periods prior to the availability of machine-made bricks. In the sixteenth century, to lay bricks with carefully double-struck joints ruled precisely in straight lines became a distinguishing aspect of high quality work. Another technique aimed at achieving the same effect of precision was known as 'penny-struck', a penny being used to score a line down the centre of the mortar joint. The converse effect was rib jointing in

Figure 12.8 A rough brick clumsily repointed using a mortar lacking the right texture and brought forward of the face-line of the bricks.

which a tool with an indentation centrally could be drawn down the line of the mortar leaving a raised central ribbon, usually rectangular rather than the half-round profile now commonly presented as reversed bucket handle. Necessarily the tool would be only as wide as the closest spacing of the adjoining bricks and, therefore, for most of its length it would leave an untooled zone of mortar either side of its track. This technique was the precursor of tuck pointing in which a central white band contrasted with the darker colour of its background. In less sophisticated work mortar might be consolidated by stippling with a brush, being finally pressed with cloth or wiped with sacking – a technique often known as 'bagged' and apparently of some antiquity.

Judgements on historical pointing may be made on the basis of examples surviving in protected areas not originally intended to be seen, although areas of special treatment of joints may not be truthfully represented by surviving material. Islamic brickwork generally follows Roman and Byzantine precedent in having been finished with flush or slightly recessed joints, except for examples of deliberate decoration with strapwork. In modern work this raised band of pointing is known as

Figure 12.9 Careful slightly raised pointing on a wall containing a wide span flat arch (Netherlands).

Figure 12.10 A late eighteenth century place brick with a 'penny-struck' pointing.

a ribbon. Ribbon pointing is not normally to be recommended.

An important decorative technique employed locally and often under the impetus of fashion is known variously as galletting and garretting (also garneting in Surrey). This was a technique with important structural advantages in irregular masonry, particularly flintwork from which it is derived. The trimming of flints for use in walling inevitably produces a large number of fragments and flints, being irregular, cause the wall mason some difficulty in ensuring that they are fully backed by mortar where they are set, as so often, against brickwork. Voids found at the back of faced flintwork are fewer and smaller where the mortar has been made more dense by galleting. The introduction of small chips of flint within the joints provided a greater fill of the surface where rounded irregularities occurred. The act of pushing flints into the joints itself squeezed the mortar further back behind and under the flints. Some gallets actually physically supported the flints before the mortar had hardened making the irregular wall more stable and ultimately the projecting gallets acted as drips and sheds to hold rainwater off the mortar surface. Galletting was, therefore, a sensible and useful technique in building flint and other irregular walling. Since much flintwork, particularly from the seventeenth century onwards, was backed by brick and often incorporated bricks as horizontal coursing (and quoins), the technique was carried into adjacent brickwork. It was used also in stonework and sometimes in brickwork itself. Small nodules of hard ironstone or firestone were incorporated in brick and stone jointing in some areas, giving the joint a very strong emphasis.

The strong enunciation of joints by the use of a fine strip of white lime putty was perhaps the ultimate refinement in pointing. In English practice this technique evolved from the use of gauged brickwork in which carefully and perfectly formed bricks laid with fine joints of 1/8" (3 mm or 4 mm thickness) in a neat lime putty or a rich putty incorporating some fine sand would be rubbed down with stone to provide an even and true surface aiming for mathematical perfection in contrast to the vagaries of hand-made textures. Tuck pointing was an attempt to simulate this precision by providing ribbon joints in a pure white pointing contrasting with the darker mortar matching the brick in colour in which the joint would be formed. In essence it was a somewhat laborious method of simulating very high quality brickwork with poorly shaped bricks although ultimately it became an end in its own right.

Bricks laid with bed or header joints of 10–15 mm in width would by this method be given a visible pointing only 3 mm wide, equivalent of the minimum bed achievable in laying with a very pure putty. To achieve this effect it would be necessary to colour the mortar to match the brick. It was particularly popular where a very dark brick was used in urban areas. The wide joint would be laid or filled flush with the surface of the brick. Sometimes the entire mortar bed would be laid in coloured material but normally a coloured pointing would be used, the surface being rich in dust from the bricks, trowelled or stippled in by use of techniques in which the tool or brush would be applied with

Figure 12.11 Rough brickwork carefully repointed with a textured mortar held back just behind the surface and 'bagged', being finally washed and brushed down while still soft.

Figure 12.12 A flint-facing removed to permit structural alterations with the removal of decayed lintels and plates and alterations to structural brickwork. Before further work began raking shores replaced the two-stage temporary support. The brick backing and galletted flint facing are clearly seen.

pressure at right angles to the surface. The craftsman then, using line and rule, would cut a precise channel 3–5 mm deep and 3 mm wide in the mortar. This would be done within 30–60 minutes of laying the pointing mortar with a tool carrying a hooked end and known as a Frenchman. This tool being run along a rule leaves a channel in the dark mortar into which a thin ribbon of white mortar is offered by sliding slices of putty across a slate (or similar) surface into the groove. The ribbon would have been cut from a mass of putty or putty gauged with silver sand rolled out like pastry to the thickness of about 6 mm. The 3 mm strips cut from this slab with a putty knife would be carefully introduced into the joint cut in the firm mortar leaving a protruding tongue of lime putty to harden off. The expense of such skilled and laborious operations reserved it for use on buildings of architectural pretension, mainly in the capital. Skilled specialists repeat this work but a good rate of working by a craftsman with adequate back-up is unlikely to exceed 2 m^2 per day, so the costs are commensurately high. To raise a band of pointing, the technique known as ribbon pointing has a very limited historical provenance, but was sometimes used also to achieve a mathematically precise effect in urban building.

Conservation judgements in pointing focus principally on the mix of mortar to be used and the style of finish. In cutting out, and after centuries of weathering, the bricks may themselves be damaged. In special work such as tuck pointing the background which will receive the final tuck must be chosen with special care to avoid the effect of surrounding each brick with a halo or dark rim. The mortar must be brought up to the face of the bricks and the pointing will, therefore, overlay damaged or irregular areas. While colour is important the crucial factor is the tone value which may vary according to light. Careful running of samples is important to achieve satisfactory surface, satisfactory colour and a tone value which allows the mortar to meld imperceptibly with the brick background. Artificial pigments available for the colouring of mortar are not necessarily light-fast and may be affected by the type of mix employed. Difficulties may arise where a mortar bed toned dark to match sooted brickwork is then exposed when the brick is cleaned (Fig. 12.4).

Since judgements must be made on the basis of sufficiently lengthy experience it is best to rely on samples known to have been in place, and therefore tested, over some years rather than those without a proven history. The type of cement employed can affect the performance of some colouring agents. It is important that the batches are consistent, in the materials supplied, the mixes made up and in the protection given during the setting period. The sample mix must be made with carefully recorded volumes of dry material from the same source and colour-matched.

Historically the mortars used for pointing may have been relatively weak – weaker than the brick and the mortar used in the walling. Sharp sand of the appropriate colour together with brick dust, white or grey cement or ground coal ash and natural pigments such as umber have been the typical constituents of conservation work and can match quite closely the original material. The sand will be fine but sharp as the mortar has to be grooved and brushed off when it is semi-dry. It is important to judge the initial colour so that it weathers towards the existing work. It will, therefore, be different (and usually lighter) initially.

12.11 Cements

The inventive turn of mind apparent in the late eighteenth century in engineering, agriculture, shipbuilding, navigation, military techniques and the science that became known as civil engineering led to the search for new cements and aimed at improvements, many being patented in America, France and England. The most successful were silicaceous limes with contents of alumina and magnesia derived by burning suitable natural rocks. One such was Roman cement, first produced by Joseph Parker towards the end of the eighteenth century by calcining and crushing nodules (septaria) dredged from the Thames estuary. These nodules are primarily iron-rich clay cemented by calcium carbonate. Calcining at about 1000°C and subsequent milling produced this and other imitative cements of a brown colour and used often in a proportion of 1:1 with sand to produce a very hard durable external rendering for brickwork. It provided a passable simulation of ashlar. The material was sometimes used with a minimal addition of sand in the production of mouldings and decorative castings.

At the beginning of the nineteenth century a number of similar materials were produced by adding chalk, limestone, clinker and ground pottery or glass to clay as a raw material, a patent of 1811 being followed by that of Leeds bricklayer W. Asdin, who took out his famous patent in 1824 for the burning of limestone, lime and clay in shaft kilns. Independently Sir William Pasley produced a similar material at Chatham in 1830. The product was named Portland cement.

Continental manufacturers soon followed with improvements in grinding before and after firing. With the raising of firing temperatures to something between 1200°C and 1400°C more powerful cements were made. The finished product generally contains oxides in the following proportion: calcium 60–65%, silicon 20–24%, aluminium 6–10%, magnesium 1%, iron up to 3.5%. Simplistically the setting process consists in the relatively rapid formation of tricalcium silicates and tricalcium aluminates coupled with calcium silicate. The silicates and aluminates react with water to form hydrated lime and a gel of silicates and aluminates from which some calcium has been withdrawn. This is available as free lime and in reaction with water the densely fibrous calcium silicate hydrates form interlocking matrices of needle-like crystals. The calcium silicate hydrates and calcium aluminate hydrates are complex families of crystalline materials and ranges of similar compounds including iron and fluorine increase the complexity of the crystal structures. A number of other inorganic compounds are formed in the setting process including some which are water-soluble, particularly sodium silicates, sodium and calcium sulphates and sodium and calcium hydroxides.

As the setting process continues these compounds can migrate into the basic crystal structures to take up permanent positions but where the content of these salts is high they can move into adjoining materials and be carried through to external faces to become unsightly and occasionally damaging forms of efflorescence. It follows that the nature of the raw materials used in the manufacture of the initial cement will significantly affect the ultimate material. The greatest care should be taken in the selection of the type of any cement employed in historic building work and where the use of cement mortars is appropriate the possibility of soluble salts being leached into ancient brickwork should be considered and safeguards put in place. This is particularly true of injections of cement-based grouts and concealed structural elements of reinforced concrete. Where, as is common, the historic building contains original cements, their careful evaluation will allow matching densities and strengths to be achieved with modern materials.

Sulphate-resisting cements are those which are low in calcium aluminates (5% as opposed to 11% upwards in Portland cement), giving the cement a greater ability to take up soluble salts and therefore more resistance to sulphate attack. The breakdown of cements by the action of sulphates is due to the growth of compounds deriving from calcium aluminate.

Some ordinary cements have been gauged with gypsum to amend their setting reactions with the consequent availability of further sulphur. These should be avoided on historic work. The commonest cement formulation, Portland cement, has a distinctive hard grey colour and a hard set with considerable tensile strength. Both these qualities run counter to the fabric of many historic buildings and it should be a general rule that Portland cement should be used only in those relatively modern historic buildings where it was initially employed and should always be introduced deliberately after careful consideration. While its colour can be modified within limits by appropriate sands and lime, some ranges of mortar colours are not achievable with ordinary Portland cement. Two other cements now available increase the colour range. Of these the most useful is white cement produced from low iron kaolin (china clay) rather than high iron content lias or shales and marls. Gas or oil are used as fuels because of their low ash content. The resultant mortar can simulate lime in colour and although less strong than ordinary Portland cement mortar is very considerably stronger than lime mortar. A tendency to surface crazing can be minimized by the use of water reducing agents, containing sodium gluconate or naphto-sulphanates, but the requirement for such strong materials is exceptional in conservation work.

The other modern product is high alumina cement. Historically some very dark cements were made by grinding slag into the product and dark mortars were made by adding well ground (and sometimes not so well ground) ash at the risk of increasing sulphate contents. High alumina cement produced for engineering purposes can simulate these materials and provide a dark matrix. Not being a historical material, high alumina cement should be used only in the rarest of circumstances where its special properties are called for and cannot otherwise be met. It must not be used in conjunction with lime. While it has the property of setting rapidly this is rarely important and the darker colours can be achieved by other means. High alumina cement has a history of structural failure and is not in general use.

A number of modified forms of Portland cement are marketed as masonry cements with particular relevance to use in sandstones where lime may have a deleterious effect. They contain inert fillers and work freely because they contain plasticizers. British masonry cements contain crushed limestone up to 40% by volume. They are a good basis for the production of renders, but have no special relevance to brickwork.

12.12 Pigmentation

Limes, sands and cements all provide permanent coloration. Sands vary from near white (silver sand) to dark red browns with a wide range of ochres, yellows and umbers between.

In addition brick dust can provide a series of stronger oranges and reds right down to carmines and purples. Dark grits and crushed granite are also available. The range of 'natural' materials is, therefore, very wide and it is hard to imagine a historical mortar or rendering whose basic colour cannot be approached very closely using these materials without the employment of artificial pigments. Where pigments are unavoidable they should be selected from ranges of pure unadulterated inorganic materials avoiding any proprietary blends which may contain impermanent stains.

12.13 Waterproofers and antifreeze additives

Mortars are sometimes gauged with proprietary materials to protect work from frost during the setting period. The material primarily used for this purpose is calcium chloride which works not so much by reducing the freezing temperature of water as by accelerating the exothermic setting reaction liberating heat more rapidly within the core of the mortar and achieving a set before the heat is lost. To introduce a readily soluble reactive salt such as calcium chloride into historic brickwork is an unacceptable risk and must be classed as bad practice.

A number of proprietary mixes available as mortar or as additions to mortar can successfully inhibit the movement of liquid water and water vapour through cement mortars to the point where mortars and renders are convincingly 'waterproof'. While there can be applications (usually concealed) where these materials can be used effectively and justifiably they are relatively rare. Specific applications might include vulnerable elements such as concrete lintels or thresholds introduced in place of decayed timber where an original fault in design should not be repeated even at the expense of historical accuracy. Calcium stearate is an additive used for this purpose. If the use is justified, commercial materials should be used according to manufacturer's instructions. There is no point in attempting a purpose-made mix.

High density is not necessarily advantageous in the production of a mortar or render which may be subject to freeze–thaw cycles. Volumetrically an optimum porosity of about 15% is sought for mortars subject to freezing. Higher porosity introduces greater weakness and lower porosity increases rigidity at the expense of an elasticity and void ratio desirable under conditions of freezing stresses. A critical factor is pore size, the resistance to moisture movement increasing exponentially with the reduction in pore size in accordance with Poiseuille's Law. To achieve the appropriate porosity air-entraining agents are sometimes employed with Portland cement. Calcium ligno-sulphate (together with surfactants) produces mixes which have a high workability, thereby reducing the need for an increased water content and at the same time containing pore structures sufficiently large to give the required flexibility of movement. Porous aggregates will assist. Careful mixing and workmanship affect the uniformity and quality of the applications. Simplistically the larger pores introduced by air entrainment give space for ice crystal growth without exerting stress on the surrounding mortar.

12.14 Instant set and additives

Pozzuolans are added to accelerate a cementicious set. They are all active silicates, the conventional additives being silicaceous materials on which a layer of active silica has been left as the result of firing at about 1000–1200°C. The extreme is a solution of sodium silicate of optimum molar ratio injected and instantly mixed into a stream of cement mortar or grout. The set is virtually instant – it can be as little as 0.5 seconds. The projected mortar is delivered from a gun, which if held more than 50 mm from the delivery point can produce a 'bounce-back' from the set mortar. Such mortars have engineering applications and are for specialist use with mortar guns. In historic structures their prime use is for repointing where water pressure or running water makes the introduction of a normal mortar impossible, as in river walls, tunnels and vaults. The finished surface is irregular but can be trimmed mechanically within the first few moments of setting. If it is held back or immediately scraped back from the brickwork an appropriate pointing can be introduced while it goes off. In circumstances where running water is the problem it is very important to ensure that hydrostatic pressure is relieved by permanent drainage. It is one thing to be faced with continually running or moving water; another to deal with many tons of water pressure on the back of a brick skin. Where water movement is slow an instant fill of joints may produce an apparently satisfactory restitution of historic brickwork, but ground water pressures building up over a long period or after high precipitation can cause sustained water pressures behind embankments and brick-facing walls which can lead ultimately to collapse.

A further and entirely different mortar can be made using two part epoxy resins. Commercially available for many specialized aspects of filling, hardening and high performance they can have important but very limited areas of use in conserving brickwork. They omit water entirely. The setting resins provide a matrix which fills the voids between the grains of carefully prepared sands, adhering powerfully to the faces and providing a strong bond with adjoining porous materials. Curing may be air-dependent and temperature control during setting is important. Durable, tough and of high tensile strength these mortars may be invaluable in small scale application, allowing special repairs to be undertaken in ways previously impossible.

12.15 Sands

Sands are predominantly quartz particles, varying in characteristic by size and shape. The availability of other minerals, particularly iron, affects the coloration. The quality of the resultant mortars depends largely upon the grading or balance of particle sizes and particle shape. The shape of sand particles is a consequence of their natural fracturing characteristics and wear in erosion. Maximum/minimum dimension relationships tend to be in the lower ratios, that is to say 1:3 up to 1:5. In other words, sand particles tend to be rounded rather than long and fibrous or flat and plate-like. The smaller grades of sand, however, may contain micaceous particles whose characteristics are normally flat. The coarser sands may contain non-quartzitic minerals but to a large extent these will have been separated out in transport and deposition. Quartz particles tend to be multifaceted sometimes having sharp edges derived during the processes of fracturing under impact in air- or water-borne situations or in conditions where rocks and crystals are stressed. Sand particles which have been subject to long exposure to mechanical movement by wind and water will have suffered abrasion which reduces the sharp-edged characteristics and smooths out concoidal fractures on the faces. The friction is usually quartz to quartz. Being of equal hardness the surfaces do not scratch each other but sharp edges striking against flat areas tend to be abraded. Ultimately the particles become round-edged. Sands with a high content of rounded particles are described as soft sands and the more angular faced materials are known as sharp sands. Particle sizes will be in ranges of upwards of 0.02 mm in diameter and an upper limit of 2 mm is generally accepted as a reasonable maximum in construction mortars, although grits in the range 2–5 mm can also be very useful in the formulation of matching mortars. Sharp sands tend to provide a more rigid mortar. In addition to the surface characteristics of the particles, grading (that is to say the proportionate number of particles distributed across the size range), and freedom from adulterants are important.

Simple visual inspection, particularly with a magnifying glass, gives a very clear idea of particle gradation and in addition allows differentiation of colour. Washing with water and filtering out the finer particles is a quick indicator of the extent of the silt and clay content. Clean, well-washed sands with a balance of particle sizes across the range and a sufficient proportion of sharp particles provides a well-balanced mortar. The objective in preparing a mix will be to fill but not overfill the voids between dry sand particles. If an overfill occurs the binding lime or cement will be able to contract on drying and shrinkage occurs. If the voids are underfilled air remains between the particles of sand giving a weaker mortar. Therefore the strongest mortar is made by choosing sands which produce the greatest possible triangulation of support between particles (i.e. a wide gradation of sizes of hard-edged grains) and the thorough filling of all the remaining voids by provision of exactly the right volume of mortar containing a minimum of free liquid which will evaporate out. In a well-balanced gritty compacted sand about a quarter of the volume is void; hence the general rule, one of lime to three of sand – a total of four. A simple test to verify the precise figure for any particular sand appears in Section 14.4.

Blending may be necessary to achieve a good balance and sands are often supplied on the basis of specific characteristics as to size and angularity. Commercial sands will normally be well washed and free of clays, silt and salts but careful verification of all characteristics is essential. Washing a sample will give an immediate indication of the presence of particles below the specified size range.

Careful analysis of sand particles and other components of mortar can provide worthwhile information about the history of the building. It must be remembered, however, that mortars vary more readily than other materials due to the supply of sands from different sources. Even from a single source the proportionate characteristics of sands may vary quite widely. Most deposits of sand have been accumulated by water or airborne deposition and the fluctuations in currents and air streams can vary the deposits widely over close geographical limits. The colour of quartz sands is usually a consequence of their iron oxide content. Often the iron oxide is concentrated and cemented on to the grain surface. In choice of sand the workability will be greater with the inclusion of soft sands. Fine work will demand exclusion of the larger particle ranges. A balance of grain size should always be sought.

A number of authorities maintain registers of available sources of sand. In the United Kingdom, English Heritage has a 'library' of samples from over 500 sources, analysed as to colour, sharpness, size, gradation and origin. This data is published in a directory which is accessible to specifiers and conservators. The directory carries a database which will make it possible to select sands to match original material used for mortars. For this purpose the original mortar must be analysed and its characteristics logged in a manner to match the database.

Colour is described using Munsell colour charts. Sizing is by sieving and grading using BS sieves and interpreted as a numerical curve. Mineral content is defined by reaction to polarized light and sharpness is expressed on a comparative granular basis. The presence of feldspars, iron oxides and clays is logged proportionately. The suitability of each sand type for use is recorded.

With this information, appropriate mixtures can be specified.

12.16 Analysis of historic mortars

Mortar analysis is one direct indication of construction history and the conclusions drawn from mortar analysis must always be carefully related to other evidence. Results must be consistent. An isolated example can be misleading and an averaging of three or five results is essential before conclusions can be drawn. Chemical analysis of mortar requires expert interpretation to determine the original constituents since some of the marker elements such as calcium oxide can derive from different sources. An analysis which provides information as to the extent of calcium oxide present fails to identify to what extent this derives from a cement or from lime and from chalk or limestone in the aggregate. Ground up chalk in the original mix will provide a high calcium oxide content in the analysis which must not be taken necessarily to confirm that a high content of putty was originally used.

The first stage of analysis will look at building techniques. Before the eighteenth century it was unusual for a building to be separately pointed, the mortar usually being finished off on the surface as the work proceeded. Pointing can, therefore, be taken as an indication of later repair, or construction within the past 250 years. Whether the pointing is original may well be determined by the discovery, in protected circumstances, of original work untouched by later hands. Alterations often helpfully seal off earlier work encapsulating it for later examination.

The nature of the finishing tends also to be characteristic of culture and period. English Elizabethan brickwork for instance tends to be double struck and the original pointing may survive in protected or later obscured positions on a building with a totally different later form of pointing as a result of repairs. Other techniques are not susceptible of broad generalizations since their pointing may vary from area to area or even builder to builder as well as from age to age. The pattern of local knowledge built up by historians and conservators is the essential background to conclusions based on techniques of this type. Some types of pointing, however, are patently modern and may have been the result of a particular fashion connected with a specific period. Ribbon pointing is such a technique, applicable to historic buildings in only the most limited of circumstances where impeccable evidence of original form is available. In some instances the techniques of repair have, themselves, become historic, just as an addition to the building may form an intrinsic part of its history. There may then be an argument for conservation in a manner which follows the evidence of preceding repair.

The nature of material used in both pointing and mortar can be important evidence in itself. Much repair of jointing has been done because of need: conversely some has been done unnecessarily for visual effect. Eroding lime mortar joints are sources of leakages into the wall and building owners have usually felt impelled to spend money on repointing because they wish to improve the structural and weathering capacities. Money is often saved by repointing only those parts most affected and for reasons of conservation the same policies would apply in current work. Where part of the structure is in serviceable condition there is a presumption in favour of its retention. Any proposal which involves repointing where the original material is still generally sound must be viewed with great suspicion. Because of the patchy nature of some repointing a façade may well carry evidence of the original finish to the mortar joints and elsewhere show evidence of a series of subsequent repairs. These in turn may be related to other aspects of repair and provide a dating sequence for the history of the building.

The nature of materials can also be helpful. Knowledge of the introduction of particular types of cement or their availability from local works will give *termini post quem*, delimiting the possible periods of construction and repair. This evidence may interact with other evidence of subsequent rendering or painting and this, in turn, with the introduction of other elements such as joinery.

Mortar analysis may be carried to the point of investigation of materials other than the cements.

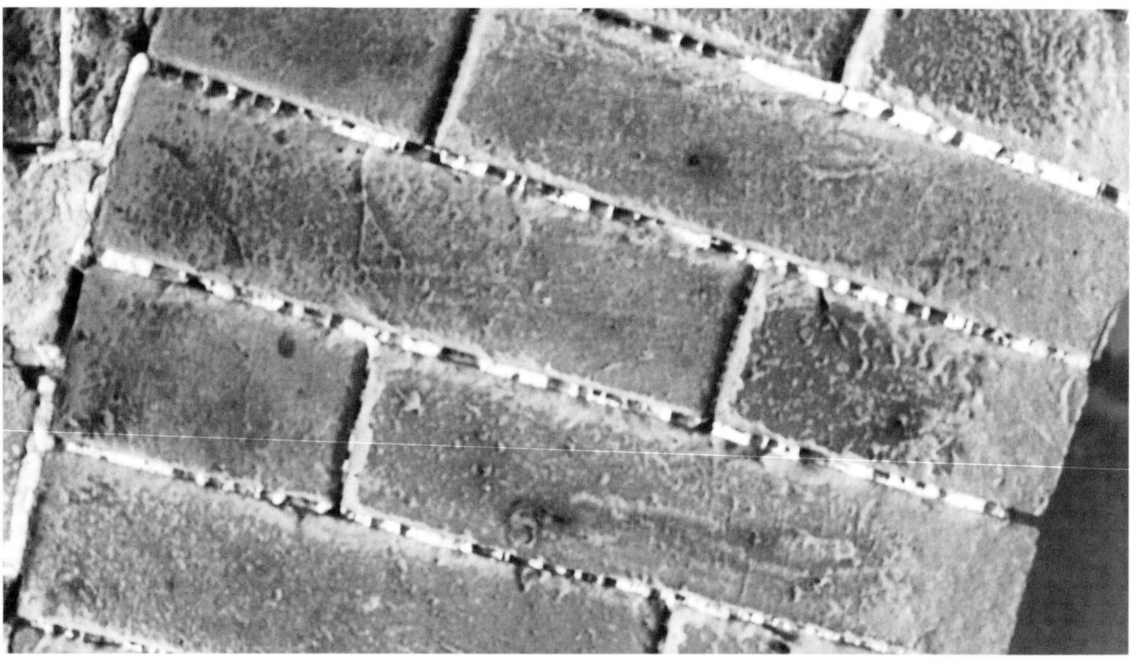

Figure 12.13 Pure lime mortar in an arch displaying 'pitting' after a century of exposure. Note the engrained on the surfaces after solution and redeposition, significantly altering the colour of the brick.

Figure 12.14 Lime mortar failing on the inner-surface of a high wall in a marine environment due to salts being carried through the mortar beds to evaporate internally. The mortar has expanded and powdered to a depth of 30 mm. This process will continue until the stability of the wall is threatened.

Cements may be dissolved out by the use of hydrochloric acid solutions, leaving a residue of brick dust, sands and smaller particles. These may be examined under a low-powered ocular microscope or under a magnifying glass with rear and oblique illumination. The crushed samples from one source may be examined with and without the soluble component. Sieve grading of the acid separated particles is useful and examination of the individual sieve fractions can identify the source of the sands originally used to a known stratum or a local pit.

Careful selection of similar sized samples will allow particle counts to be made in which averaging will give reliable indications of the distribution of particle size and particle nature. Sharp and rounded sands can be differentiated, coloured sands and fired particles likewise and a typical fingerprint can be established which will be helpful in determining which mortars come from an identical or similar batch or mixing. While a similar 'fingerprint' is very good evidence of contemporaneity, variations in 'fingerprints' are much less helpful in this regard. Many different mortars can be incorporated into one phase of construction but accidental repetition of the same mix of components at different historical periods is much less likely. This argument can be extended to present conservation in support of the notion that a mix used in current work should not reproduce faithfully the original material but should be comparable and compatible but different to allow it always to be distinguished from the original work.

12.17 Mortar design: testing and research

Essentially there are three methods of approach to the complex problem of design of mortars. Specifications can engage aspects of all methods and usually do.

The primary technique involves the discovery of a material which has performed well. An original mortar will survive in the core of a wall. Original pointing may be found in the shelter of an overhang, under an arch or covered by later alterations. Analysis of the constituents, methods of preparation and the circumstances of application and use can, therefore, be expected to give an equally good material. However, while aggregates may be defined with some certainty and analysis will determine the nature of the binder, methods of preparation and application are unlikely to be accessible and a degree of supposition, therefore, enters the equation.

A more certain technique is the preparation of sample mortars and their subsequent testing. Samples will inevitably be made on the basis of prior knowledge and with available materials. They may be tested mechanically, by accelerated weathering and in field conditions. While poor samples may fail in the relatively short term the successful material will, by definition, be distinguished only after many years of field trials. Satisfactory results are, therefore, long delayed.

The final method is theoretical design. The performance of components being known, a theoretical blending of suitable factors will determine mixes and methods of preparation.

The common factor in these approaches is the complexity of the organization and the considerable time and expense involved. Much may hang on the results, the difference in repointing cycles of between ten and fifty years representing a large amount of money. A badly designed mortar can greatly diminish the visual quality of a historic building. A poor mortar can result in water penetration, damage to brickwork, staining and early failure. The conservator, however, can neither assemble the resources to conduct long term and meaningful experiments nor is likely to have available analytical results for the historic structure itself. The problem is to provide a relatively swift instruction using materials that are readily available. Almost inevitably this will mean that decision-making is based on background knowledge and experience coupled with close visual examination and possibly some field tests, minor laboratory analyses and possibly manufacturer's advice. The problem is immediate and the consequences far off. Only a body with long term objectives and a wide brief has the ability to test, analyse, consider and reconsider. The busy conservator can respond only to an immediate need by a blend of personal and others' knowledge and experience. It follows that the necessary research will be carried out only by a trade association, institution or government body, perhaps in conjunction with an institutional building owner whose interests are long term.

To visualize the statistical complexity of the problem a wide variety of components and methods must be remembered.

- Sands and aggregates vary as to particle size, particle shapes, colour, reactivity and secondary components.
- The binders may be limes, which in turn can range across many varieties of hydraulic through to pure 'air' lime and can be given a set by different types of pozzuolan.
- Cements vary through several principal types of strength and colour ranging from black to white and, in some parts of the world, gypsums of varying characteristics may also be used as binders.
- Colorants and other entrained additives are possible inclusions in a wide variety of ranges and a fundamental ingredient is water whose purity and extent of use are critical.

Figure 12.15 Mortar trials, Hadrian's Wall (courtesy of English Heritage).

Figure 12.16 Penetrometer used to test stiffening rate of fresh mortar (Smeaton Project: Phase I) (courtesy of English Heritage).

- Methods of preparation vary through hand mixing and several forms of mechanical mixing, with standing or mixed time and preparatory times being important.
- The circumstances of laying include the wetness of the adjoining material, the size of joint or filled void. The adjacent materials may be important and the behaviour of mortar is powerfully controlled by typical and special weathering conditions.

The consequent variants in mix can run into millions: hence the need for reliable guidance.

A major analytical project has now been running over a period of some years under the leadership of English Heritage in association with the Building Research Establishment and, in initial stages, with Bournemouth University and ICCROM (Rome). It is known as the Smeaton Project in deference to the English engineer who conducted some of the earliest analytical surveys of strengths of mortars for the construction of the Eddystone Lighthouse. (He settled on a hydraulic lime mortar given additional pozzuolanic set by the use of imported crushed volcanic ash.)

The Smeaton Project grew out of experimental work begun by English Heritage in 1986 to identify suitable mortars for use in the conservation of Hadrian's Wall, the Roman wall built in the second century AD by the Emperor Hadrian between the Tyne and the Solway Firth across the north of England. The wall was originally constructed in relatively small rectangular stone set in a mortar typical of Roman practice – lime with crushed tile and kiln waste, crushed sandstone or local sand and occasionally traces of tallow or similar fats. These mortars have survived in the core.

On Hadrian's Wall, conditions of exposure vary between 'quite severe' and 'severe', and frost cycling is high. Upwards of six cycles may occur in a day. In earlier restoration work cement-based mortars had been used in repointing and rebedding. While themselves sufficiently durable to withstand the extreme exposure to which the wall is subjected, these mortars contributed to the deterioration of the wall in a number of ways that are typical where cement is used on contact with weaker and more porous materials.

Thus, a simple comparative testing programme was initiated to try to develop more appropriate conservation mortars than would be compatible with the original masonry but still able to withstand the severe weather conditions at the wall. This largely site-based testing programme involved the preparation of a large number of mortar test cubes based on a wide range of binders from pure limes to Portland cements. The mortars were mixed in varying proportions of binder:aggregate with and without pozzuolanic and air-entraining additives. Large test cubes (150 mm) of each sample were placed on exposure racks in two locations near the wall. Sets of smaller cubes (50 mm) were subjected to the BRE salt crystallization test for durability. Subsequently, selected mortars were used to consolidate small trial sections of the wall.

An evaluation of both the test cubes and the trial wall sections raised a number of questions and indicated some surprising trends. The

Figure 12.17 Sample cubes after fifteen cycles of the salt crystallization test (Smeaton Project: Phase I) (courtesy of English Heritage).

overall framework of the Smeaton Project was designed to investigate some of the phenomena observed at Hadrian's Wall in a more controlled laboratory setting.

The Smeaton Project has passed through a series of phases and publications, during which time the range of available materials has increased considerably. Thus, while initial phases of the research concentrated on the effects of set additives, particularly brick dust pozzuolans and on lime-based mortars, more recent work has begun to examine the characteristics of mixes based on the numerous domestic and imported hydraulic limes which have recently entered the marketplace. The conservator is referred to the numerous publications on the project (see Bibliography) for a detailed discussion of results. However, a number of generalizations are possible.

Perhaps the most notable result of Phase I was the confirmation of earlier indications that the addition of small quantities of cement to lime:sand mortars (less than a proportion of 1:3:12 of cement:lime:sand, i.e. cement in proportion less than one third of the volume of lime) is likely to have a detrimental effect on the strength and durability of the mortar. Mixes richer in ordinary Portland cement do have a strengthening effect on mortars, however, due to the extent of cementicious action.

Information was also gained regarding some of the properties affecting the behaviour of lime mortars modified by the addition of crushed brick or tile. Low-fired brick dusts (fired at a temperature less than 950°C) seem to have the most positive effect on the strength and durability of lime:sand:brick dust mixes. Clays fired at higher temperatures of 120–1350°C (such as High Temperature Insulation or HTI powder) proved less consistent and less satisfactory as a pozzuolan.

As regards particle size, results indicated that brick dusts in the lower size particle range (particles < 75 microns) can act as reactive pozzuolans assisting initial set and increasing durability, especially if the dust forms a significant proportion of the mix. The larger particles (> 300 microns) of brick in mortars seem to behave more as porous particulates or air-entraining additives which again assist durability. The air content of the aggregate aids carbonation and improves resistance to frost and salt crystallization.

It is not possible to be conclusive about practice implications resulting from this research until results have been tested on a larger sample pool and field trials have been carried out. However, some general indications based on observed trends can be summarized as follows.

12.17.1 Cement gauging of lime mortars

The addition of small quantities of cement to lime:sand mortars is likely to have a negative effect on strength and durability. In practice, this means that cement gauged mortars with less cement than a 1:3:12 (cement:lime:sand) mix should be avoided. Stronger mixes (such as 1:2:9 and 1:1:6) do improve durability but they may be unacceptable in terms of increased hardness and decreased water vapour permeability for soft, porous masonry. Similarly, pure lime mortars may not have adequate durability in very harsh climates. As always, decisions must be made based on the nature and condition of the masonry and on the environmental conditions which it must withstand, as well as an informed understanding of the original material and technology.

12.17.2 Brick dust gauging of lime mortars

The addition of brick dust to lime:sand mortars does improve strength and durability, especially if the brick dust is a significant component of the mix. While low-fired brick dusts seem to have the most positive effect it is not possible to say whether this result is due strictly to firing temperature or is also related to clay type. Both large and small particles seem to play a part in the improvement of strength and durability, although these mechanisms are not yet completely understood.

Thus, if brick dust gauging is considered desirable to enhance the durability of a lime mortar, best results will probably be achieved with a low-fired brick dust in a range of particle sizes from 38 microns to 1 mm. Mix ratio will depend on the characteristics of the lime, the brick dust, and the other aggregate present in the mix. Proportions ranging from 1:2½:½ to 1:3:1 have proved successful in laboratory trials but site tests are required to achieve desired characteristics of colour, texture and hydraulicset in particular situations.

In selecting an appropriate mortar for laying and/or pointing, the type of brick and the overall performance of the wall are fundamental but it is only in the case of very high strength brickwork that stronger mortars will be justifiable in which case the careful provision of movement joints is essential.

The emergent published research and the careful use of registers of available sands coupled with manufacturer's and trade association information on the widening range of limes and hydraulic limes available now provides a basis for careful and effective mortar design.

13

Case studies

The sequence of case studies presented here illustrates a series of quite different problems. Medieval brickwork in Venice – the crypt of the Church of San Marco; the repair of sixteenth century brickwork and terracotta at Laughton Place in Sussex, followed by the renewal of sixteenth century chimneys in new material to authentic designs at Hampton Court, London; the repair of brickwork of the industrial era at St Pancras Station, London; and finally the repair of vernacular timber framing and brick construction in Baghdad.

13.1 The Crypt of San Marco, Venice (Architect, Dr Ettore Vio, Proto of San Marco)

Between the years 829 and 832 a great church was reared in Venice to house the Relics of St Mark the Evangelist, transported from Alexandria to the young republic on an island in circumstances variously described as piracy on the one hand and transmigration of the Christian community on the other. Two centuries of academic debate have culminated in the recent theory that the crypt of the church which houses the relics in a stone sarcophagus carried on massive columns is a substantial part of the original building of the ninth century, extended eastwards by enlargement of the central apse of the three originally built. This enlargement, in a major remodelling of 1070–1092, was a complex operation making use of earlier salvaged stone and brickwork. In consequence virtually the whole of the visible material in the crypt is of the ninth century or earlier, although the actual workmanship is almost entirely of the eleventh century.

Nature of the construction

The City of Venice stands on a deep bed of clay overriding sandstone. The clay carries a superficial deposit of silt and vegetative material of varying depth. It was common practice in the city to provide major monuments with foundations of short, close-packed timber piles overriden by transverse baulks of timber at a depth that would ensure their permanent saturation. In this condition the timber would survive indefinitely. The more important buildings were then provided with a foundation of stone carried up to a height above which water would rise only rarely. From this level upwards the construction was usually in brick. In the case of the Church of San Marco this type of construction was used in both major phases of building, the stonework being, in part at least, squared limestone imported from the mainland where various Roman settlements had been abandoned after the barbarian incursions which gave impetus to settlements on islands such as the Venetian group. In some instances marble was used rather than limestone.

The purpose of the close-set piles was as much to compact the clay layer as to carry the load a further metre or two into solid ground. Densification reduced water content and gave greater load bearing capacity. The use of stone at water surface level provided a structure less dependent upon the stability of mortar than would have been the case with brickwork. As water surface level is the most vulnerable zone the use of stone above the piling is understandable. Such stone footings have long been visible in the crypt and marks on individual stones give evidence of their former use elsewhere. There is also documentary evidence that Doge Giovanni Participazio had

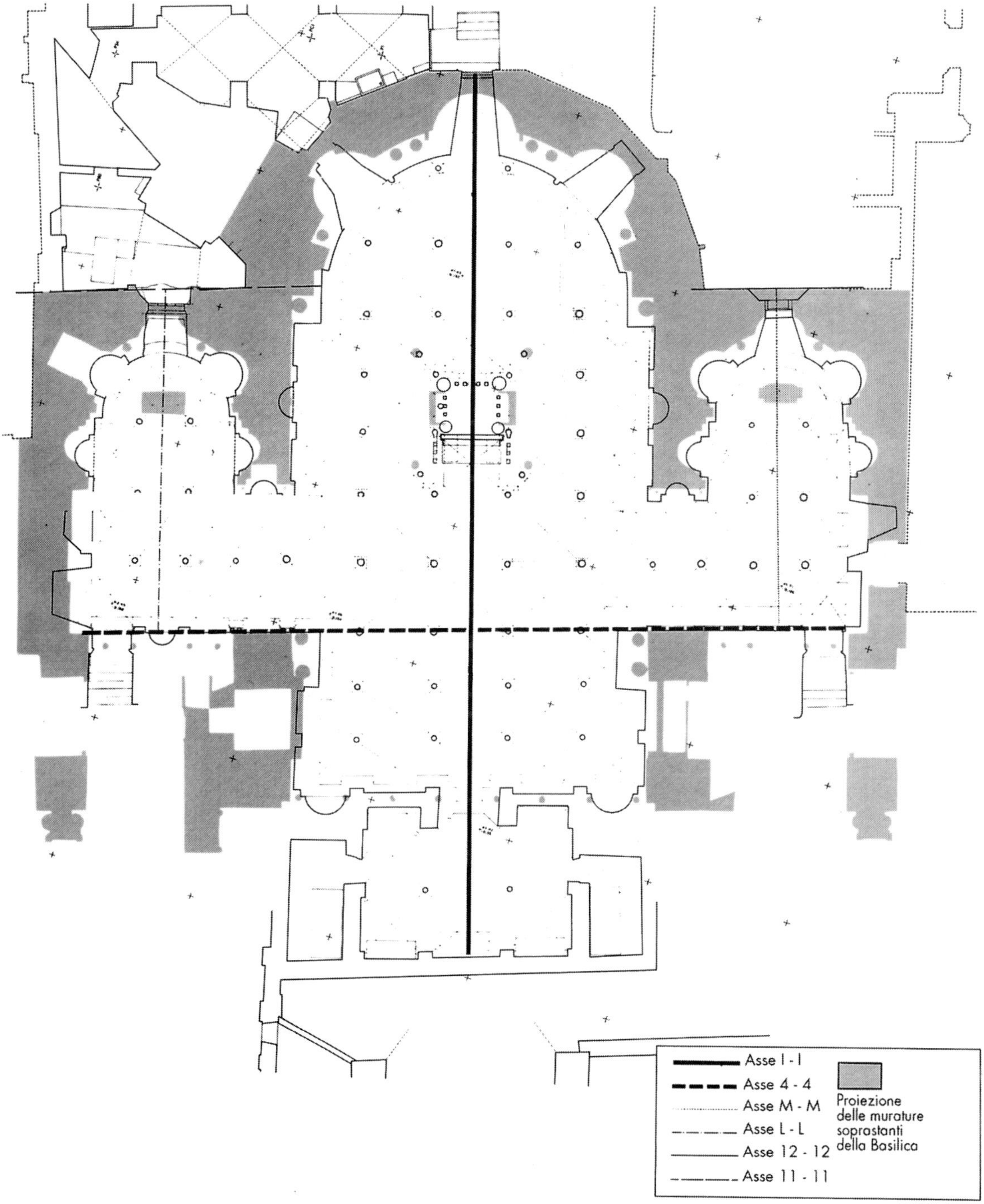

Figure 13.1 The Crypt of San Marco. The plan is shown in firm line with the plan of the church above being shown in solid tone.

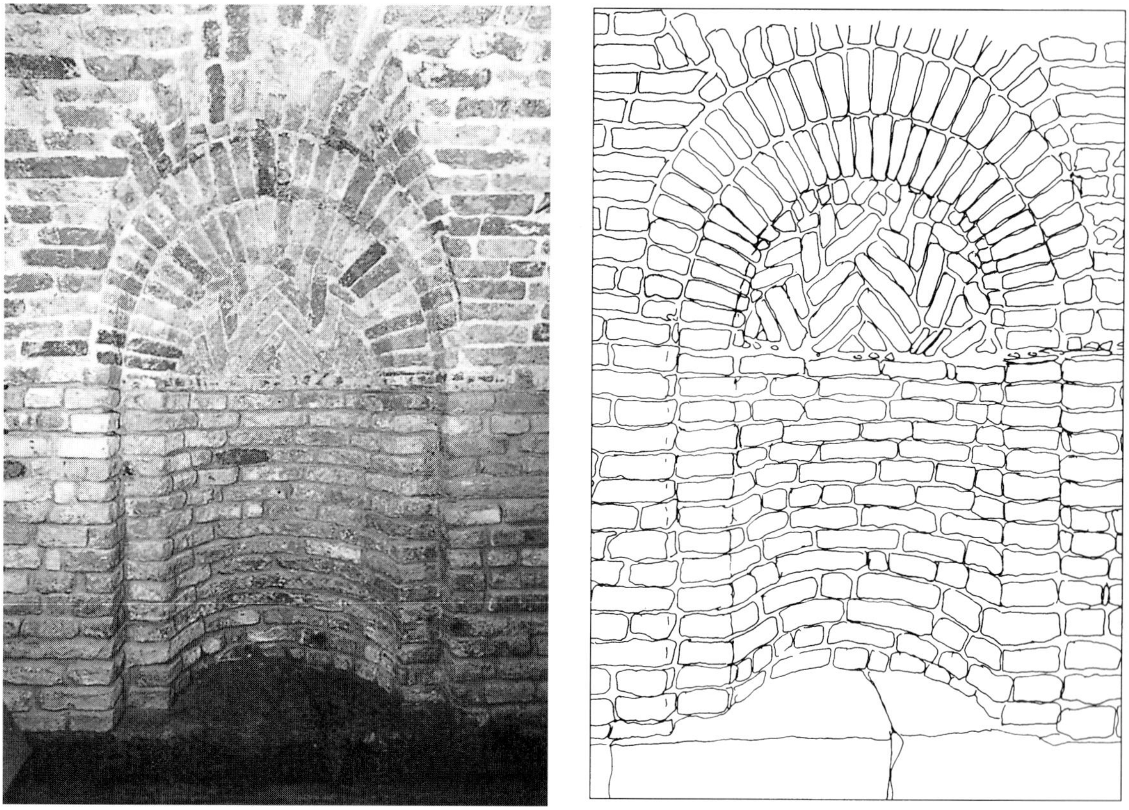

Figure 13.2 The Crypt of San Marco. An apsidal recess in the crypt with a drawn representation on which detailed brick-by-brick instructions may be imposed.

these salvaged stones available to him for the work in the ninth century.

The bricks now visible in the crypt are of rough formation and many of them carry signs of physical damage suggestive of reuse. They are of moderate porosity and hardness and, not surprisingly, of local origin. They are variously burned to a range of colours from straw through yellow-buff to red. Three brick types may be identified by differences in size, colour of inclusions, colour of matrix, nature of voids and shaping. They represent different sourcing at different periods, and some salvage also. Although there is evidence of Byzantine influence in the design of the building, the laying of the bricks in the crypt (and throughout the church) is in the Italian manner using normal mortar beds, rather than in a Byzantine manner with deep mortar beds which reduces the number of bricks needed The construction of the crypt is multi-columnar, a regular pattern of short, monolithic columns with simple rounded capitals carrying a system of brick cross vaults which does not depend upon separate arching between the columns. The cross vaults are not square but rectangular and are of consistent dimensions in the outer apses and in the western section of the crypt. In the principal apse, however, the rhythm is considerably modified. On archaeological examination it became clear that the original profile of the crypt as formed in the ninth century had been distinctly altered in the refacing undertaken in the eleventh century confirming the later alterations. Implicit confirmation of the eastward extension of the central apse derives from the positioning of the catafalque of the Evangelist at a position which would have been at the focus of the ninth century central apse. A comparison with contemporary churches such as the Cathedral at Aquileia confirms the

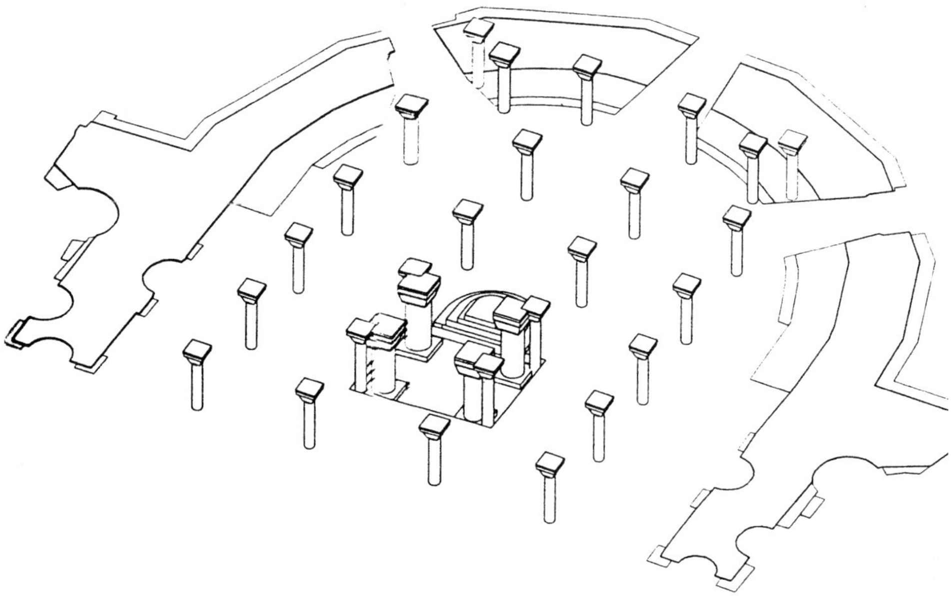

Figure 13.3 The Crypt of San Marco. An isometric perspective showing the columns and wall. While the walls, set against solid ground, are subject to continual saturation of the brickwork, the vaults are isolated and absorb moisture from the air. (Courtesy Dr Ettore Vio.)

practice of placing relics in the focal position in the apse.

The conservation problem

Although the crypt is shallow and the floor surface is less than two metres below ground level, it has inevitably suffered from water ingress and salt penetration. The water of the Venetian lagoon is moderately saline although continually replenished by the rivers which flow into it. Consequently the walls and parts of the vaults have built up uneven and sometimes substantial volumes of salt repeatedly apparent as efflorescence. Almost inevitably attempts had been made over the centuries to cure this problem by providing a surface coating of plaster and cement of different compositions applied at different times, the last such campaign having been carried out under the architect responsible for the major investigative and repair works following the Second World War, Proto Sig. Forlati. The problem therefore faced by the administration responsible for the recent repairs under Proto Dr Ettore Vio was to overcome the continuing ingress of water at low level and the evaporative effects at higher levels.

The techniques of waterproofing, surfacing and discharge of below floor level water are of no direct concern here since all the brickwork is in the vaults and in the wall facings above floor level.

Investigations

The entire structure was examined with great care by direct physical investigation which had been preceded by non-destructive analysis.

A number of techniques were applied including ultrasound, radar and magnetometry, of which the last two have direct relevance to work on the brickwork in the crypt. Deep penetration radar was used to determine the foundation depths.

Under the direction of Sig. P. P. Rossi ultrasonic testing, which differentiates between the density of materials by speed of travel of the sound wave, was used to determine the mass and extent of brickwork and in one instance (in a pier in the church itself) demonstrated the existence of an iron chain in the core of the brick structure.

Three physical methods of investigation were employed.

1. The extraction of samples of mortar for the chemical analysis of the original lime, which proved to be an eminently hydraulic lime with rather more than 3 parts of sand to 1 part of lime putty.
2. 60 mm diameter borings were made through the lower walls and foundations on diagonal downward alignments, demonstrating the continuity of the footing courses of stone. Petrographic analysis of the cores by Prof. Lazarini differentiated between the types of stone, all of which proved to be durable limestone and marble including some elements of a very pure marble which may have derived from the Proconesian quarries in the Marmara, near Constantinople.
3. With this assurance of the stability of the footings, tests were carried out with hydraulic flat jacks to ascertain loadings in the brickwork. Horizontal courses of mortar were removed and micrometer strain gauges were used to measure the deformation and the hydraulic pressure required to reinstate the internal loading on the wall.

On the basis of the information thus obtained static loadings were calculated and modelled mathematically by computer. The structural stability of the basilica having been verified physically and analysed in theory it was possible to proceed on a basis of structural certainty. The remaining problem was to overcome water penetration and salinity in the brickwork.

Figure 13.4 The Crypt of San Marco. A pen-reinforced photograph of vaults and walls. The mixture of limestones in the lower courses gives way to brickwork in the upper levels and to brickwork entirely in the vaults. Ground water is impregnated with carbonates. (Courtesy Dr Ettore Vio.)

Refurbishment of the brickwork

In the crypt several different conditions of brickwork existed but the variations were not such as to justify radically different methods of treatment. Areas of weakness in the brickwork were identified and voids were located.

The intervention

The objectives of the intervention were:

1. To provide free, below-ground drainage for liquid water and thereby maintain a zone above floor level in which there are no hydrostatic pressures.
2. To retain the existing structure without alteration to the historic materials.
3. To avoid the continual formation of efflorescence while recognizing the impossibility of creating a damp-proof barrier behind the brickwork in areas beneath the church.
4. To provide a surface of sufficient porosity to maintain a gaseous continuity between the inner pores of the brickwork and the internal atmosphere of the crypt, while avoiding the transmission of water-borne salts.
5. To provide a visually acceptable finish.

Having verified the adequacy of the structure in relation to the loadings it carries (by the use of flat jack technology) the voids behind the brickwork, identified by borings

Figure 13.5 The Crypt of San Marco: Conjunction of vaults and walling. In this position the ground-wetted wall comes into contact with the suspended vaults. (courtesy Dr Ettore Vio.)

and by non-destructive examination, were filled with a resin-rich cementitious grout injected under pressure, the resin being included to ensure that the grout itself remained impermeable. Injection and verification of the successful pressure resin grouting was carried out through fine cores taken with diamond cutting heads and verified by endoscopy.

All the superficial renderings, largely of cementitious material, were removed from the brickwork, the joints were cut out to a depth of 50–60 mm and where necessary brick repair was carried out using matching compatible material. The extensive impregnation of salts was analysed, the results confirming that the salts were primarily sodium chloride. Over depths of 70 mm salt concentrations reached 2 kg/m^3 on the basis of analysis of material extracted by core boring. Over a 40 day period persistent application of sepiolite poultices applied as renderings reduced the salt content by as much as 30%. The sepiolite was applied as a pumped cream, building up to a 5 mm coating unrestrained by containers and permitted to fall away when dry. The extent of the work varied from place to place, the lower levels being more heavily saturated with salt due to greater water penetration under hydrostatic pressure.

Because of the disparities in impregnation and in the ratio between salt and water in the brick at varying heights and in varying positions a careful campaign of temperature and humidity control was undertaken monitored by thermography of sample areas of brickwork. The work was carried out during the summer and autumn, both for reasons of air humidity and water levels in the lagoon. The areas being controlled were enclosed within temporary plastic sheet barriers in which daily thermal cycling was relatively minor. Thermal mapping in relatively constant conditions then became possible, with the gradual long term reductions in humidity zones of excessive drying around the intrados and extrados of wall arches contrasted distinctly with zones of persistent humidity in the lower levels of brickwork and it was, therefore, possible to arrange dehumidification to provide balanced drying and therefore a greater uniformity in reduction of saline content.

Generally four to six applications were sufficient to remove 25% of the salt content. As many as ten successive applications were made.

In principle the authorities had agreed on the use of synthetic resins as moisture inhibitors and consolidants. Material samples, 50 mm × 50 mm × 20 mm were provided to give data on water absorption and these were subject to exhaustive tests on absorption and crystallization in conditions equivalent to those anticipated in the partially dehumidified brickwork.

The samples revealed ranges of porosity from moderate to substantial. The absorption coefficients varied between 0.667×10^{-2} through 1.26×10^{-2} to 2.36×10^{-2} grams/cm^2 × seconds giving a range of approximately four times in absorptivity. The samples were then used for tests on absorption of resin

Figure 13.6 The Crypt of San Marco. The suspended vaults. The rugged and irregular brickwork was originally plastered. With the removal of sequential surface coatings salt deposits have been removed. Exposure of the brick requires elimination of the passage of salt-laden solutions. (Courtesy Dr Ettore Vio.)

polymers. The materials came from the Rhodorsil range, being RC70, RC80 and RC90 and RC4673 in a white spirit carrier. All are ethyl silicates (RC 70) or polymethyl siloxanes. In addition Rhodorsil RC100 was tested. This is a silane carried in an aqueous solution at 10% of active material. In this condition it has a pot life of some hours, sufficient to allow impregnation.

Further tests were carried out with Tegosivin HS, alkyl-alkoxy-silane in a mixture with ethyl silicate. Comparative absorption curves were studied extensively to determine aqueous absorption before and after treatment. The range of samples was tested as to absorption, crystallization of materials and artificial ageing with the combined objective of determining the effective water repellency and consolidants over extended periods at different densities of acceptance. In the ultimate, a series of further tests with modified primary materials showed that the optimum range of consolidation and liquid water resistance lay with a mixture of ethyl silicates and polymethylphenyl siloxanes under the proprietary designations of RC90 + RC11309 and in other cases RC100 + RC4673. This result of laboratory testing had then to be subject to site investigation in which circumstance recipient materials were much more varied and conditions less controlled.

Parameters considered were:

- water absorption due to capillarity;
- effective protection in relation to absorption;
- drying behaviour;
- ageing (by artificial tests).

Tests before and after treatment included:

- water absorption;
- drying indices.

Methods of application were, therefore, tested in site trials prior to general application. On completion of the site trials brickwork repairs were undertaken and necessary services were introduced into joints chased and modified. Matching salvaged brick was used in renovations and repairs and service concealment. With these exceptions the whole of the now visible brickwork is that recovered from historic periods.

The work proceeded with the extraction so far as possible of the soluble salts with the help of air heating in a spring, summer and early autumn programme. The removal of mortars from joints materially assisted the salt removal. On the removal of the sepiolite poultices pointing was made good and finished with recessed joints. The mortar was made from an eminently hydraulic lime with fine sand and ground marble aggregate.

Finally, after allowing due time for the setting of the mortar (not less than 3 months), a regime of dehumidification by artificial methods brought the mortar and brickwork to a stage at which the consolidant could be introduced.

Ambient conditions were achieved in which relative humidity was in the order of 70% at temperatures in the order of 25°C. In these circumstances the moisture content of the bricks yielded values between 18 and 21%.

The application of the specially formulated, spirit-based consolidant/water repellent followed careful hand cleaning of the brickwork and was the final act in the programme of work. The application was by spray. Work was carried out on a carefully calculated basis to deliver the appropriate amount to each superficial area ensuring adequate penetration into the brick without destroying its porosity and without creating a superficial hardened layer which could ultimately break away. In initial tests the volume of applied material was calculated to minimize colour change in the brickwork although ultimately the natural colours of the brick have been enhanced (in the same way as dampness tends to heighten colour). On completion the vaults took on an attractive overall warm soft brick red.

It will be recognized that these vaults are an unusual and special case in which funds and technical skills were available to pursue the full range of scientific enquiry and analysis. It will probably be many years before the decisions made can be vindicated and full technical conclusions drawn. Past generations have made many unsuccessful attempts to solve the problem with notable lack of success. It would be fair to conclude at this point that a similar operation attempted without equivalent technical backup would be inherently risky to the point of unacceptability. The use of sophisticated chemicals carries with it inherent risks, heavily underlined by the Society for the Protection of Ancient Buildings (UK) in its comments on the use of resins in stonework. Such risks are not to be undertaken lightly nor should one carefully calculated example be copied blindly.

One final factor in the equation, as important as the care in the intervention is the physical regime under which the monument is maintained. In the Church of San Marco a system of background air control has been introduced so that the temperature and relative humidity can be maintained with a minimal differential between the ambient air and air entrained within the masonry. The objective is to minimize vapour movement out of the brickwork.

13.2 Laughton Place, Sussex – (Architect, John Warren for the Landmark Trust)

This case study concerns a post-medieval tower restored from dereliction.

On the flat farmlands just north of the Sussex Downs, where the river Ouse makes its way down to Lewes, a solitary tower stands remote and romantically isolated. With its battlemented top it looks like the tower of a church which has lost its nave. This is all that remains of the important medieval manor of Laughton. On closer inspection, the site proves to be surrounded by an intact moat where fragments of other sixteenth century buildings are to be found among the trees.

The importance of the tower lies in the terracotta brickwork ornamented with designs of Renaissance derivation which are used for string courses, plinths, mullions, window heads and friezes. Their quality puts the building among those few great mansions such as Layer Marney in Essex and Sutton Place in Surrey, as well as vanished mansions like Nonsuch, in which this particular form of detailing was used in a brief florescence in the early sixteenth century. The date of the tower is in no doubt: *L'an de Grâce 1534 fu cest mayso faicte.* The attribution in medieval French cast into the terracotta boast once mounted on the face of the building may hint at the continental derivation of the craftsmanship in terracotta which became fashionable and equally quickly disappeared within the space of about 40 years.

The owners of the manor and builders of the tower were the Pelham family. The tower itself was probably an addition to the solar wing of a timber framed hall house. The construction of the tower in 1534 was evidently the first stage in an enlargement of the mansion, whose appurtenances included extensive constructions around the moat. Clearly the intention was to build on the scale of other great Tudor mansions. A drawing on

Figure 13.7 Laughton Place. Solid block terracotta used for a string course incorporating the 'Pelham buckle'. All the terracotta used in the sixteenth century work is solid moulded block, forming mullions, string courses, spandrels, framings and plinths. On the perpends of the upper courses of brickwork can be seen the double-struck pointing copied from original work.

an estate map of 1641 suggests that an earlier timber building survived and probably continued into the eighteenth century when in 1733 the building was heavily remodelled to include the tower as the centre-piece of a very early Gothic Revival mansion – sometimes described as a Rococo Gothic farmhouse. The farmhouse survived in increasing delapidation through the Second World War and was demolished about 1950 leaving the tower standing alone. The principal alteration it had suffered was the construction of a steel and concrete upper roof to enable it to serve wartime purposes as a defensive point and lookout.

Significant to the conservation were a series of vertical structural cracks running from top to bottom of the tower on its northern and southern sides and the nature and importance of the terracotta work, more of which came to light as investigation proceeded.

The structural failures had evidently been long standing and were the result of differential movement between the stair tower and the main block of the tower itself.

Structural problems

The tower consists of two elements – the four-storey pile of single square rooms with timber floors, and attaching to one corner of the pile the massive buttressed tower containing a brick spiral staircase. Calculation showed that the average density and, therefore, the ground loadings of these two parts of the structure were very different. The foundations are relatively shallow and stand on marshland deposits over Gault clay. In consequence the pressure bulbs are unequal and differential settlement had occurred at an early stage in the construction – probably before it was even finished. A consequential shear crack passing vertically up the south wall through the windows must be dated to within a decade or two of building, because the windows were blocked with identical masonry in order to provide greater support. In all likelihood this blocking took place in the later stages of construction, and the problem may have been an inhibitor to further investment.

At the point of survey for restoration the lateral restraint within the tower had been lost by the decay of the timber floors with the exception of two principal girders inadequately seated on to the brickwork. Vertical cracks on the northern and southern sides of the tower demonstrated the detachment of eastern and western faces which left its stability entirely dependent on the tensile strength of the steel in the wartime concrete roof supported on rolled steel joists. These steels and the reinforcement were in a state of advanced decay.

The new building owner – Sir John Smith of The Landmark Trust – decided to restore the tower without additions other than minor alterations at the junction of the stair tower and the main pile, essential for access and sanitation. No new openings were to be formed although one blocked original doorway was to be reopened.

Two aspects of the extended and complex restoration will be dealt with here: firstly the temporary and permanent achievement of stability and secondly the restoration of the terracotta and brickwork.

Figure 13.8 Laughton Place. The blocked south window uncovered. Originally limewashed and built-up due to settlement at an early stage, fractures in the terracotta mullions are clearly visible.

Stability

Examination of the cracking in the structure demonstrated that initial settlement and structural failure on the southern wall, and probably also on the northern, had occurred within a few decades of construction and perhaps within a few years. The blocking of the southern windows had been achieved using surplus brick from the building operations. These windows had subsequently been plastered over on both faces and their existence was unknown due to the later plaster although some inconsistencies in the pattern of cracks suggested irregularity in the bonding of the wall. However, the blocking had had the desired effect, having been so successful in stabilizing the wall that consequential shear cracks occurred in parallel rather than through the original weak zone represented by the ground and first floor windows. It was possible to identify the effects of the eighteenth century additions and to determine that the structure was at that stage already stable.

In this eighteenth century work the original house was removed. It was probably a light structure, being timber-framed (although there is no certain evidence of its construction). The eighteenth century additions, however, were all in brickwork and calculations suggested that they imposed a load on the adjoining subsoil rather less than that of the tower pile and very considerably less than that of the stair tower. However, some further movement not attributable to the differential between the pile and the stair tower appeared to have occurred at that time. Cracks which had then opened had been made good. After the Second World War, only about twenty-five years before the restoration, these additions had themselves been removed – the mansion

Figure 13.9 Laughton Place. The corset of scaffolding restraining further movement. Loss of restraint at floor levels due to timber decay had allowed the tower to crack vertically on opposing faces. Raking shores were rejected in favour of a scaffold from which uniform lateral restraint was delivered until internal tie-rods were in position.

being in decay – and the subsoil had been relieved of their relatively moderate load. Any recovery of the subsoil would appear to have been in the humus-rich overburden rather than in the underlying clay. There was some evidence of this recovery but it was not regarded as sufficient to justify underpinning a structure which at foundation level appeared to have stabilized. Consequently it was decided to rely on internal restraint rather than underpinning, which would have involved short-bored piling.

To provide temporary restraint to a structure which might fail if interfered with, reliance was placed on a corset of scaffolding. An engineering solution involving massive raking shores of scaffold work was rejected in favour of the construction of an isolated, external, cross-braced scaffold separated from the tower itself by 100–150 mm and entirely disconnected from it. Horizontal screw jacks were then fixed to the scaffold and moderate pressure was brought to bear evenly on opposing sides of the tower. Wide pads were provided at the pressure points.

This solution was economical in scaffolding and, much more importantly, it was designed to provide even stressing against the walls. Any pressure on the scaffolding deriving from movement of the masonry was immediately reflected in an equivalent pressure on the opposite side. With this scaffold in place it was possible to work on the masonry and remove the internal steel and timber girders.

Permanent restraint was provided by reconstruction of the floors with equivalent weight timber girders and joists. This floor structure was reinforced by a grillage of stainless steel tie rods, the ends of which terminated in the masonry.

Terracotta and brickwork

Examination showed that the terracotta had been purpose made for the Pelham family and in all probability on the site. Although the moat itself is medieval and therefore predates the tower and other constructions around it, pits and depressions in the vicinity suggest access to the underlying clay and a stratum of suitable sand lies not far distant to the north. A few miles south plentiful supplies of chalk provided a non-hydraulic building lime for all local building purposes. There was no indication of the use of brick dust as a pozzuolan. There are some strata in the chalk, however, which would provide a moderately hydraulic lime which might have been used in part of the structure, particularly in areas where an early set was required, such as the staircase. These strata contain clays and silts and provide a hard chalk used also as building stone.

Comparison with other terracottas from this same period, notably at Sutton Place, Surrey, shows Laughton as having been unusual in having terracottas which fired to a deep cherry red and, when well or overfired, to a clinker blue. Almost all other contemporary terracottas are brownish and flesh pink. Analysis of fragments of terracotta recovered from

Figure 13.10 Laughton Place. Original terracotta mullions, including one replacement salvaged from original fill. The builders evidently had a surplus of terracotta components some of which were built into voids and were finally recovered for use in the repairs.

Laughton moat were carried out at the Department of Archaeology, University of Southampton by Dr David Williams on a comparative basis with other contemporary examples. The following summary derives from the analytical survey by Dr Maurice Howard. The samples are now filed in the thin section collection of the Commission of Historic Buildings and Monuments.

This summary concluded that the Laughton examples were not the production of a site common to the others: 'The Laughton fragment was found to be made of an isotropic clay matrix with sub-angular quartz grains, mostly under 0.3 mm in size together with a little iron ore silt stone and flecks of mica. The quartz grains were fairly standard in size throughout the samples from all the sites, suggesting a common practice in the way the raw materials were used. The isotropic nature of the clay matrix in all the samples points to a firing temperature in excess of 800°C. The clay chosen for Laughton was by far the most red of all the clays ...'

The terracotta units were evidently formed of refined brick clay almost certainly achieved by passing clays in water suspension across ponds designed to allow for sedimentation and later extraction. The units were made solid. There was no indication of any attempt to provide thin-skinned terracotta as in later practice. Consequently the terracotta units were built into the structure in the manner of bricks and were expected to be load-bearing in context. They were, therefore, generally confined to sizes broadly typical of brick and large floor tile although some larger units were found in the excavation of the moat. In consequence the components were designed to fit in much the same way as 'small stone' masonry, the mullions, for instance, being formed as matching internal and external units embedded back to back.

There was a common vocabulary of Renaissance features in the designs implanted on the terracotta at this period including coats of arms, foliate arabesques, vases, pendants, pelicans, ribbon work with occasional putti and other devices of recognizably continental Renaissance origin. The embossed designs were principally applied within the cavetto of mouldings. The unique feature of the work at Laughton is the consistent use of the 'Pelham buckle', a device based on a circular buckle with a vertical central bar, which the Pelham family had made their symbol.

In addition to its use for window and door surrounds, terracotta was employed in the capping moulds to plinth and string courses. In addition it formed the crowning feature of a corbelled course to the parapet of the tower. The design was a low cusped arch executed in each corbel brick providing an appropriate transition from the main wall to the upper

Figure 13.11 Laughton Place. The surviving merlon (central) abutting the turret, which gave evidence for the restoration of the battlements. The parapets had been significantly altered during the construction of the mansion, but one merlon survived, providing evidence of the height, size and detailing of the battlements.

structure. The merlons of the battlements had entirely disappeared, however, with the exception of a transition between the stair turret and the tower, where some damaged fragments remained in situ. These were enough to determine the profile and construction of the original battlements. Geometric analysis showed that only one possible combination of merlon and embrasure would accord with the extant chimney stack serving the two upper floors and the battlements were, therefore, restored on this basis.

During the work evidence of other earlier openings was found at all levels. Much of the structure had been so mutilated in the forming of additions that it was no longer possible to leave it as an external skin and where a new skin was provided the internal evidence was

Figure 13.12 Laughton Place. Replacement terracotta before incorporation. Blocks were made from clays similar to those used originally and to the original profiles but omitting the embossed designs to differentiate between original and replacement work. In this photograph the white powder is merely a drift of fine blown snow.

Figure 13.13 Laughton Place. Repair of fractured blocks. A block showing fibre-glass dowelling and a repaired door head. Commercial epoxy resins were employed, the strengths being reduced to approach compatibility with the brick.

repeated in the construction so that the new facing retains the arching, the straight joints and, wherever possible, the original brickwork to repeat the historic evidence. A remarkable feature of the work at Laughton is the hood mould used consistently over the windows. This is unusual but not unique in contemporary work. In later smoothing over, however, most of the hood moulds had been removed and these have been left exposed in their mutilated state.

Some terracotta units had been discarded during the original construction and these were recovered and reused wherever possible even where they had been discarded because of over-firing. The brick manufacturer (Swanage Brickworks) was selected on the basis of the ability to provide hand-crafted material matching in colour and texture the original bricks and willingness to provide the special replacement for the terracotta work. These were formed to the original profiles and sizes, but all moulded and embossed designs were omitted. It is, therefore, possible to identify the replacement terracottas at a glance.

A number of original bricks were found in crushed and sometimes in fragmentary state. Overloaded mullions and shear cracks running through window dressings were the cause of fractures which justified individual brick repair. Wherever possible they were salvaged, matched and if necessary small

Figure 13.14 Laughton Place. Fractured blocks repaired and replaced. Where serious fractures had occurred the units were removed, insertions were cut where required and the block was reassembled as a coherent unit before being built back into the structure.

missing pieces were cut from appropriate material as inserts. The fragments were then joined using polyester or epoxy resins heavily filled with brick dust. A number of important junctions were formed using dowels of fibre glass set in polyester resin, the drilled material providing a filler for the bonding.

Analysis of the original mortar by hydrochloric acid reduction determined that a mix ratio of 1:3 had been employed and that the facing work had used a fine white sand. An equivalent sand from the Wealden beds in Surrey was used in the new work, which was entirely carried out in lime slaked on the site.

A pit penetrating the water table was dug at the commencement of the works and the slaked lime was matured for three months. The profile of the original double-struck pointing was evident in the unweathered pointing in the soffits of the arches to the flying buttresses of the tower. It was, therefore, possible to repoint the brickwork and to finish the new work in a manner which exactly reflected the original construction. The same pointing was carried across the single remnant of the eighteenth century house, this being a pedimented portico which had been applied to the eastern face.

This surviving fragment of Laughton Place is important for the rarity of its terracotta and as an example of a building type which is sufficiently unusual to have caused considerable discussion as to its purpose. The terms solar tower and prospect tower are both used. The solar tower explicitly added to the accommodation at the private end of the hall and the prospect tower provided surveillance and some visual dominance. Laughton appears to have served both functions.

Sir William Pelham (*c*.1486–1538) came to hold a manor which had been intermittently crown property until, in the fifteenth century, it came to Sir John Pelham. William attended Henry VIII at The Field of the Cloth of Gold and at a meeting of the royal houses of England and France in 1533. His enlargement of Laughton Place is consonant with the creation of other great houses of the sixteenth century where the Francophile tastes of Henry VIII's court and the influences of the court of François I are perceptible in a stylistic overlay. Sir William died in 1538 having achieved no more at Laughton than the tower and bastions on the moat. Having lost its medieval aisled hall and its eighteenth century enlargement, it stands a monument to one short and important period of building activity at the end of Sir William's life.

13.3 Decorated chimney shafts at Hampton Court (Architects: Feilden and Mawson)

The early history of brickwork in England is closely integrated with the reduction in the risk of fire and therefore the building of brick chimneys. Long before timber framing had gone out of fashion the heavily-daubed timber and wattle flues of medieval England were succeeded by the great brick chimneys of the sixteenth century. These developed as an art form of their own – high clustered columns, patterned with diapers, swirling ribs and

Figure 13.15 Hampton Court. Chimneys of the Great Kitchen. The rebuilt chimney cluster replaces metallic flue pipes which had been substituted for the Victorian stacks known primarily from engravings. No proper record had been kept from their demolition. (Courtesy Feilden and Mawson.)

chevrons, the shafts set on massive bases and terminating in over-sailing heads like capitals. These great shafts were effectively giant chimney pots. Their patterns were later imitated by great terracotta pots and the resulting roofscape became a signal feature of what is known as the 'Tudor style'. The evolution of the clustered chimney must have a fundamentally aesthetic explanation. Initially there was clearly less material to be found and less labour involved in the grouping of stacks formed individually and, for a given amount of material, greater stability if they were set diagonally. The growth of this idea into the fanciful and elaborate chimneys that rose above palaces such as Nonsuch and Hampton Court goes with stylistic development, showmanship and pretension. They have become much-loved architectural features and have many times been rebuilt in the early buildings and imitated in revivalist work. At Hampton Court not a single one of the many clustered chimney stacks is original. Victorian rebuilding and subsequent repair are the source of the present admired roofscape.

Among these the cluster over the great kitchen chimney is an entirely modern creation replacing a utilitarian structure with a replication based on a drawing made 100 years earlier of chimneys which had themselves been repaired or rebuilt perhaps several times. The life of chimneys of this type is generally less than 100 years in use as a flue or perhaps two or three times this length of time as a decorative object carefully maintained.

In the Victorian rebuilding the chimneys were faced with red rubbing bricks of high quality. These were red or orange in colour and cut to shape by hand, bow-sawn, using cutting boxes and templates and rubbed back on completion to take out irregularities. They were laid in lime putty with joints of 6 mm to 9 mm on average. Mortar was finished flush and the effect is clean and pristine.

In the newly built chimneys modern rubber brick material was used, made with brick earth with a 75% clay-silt and 25% sand content. This inversion of the normal proportions is said to give a stronger brick which is more resistant to attack by wind and frost and to mechanical damage on the arrises. The material is cut using a bow-saw in cutting boxes which are made to the finished profile of the brick so that the subtle and ever-varying curves of the outer profile can be followed accurately as the brick is sawn. Traditionally the bricks are cut using a bow-saw but on this occasion because of the harder nature of the more frost-resistant bricks selected, they were cut mechanically to shape by a skilled bricklayer working to templates who then hand-finished them using a file or chisel. A traditional form of brick was thus achieved using modern techniques of preparation, firing and shaping. In the ultimate these methods of preparation will distinguish the component,

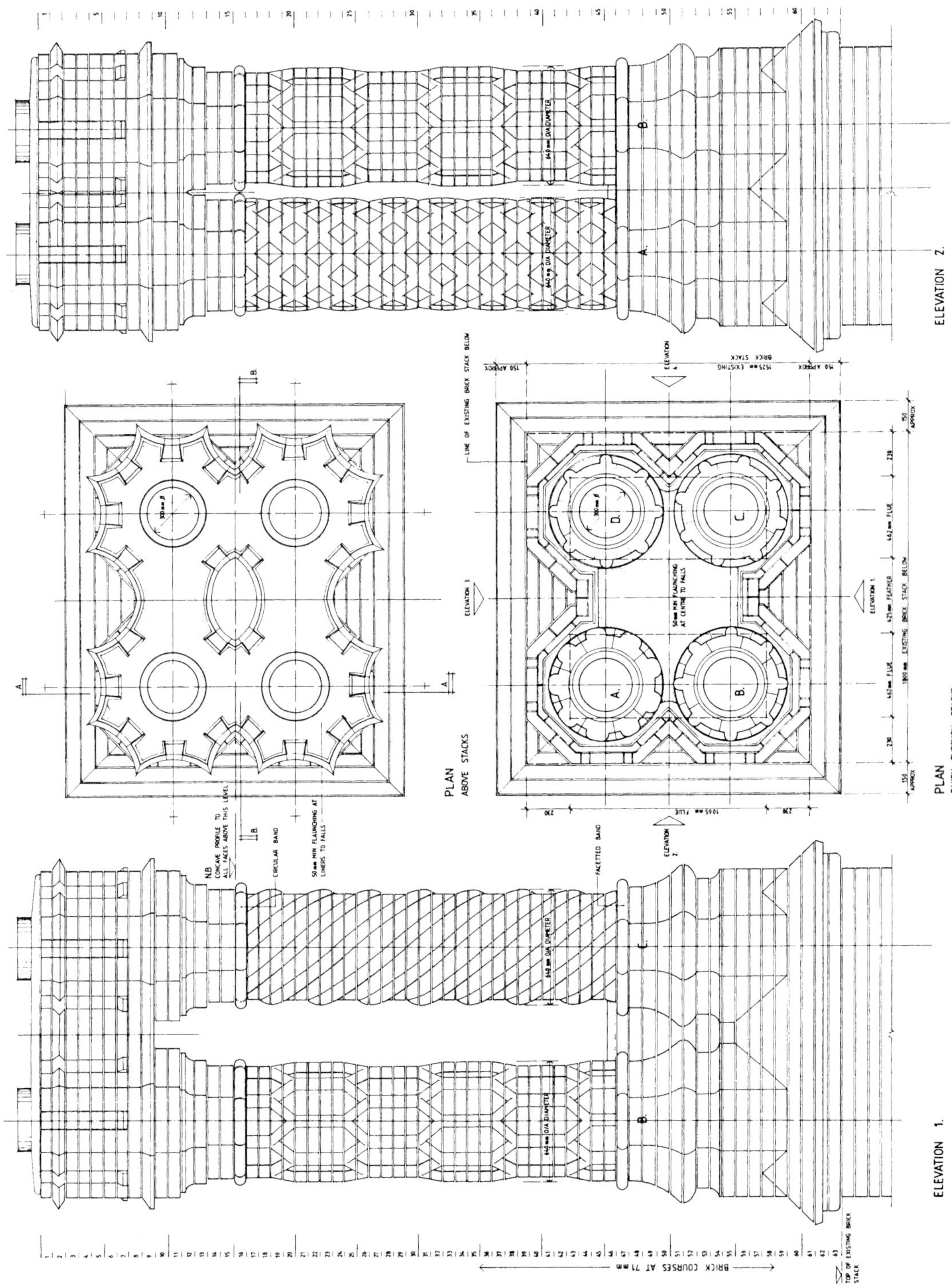

Figure 13.16 Hampton Court. Reconstruction drawings for the kitchen chimneys. The elevations shown represent a best assessment of the Victorian stacks on the basis of detailed pictorial evidence and comparison with other examples on the building. (Courtesy Feilden and Mawson.)

Figure 13.17 Hampton Court. Cutting boxes and templates. The templates used for accuracy in laying are seen above and an empty cutting box is central below with filled cutting profiles to the left. (Courtesy Feilden and Mawson.)

Figure 13.18 Hampton Court. Brick cutting boxes in use. The profiles for brick cutting are seen on the conclusion of shaping. Each cut brick is slightly oversize, as determined by the 'box', to allow for rubbing back to the exact profile. (Courtesy Feilden and Mawson.)

and enable it to be identified for what it is, although in use it will provide the precise shape, colour and texture required. The quality achieved in this way is described by the architect responsible for the scholarly recreation of the great kitchen chimney stack at Hampton Court, John Hyett of Feilden and Mawson, working under the direction of Alan Robson, architect.

The following quotations are the words of John Hyett, taken from his report on the project in the *Transactions of the Association for Studies in the Conservation of Historic Buildings*, 1992, 17, pp. 1–14.

> Attempts have been made in recent years to replicate Tudor designs, using pre-shaped cast bricks for economy and performance. Cast bricks, being harder and more vitrified, are classified as having a higher performance than soft rubbing bricks and are less prone to weather damage. Being harder, the cast bricks cannot be rubbed. Consequently, correction cannot be made in the completed works. The result is, therefore, not smooth and regular, and a typical swirl pattern built in this way with thin mortar joints does not run smooth, but rather jolts around in a series of staccato jerks. On the whole, the result is unsatisfactory and particularly so when the offending stacks are to be viewed adjacent to or juxtaposed with smooth Victorian rubbed brickwork.
>
> Some brickmakers have devoted much time and energy to perfecting cast brickwork. Precise wooden moulds crafted by pattern makers are generally used into which the clay–sand mixture is thrown. Allowance is made for shrinkage and great care taken in air-drying to minimize distortions in firing. It seems that the more deliberate the attempt at ensuring accuracy in the size and shape of cast bricks, the less satisfactory is the end result.
>
> Replica 'Tudor' chimneys aside, there are now many examples of precision brickwork which are in a way lifeless, lacking the subtlety which comes from the craftsman's handiwork at stages in the process of making and laying ...
>
> Tudor stacks often have the appearance of having been built solid and then carved in situ. It is thought that the technique was to cast the bricks to bald geometric shape in the solid, to allow the green brick to dry and harden in air, to scribe the outline of decoration and shaping, and then to cut the green brick with a sharp knife to the required shape. On firing, the knifed surfaces assume a polished appearance, by contrast with the sand-faced surface of the moulded solid body shape. A person employed in hand-cutting the green bricks will have become quite expert at cutting and shaping, which accounts for the quality of surviving examples of Tudor work. There is nothing of the mechanical in Tudor work for the reason that the Tudor craftsman was evidently reluctant to abandon hand-drawn lines and sections. The emphasis was on proportion, curves of ever-changing radii, and the appearance of light and shade on deeply incised mouldings. The bricks were carved in the green state partly because it was much easier than cutting the baked material, for much the same reason that, for example, green or unseasoned oak was used in preference to seasoned oak in timber framing. Tudor tools were of inferior iron and were not hard or resilient enough to deal with fired bricks, unless the work involved simple slitting to form closers and batts. The evidence of hand craftwork is further enhanced by the firing process and, when the bricks are laid, often with thick bed joints, beautiful and subtle effects of light and shade can be seen in the work.

Setting out a major chimney stack is a skill in itself and like the cutting of the bricks makes use of the practical application of dimensional constraints. Again, in the words of John Hyett:

> The finished elevation drawings were a guide to setting out and template manufacture. The actual setting out was at full size, commencing with the plinth and heads. Attention is not paid to any specific dimension. It is much the best thing to put aside numerical dimensions altogether, relying instead on proportion and geometry. The bifurcating bricks in the shafts needed ten templates each, as did the helix pattern bricks, to show the stepping off necessary around the base circle circumference for the travel of the pattern per course. This is the same, regardless of the height of the units.
>
> The full height of the stacks was ascertained from the proportion existing between the total height and the total width from comparison with similar chimney types at Hampton Court. There are 64 bed joints in the construction; allowing a nominal joint size of 3/16" per joint gives an allowance of exactly one foot for all mortar beds.

> Dividing the remaining distances with dividers gives the actual height of each cut-brick unit. Provided that the increment is clearly marked on the template, and that it is cut, planed and sanded to that size, it need never be physically measured. This is advantageous because it could, of course, turn out to be a very awkward decimal or fractional size.
>
> Similarly, with the amount of oversail to the heads, the proportion of the shaft to head diameter is all important. Oversail is confined to three courses in the head; too little and the individual heads will not meet, too much and the structure becomes rather weak and extremely difficult to lay.

The mortar used was formed from an eminently hydraulic lime which develops an initial set in about four hours with the addition of a small amount of white cement providing a jointing just weaker than the bricks themselves so that the whole body of the chimney stack acts as one in the thermal and moisture cycling to which it is subject. John Hyett describes how the work was programmed:

> The single most important factor to be considered in programming the works is the recognition that cut and rubbed brickwork is necessarily a wet process, owing to the extreme porosity of the bricks and the need to control suction when laying. Buckets full of clean water are required for cut-work immersion and for wetting down the top of the brickwork before laying commences.
>
> The wetting down is done first. The bricks are laid out in their final position before being immersed, about six at a time, and left until no bubbles emanate from them. A bed of mortar is trowelled on to the very damp breastwork, the brick is removed from the water, held over it until the excess has run off, then laid straightaway in the finish position. The work has to be pointed as it is built. It is normal to lay first, then check with a template as necessary prior to pointing up.
>
> Because the bricks must be laid in a fully saturated state, it follows that drying occurs gradually over a prolonged period even in good drying conditions. This work should not be attempted if there is any risk of frost.

In conservation there must always be circumstances in which simulation is necessary. The chimneys of Hampton Court are a most persuasive example of the relevance of careful reconstruction of architectural elements lost by time, weather, human activity or inadvertence.

13.4 St Pancras Station, London (Architects: The Conservation Practice. Partner in Charge: Barry Stow)

When the Midland Railway reached to London it made its terminus in the parish of St Pancras, where between 1868 and 1874 the Directors built a railway station whose iron train shed spanned an unprecedented 245 ft. Against this daring and dramatic feat of engineering the Company built what still may fairly be described as a breathtaking neo-Gothic headquarters and hotel, displaying the products of its hinterland in an extravaganza of Nottingham brick (Gripper's patent) and Tucker facings, set off by shafts of granite and dressings of Mansfield Ketton and Ancaster stone. Whereas the train shed which it abutted conceded little to design as taught in the schools of art and preached by Ruskin, this massive Gothic pile was full of allusions to Lombardy, Venice and the Île de France, embellished with details from Westminster Abbey, Salisbury and Amiens Cathedrals.

St Pancras may stand for all those showy and romantic Gothic piles being built across mid-Victorian England. In its curves, its asymmetry, its great tower, high pitched roofs, multiple dormers and lavish adornment it is an extreme in which taste and fashion reached a point of exuberance from which it could only retreat.

After the neglect of a century during which is narrowly escaped destruction by incendiary bombs and, subsequently, demolition, changing aesthetic values and the arrival of the Channel Tunnel railway have produced a programme of refurbishment in which the entire external face has been overhauled, cleaned and made fit for further use.

The Midland Hotel had been maintained in a usable condition for the whole of its life although at the latter period it had fallen out of economic use and was held on a 'care and maintenance' basis. This was not sufficient to ensure total avoidance of structural damage and, in particular, lead flashings behind parapets and

Figure 13.19 St Pancras Station. The main façade of the building: the scaffolding is being withdrawn as work proceeds from the top downwards.

Figure 13.20 St Pancras Station. The cleaning and repair provides a matching and even surface but some staining from the stone coursing is already becoming apparent.

gables and the soakers and valleys which were a necessary part of the structure, had failed, allowing water damage to woodwork and particularly saturation of brickwork on gables and at the upper levels of walls. In addition earlier acidic cleaning on the west face (with 'Neolith') had removed the fireskin of the facing brickwork over substantial areas. The building had, in any case, been subject to the ravages of a smoke polluted environment with sulphate depositions over the whole of its life, being in a densely built-up area of the city, downwind of a wide zone of long-established development. The problem was no doubt augmented by its proximity to two great railway termini and a major gasworks.

Initial analysis involved careful examination of the structure which, with one minor exception, proved to be free of serious defects. Examination of all the weathering features including copings, roofs, leadwork, stone dressings, window surrounds and the stone, the brick and the pointing. The main structure was shown to be entirely in mass brickwork laid in a hydraulic lime mortar with varying quantities of ash as a pozzuolan. The core bricks were of Gripper manufacture (Nottingham) and the facings were a red brick from the vicinity of Leicester identified in the frog by the manufacturer's name, Tucker. On analysis they proved to have come from a number of slightly different clay beds and probably to have been made in different yards. These facing bricks were the most important component of the restoration. They were double-pressed, hand-made bricks from Etruria marl.

Figure 13.21 St Pancras Station. The cleaned brickwork reveals earlier repairs in which a dark mortar was used to match the previously dark-stained mortar.

In arches, alternating with stone, red rubber brick was used and this may be the only visible material which had not been procured from a Midland source – the likely origin being East Anglia or possibly south of the Thames.

The stones used were principally Ancaster and Ketton in exposed positions at high level. Both these stones abound with calcium carbonate and are classified as limestones with proven records of durability. The principal stone used in dressings, however, was a sandstone from Mansfield which is rich red in colour and no longer available. In substitution two stones from central England were used.

The red Tucker brick is a semi-engineering slop-moulded kiln-fired brick which varies slightly in colour dependent on the firing. The colour variations were carefully mixed in the original work and the attachment of the outer skin, as was not uncommon, was relatively infrequent, many of the visible headers on the outer surface being 'snap' headers which maintained the pattern of bonding without close integration. There is no obvious explanation for the employment of this technique which was not uncommon in the nineteenth century and is found, for instance, in fortifications such as the Martello Towers.

Being water-released these Tucker facing bricks have a moderate fireskin. In areas where the walls have been saturated this had 'blown off'. This, coupled with the earlier effects of acids and the loss of pointing was the principal defect on the elevations.

Initial approach

Damage analysis was undertaken after trial cleaning. This allowed comparison between the initial and exposed condition, demonstrating causes of failure. These were broken down into naturally occurring and those arising from use maintenance regimes or other external acts. These categories in turn were broken down by nature of the defect and the agency giving rise to the problem.

The results of this process revealed the nature of the stresses to which the building would be subjected. Some factors, such as coal tars and repairs intended to match soot-blackened brickwork could be discounted as these conditions were not expected to recur. Specifications were then written on the principle that new brickwork should match the original, in its 'as-laid' condition, in character, bond, colour, texture, size and material. Care was taken to eliminate the defects in detailing which had given rise to unnecessary loss or decay.

Initial cleaning was carried out with stiff bristle brushes, mechanically driven, hard aggregations being removed by chisel (air powered). On brickwork a granular abrasive under low air pressure was applied to persistent discoloration, using glass powder (0.1–5 mm) and stone powder (0.005–3 mm) in wet conditions.

Restoration

Restoration of leadwork, blocked downpipes, gutters and other metallic elements was on a straightforward like for like basis with improved methods of combating blockage and decay. Where appropriate, stainless-steel was used in substitution for forms of iron capable of rusting.

In the selection of brick care was taken to match both the slop-moulded face, the level of firing and the variation in colour, typical of the original bricks. Clays from similar seams now worked at Ibstock in Leicestershire were tested and ultimately the right mix of clay was achieved. However, modern techniques of manufacture and firing produced a surface more even in character than the Victorian brick. It proved difficult to match the sharpness of the arrises originally achieved. With the recovery of a discarded brick press, however, manufacture of the very considerable numbers of bricks required was taken out of the normal stream and a special production run by the Ibstock Company provided a satisfactory colour mixture. The slight variation in size consequent on firing at different temperatures exposed a skill in the Victorian bricklayer not now normally available. The original craftsmen had accommodated variations in size brick to brick, by selection and adjustment so that there was a continuity of alignment which was not precisely mechanical but satisfied the eye. In emulating earlier work over larger areas careful mixing and careful laying achieved a matching result. The new bricks are identifiable by careful surface examination and by frog mark.

Rubber bricks, which had been used in the 'Venetian' arches of the façade, were cut out where they had failed and new material supplied by Messrs Lamb was supplied, sawn and rubbed to the size in matching density and texture. This brick is manufactured by hand from fine brick earth with a high percentage of ferruginous clay. Finishing was by carborundum stone to ensure an ultimate perfect fit.

The mortar analyses carried out revealed considerable inconsistency. All the mortars were lime-based and it appears that the original lime putty was slightly hydraulic. The principal variation was in the addition of a pozzuolan which appears to have been brick dust, whose effectiveness depends upon the active silica deriving from the original sediment in the coal strata. No cementitious material was found. Calcium sulphate revealed in the analyses was taken to be a pollutant deriving from airborne sulphur and possibly from the ashes used which probably derived from the locomotive ash pits and in consequence were the product of coals of a wide range of sources.

In the restoration work eminently hydraulic continental limes (from Belgium, Italy and France) were employed with a carefully selected sand containing substantial grits. In the replacement pointing and the pointing in the new work the same lime was employed with a silver sand and additional grits. Some areas had been repointed in a dark colour to match the original stained condition. These were cut out and repointed to match the remainder of the wall.

Other bricks used in the façade were sand-faced reds and blacks in small quantities appearing in quoins and decorative elements. These were matched by equivalent modern brick.

Additional security has been provided at selected and important points in the junction of the outer skin and the main mass of building by the introduction of short stainless steel rods acting as ties set into pockets drilled in the structure and bedded in lime mortar.

No deleterious effect was observed in the conjunction of limestone and brickwork, and the sandstone originally used has failed by acid weathering and soot incrustation rather than by alkalinity. The higher pH value has probably had an advantageous effect.

In final cleaning of the brickwork an initial brushing with stiff bristles was followed by removal of clinker with sharp chisels. The surface was then carefully dressed with a low-pressure air spray with a water mist using stone or glass powder with particle sizes up to 0.5 mm at pressures up to 40 lb per square inch. In this process operative control is the key to success. For stone cleaning a fine pencil abrasive was used with ultra-fine particles on carved sandstone.

The original Mansfield stone not now being accessible, other stones from the same area were used as replacements – these being Red Wilderness and Corsehill, which are both of similar colour and texture to the original but will be identifiable as replacement work. The replacement mortars in association with the stone were colour-matched using carefully selected sands and some earth pigments to provide a tone value close to the original which may be expected under modern conditions to weather to be consistent with the rest of the façade.

Records of the work now available will provide future restorers with direct access to the materials used and will simplify future maintenance. Any building, whose quality depends upon a high level of finish, as in the Midland Chambers, will inevitably need careful restoration at regular intervals. The conditions of air pollution during the forthcoming century are likely to be very different from those of the 120 years which preceded the recent restoration and careful maintenance avoiding the saturation of brickwork and the avoidance of corrosive cleaners such as had been used will also lengthen the maintenance period. It may be that two centuries pass before this façade requires overhaul as extensive as the recent refurbishment.

13.5 Restoration of courtyard houses in Baghdad (Architect, John Warren with Dr A. Walls)

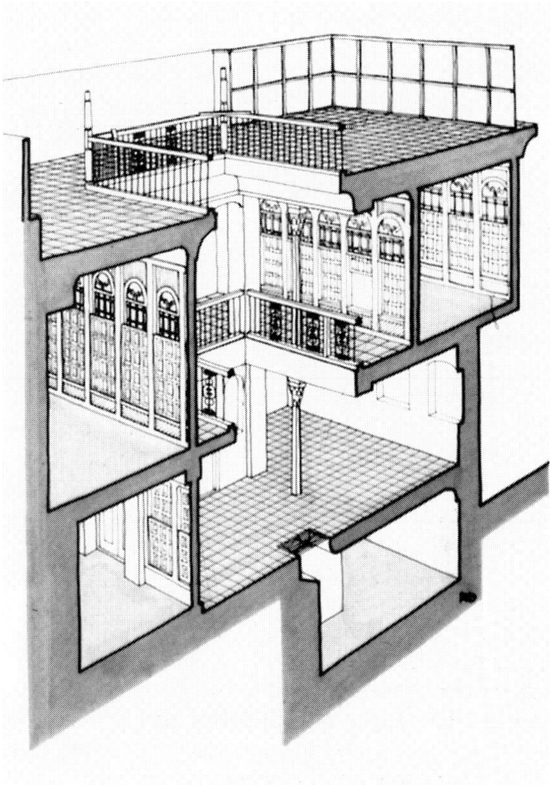

Figure 13.22 House in Kadhimiye, Baghdad. A cross-section showing the lower brick construction giving way to timber. The basement is partially below adjacent street level over which a timber 'shenashil' projects. Party walls are in solid brick.

General description

An evolutionary tradition over several thousand years brought the nineteenth century houses of the cities of central Iraq to a remarkable pitch of vernacular achievement. Their extensive basements and semi-basements were constructed entirely in burned brick with elaborate vaulting incorporating muqarnas and intersecting liernes. The ground floors were of solid brickwork and on the internal faces of the courtyard friezes of pierced brick mosaic vied in elaboration with brick mosaic panels, door-heads and ornamental spandrels. In the upper levels of the houses timber-framing was predominant infilled with timber panels, mud brick or burned brick, faced in plaster. The roofs were built up to fall with earth which would be compacted after every heavy rain except where it was topped with brick paviors to provide a wearing surface.

Materials

The bricks and tiles are made from a relatively lightly burned yellow brick locally known as Farshi (Persian). According to degree of firing it was selected into two or three categories, the best fired of which was always used for structural purposes, particularly in the basements and the less or more lightly fired was used for the extensive cutting and shaping which provided the components for the mosaic pattern decorative friezes and inventive detail that meant so much to the inhabitants of these inward-looking, enclosed homes. This soft brick, like a 'rubber' brick, could be sawn, carved and chiselled into shape, slabs as fine as 30 mm being readily cut out with bow-saws.

Mortars used were gypsum (juss) and lime, in varying combination.

Nature of the problem

Two principal faults dominated the conservation of the brickwork. Structure in contact with the ground was subject to very severe salt of impregnation from sulphates, nitrates, carbonates and other pollutant materials. The land on which the houses had been built was the product of a long period of previous habitation and, therefore, contained substantial organic remains, reinforced by the effluent from localized sewage disposal. With the introduction of modern services, water supply and drainage systems had been constructed. With ageing there had been extensive leakage of potable and foul water producing an artificially high water table under the historic areas. This water table was deeply impregnated with natural salts, reinforced by the products of decay. In consequence almost all brickwork in contact with the ground produced a continual bloom of efflorescence and was subject to breakdown with crystalline expansion.

The prime agency of decay in the upper levels of the building is the termite. In the warm, moist soils of Mesopotamia many varieties of termite (or white ant) flourish. The upper storeys of the houses constructed of poplar wood have a relatively short life, the timbers being hollowed out and often completely eaten away beneath an apparently sound surface. There is a mixed variety of consequential structural failure.

Figure 13.23 Traditional Houses in Baghdad. Failure of cut brick spandrel panels. Serious structural movement caused the breakage of panels of sawn decorative yellow brick, which were taken down and reset on a new bedding with replacement inserts as necessary. Paper templates of the joint patterns were made for each panel.

Figure 13.24 Rusafa, Baghdad. Semi-basement vaults of a house showing solid brick construction. The soft yellow Farshi (i.e. Persian) brickwork, laid sometimes in gypsum mortar, is elaborated here into liernes carrying a low patterned cut-brick dome.

The remedies to these two problems are outside the scope of this study other than in general terms.

In principle, the problem of saline deposits is susceptible of a neighbourhood solution in which all services are renewed to eliminate leakage. In some areas ground barriers of frozen material were necessary to provide stabilization, the freezing being accomplished by the introduction of a pattern of tubes through which refrigerated brine can be circulated. Severe salt impregnation is susceptible of treatment with sepiolite or attapulgite clay poulticing or alternatively the use of cellulose poultices and in many instances the deep raking out of joints and repointing provides a mechanical method of removal of pollutants.

Brickwork in the upper levels of the structures had cracked and settled, panels bursting out of their timber framing as loads changed with loss of structural support as timber failed. Each panel was given temporary support by the widespread use of screw-jacking with occasional wedging. Manual methods were used throughout to restore the structure to its original levels while maintaining a system of temporary support throughout. The principle applied was that all new temporary loadings should be applied as broadly and uniformly as achievable and in this manner entire floors could be raised, defective timbers removed and replaced with timbers impregnated against further attack.

Re-establishment of the structural positions of the original construction was doubly im-

Figure 13.25 Kadhimiye, Baghdad. Repaired panels of brickwork reinstated. The panels have been taken down bodily, reassembled against paper templates, and repositioned as a single assembly.

Figure 13.26 Baghdad. An elaborate cut-brick mosaic portal. In this relatively undamaged example the background is intact and the repair consists of inserting replacement tesserae into voids from which failed pieces have been removed or lost.

Figure 13.27 Kadhimiye, Baghdad. Comparative patterns of brick mosaic in portals. There is a wide range of geometric patterning in brick mosaic. The principal types, all of fundamentally similar construction, are illustrated here. (With acknowledgement to students of Prof. Dr Mohammad Makiya.)

portant because in the complex interlocking pattern of the courtyard houses of Islamic cities party walls are found on two or three sides of the structure and any failure within one building is reflected in its neighbours.

The materials used were limited to the locally made brick originally burned traditionally using wood. The bricks were made from a clay with a sufficiently calcareous content to fire to a straw colour. These Farshi bricks had been burned at relatively low temperatures and were of a softness which allowed them to be easily scored and fractured. It was this characteristic that allowed the use of the one material for cutting for mosaic work, carving for bas relief and as a general brick for structural purposes. Much the same material provided the body for pottery and where elements of colour were required it was the underbody for pigmented glaze, typically blue and green glazes used as pinpoints in brick mosaic. The same clays are now fired using oil fuels. The more lightly fired of the modern products are a close match to the original material in colour, density, porosity, texture and compressive strength.

This provided the material from which mosaic components or tesserae were cut to make up the intricate geometric shapes required in spandrel and frieze. Selected tiles were sawn on their length to provide slabs as thin as 30 mm and these were then cut to shape using saw and chisel, the geometric profiles required being set out on a paper template. Some, involving star shapes with acutely pointed arms, might demand three or four attempts to cut them before a successful result was achieved. The sawn face of the brick would be rubbed down against another brick prior to the cutting being undertaken.

The traditional method of assembly was to lay the cut bricks on a template in panels face down in a bed of sand contained within a former made to the profile of the lunette or spandrel. A slurry of modified gypsum/lime plaster was poured in securing all the components. Finally the recessed joints would be pointed up in lime mortar and the entire panel bedded in position. This same technique was employed in restoration, the template being formed of paper cut with openings into which the cut bricks fitted. The cut-out sections of the template acted as controls over the shaping of the components.

Larger and often modelled bricks of similar composition were applied direct into a backing of gypsum plaster (modified with a little lime, 1:3:8 lime:gypsum:sand) on the basis of a design pricked out on the backing and demarcated with pencil lines. The regular geometric motifs of complex string courses would be achieved in this way, the decorated course or panel sitting in the same plane as the outer face of the regular brickwork below. The finished quality of the panels depended on the care with which the Farshi brick was worked by the cutters (in this case Muslim craftsmen from Bengal) who finished the faces with fine axe or chisel.

The specifications laid down the type of brick to be used by comparison with specific samples held and retained for reference. Likewise standards of workmanship were agreed and referenced by sample. The skilled operatives were Bengali and Iraqi, only senior administrators understanding the language of the contract – English.

Operating methods applicable to closely integrated societies, where trade practices are highly codified, may not be applicable where there is no common language between operatives and administration and where the methodology varies widely in the absence of common background and continuity. It does not follow that good results are unobtainable in such conditions. Direct methods of instruction based on performance requirements and sample work can cross the boundaries of convention and overcome the vagaries of translation, the initial analysis having been carried through conventionally. The conclusions may then be translated into practical artefacts to which the operatives relate by way of performance and workmanship.

14

Good practice

Centuries of experience and a great deal of research have produced a vast reservoir of knowledge in the manufacture of bricks and mortars, in the behaviour of brickwork and in the associated technology. Much of this may not be directly relevant to historic buildings and the work of the conservator, but the principles and practical information available in the guides, practice notes and standards relating to brickwork do contain much that matters in the proper conservation of brick structures. In principle published guidance is the norm from which the conservation of historic brickwork will depart deliberately and only when a careful examination of the circumstance dictates.

14.1 Before work starts

Except in the case of a structural emergency there should always be time for careful project planning. In normal circumstances the owner of a historic building will be advised of its condition at regular intervals of perhaps four or five years, having commissioned regular routine surveys of the fabric. From this advice it will be possible to plan conservation work sufficiently far ahead to ensure that serious and consequential damage does not occur and to arrange that the conservation works necessary are carried out in a logical order and economical sequence. Scaffolding, once erected, should be used to carry out all tasks in the relevant area and this may involve bringing forward some operations and delaying others. Assessment of need should be made with care. Over-enthusiasm on the part of maintenance engineers, building owners and conservators will result in unnecessary renovation and a perpetually pristine condition. With brickwork this is generally not necessary as the material tolerates ageing and often looks the better for it. On the other hand an incipient crack may suggest immediate action to remove a load, deal with a failed support or provide for movement, and nothing is to be gained by delay. In situations of gradual ageing such as the failure of roof tiles a very useful guide can be the rate of annual expenditure in replacement. If the costs of replacement do not exceed the bank interest to be gained on the money which would be required for complete reroofing the building owner is better served by retaining his fund in a financially productive use. Only when the annual costs of replacement reach the same figure as the yield on the investment for reroofing does wholesale repair become justifiable. This simplistic equation assumes that loss of the occasional failed tile does not present a risk of damage through water entry. Where such a risk is apparent repair takes on a different urgency.

Always, before work starts, a careful record of condition will be made and used in planning. Ultimately the record goes to the archive. Photography is a very important tool in making records.

14.2 Selection of materials

Many conscientious conservators take the view that materials to be used intimately in historic building repair should be sourced, handled and applied in as nearly as possible the same way as the original materials they replace. Compatibility of performance and appearance and integrity in conservation are the driving criteria. There are, however, situations in which the work required must, by choice or force of circumstance, be in new materials. Many traditional materials are not covered by the manufacturing guarantees and standards applicable to modern materials. Second-hand bricks and tiles are, for instance, a very variable commodity which frequently have been badly handled in demolition, recovery and transport. Careful physical examination is essential and, as with most fired clay products, the 'ring' when lightly struck may be a necessary test for each tile. Bricks should be freed of previous (lime) mortar, vegetable growth, salts, painted and plastered surfaces and visible defects. Absorption and other qualities may have to be tested. They must be categorized as to frost resistance and weathering capability. The required characteristics of colour and surface texture must be available if matching is the objective. Tiles must be free of lamination and of rusted pegs, which may sometimes be left in a peg hole having been broken off with a core of unoxidized iron dangerously remaining. Sands, limes, cements, tiles and bricks can all be purchased with reliable standards of certification and in conformity with British or other international standards. Lime is best purchased in tub unless a specific burning is made to match an original material, in which case on-site slaking is an entirely satisfactory traditional method. A slaking pit may be used.

14.3 Brick strengths, weights and performances

14.3.1 Weight

Clay brickwork weighs upwards of about 2 tonnes per m^3 (1.5 tons per cubic yard) for common and hand-made bricks, rising to 2.4 tonnes per m^3 (1.8 tons per cubic yard) for engineering bricks, where normal mortars and joint thicknesses are employed. This information assists the conservator in calculating foundation loads and the effects to be anticipated from new brickwork.

14.3.2 Strengths

It is frequently presumed that bricks are a fairly consistent product in terms of strength. This is simply not so, a factor of some 15 times relating the weakest to the strongest. A minimum load bearing capacity under BS 3921 is 5.2 N/mm^2 (725 lb/in^2). The strongest brick is comfortably in excess of 100 N/mm^2 (15000 lb/in^2) and most bricks fall in the median range between 20 and 50 N/mm^2 (3000–7250 lb/in^2). Bricks classed as engineering generally lie in the range of 50 N/mm^2 (7250 lb/in^2) – 70 N/mm^2 (10500 lb/in^2). In calculation the tensile strengths of brickwork are disregarded. They exist, however, although they are both difficult to measure reliably, and often inconsistent in historic brickwork.

Absorption has been taken as a guide to frost resistance because porosity does not necessarily describe pore size or nature. Absorption is now regarded as too crude a method to be entirely reliable in this regard; nevertheless it can be a helpful indicator. Low absorption, that is less than 7.5% by weight, is generally taken to indicate a frost resistant brick. Some bricks with higher absorption, however, if they are well burned, will be equally resistant and some with low absorption fail under repeated frost cycling. Class A engineering brick tends to have an absorption in the order of 4.5% max. by vacuum or boiling methods and class B bricks go up to 7% absorption of water.

14.3.3 Quality

Bricks are generally classified in four grades of structural quality of material or matrix, namely:

(i) high quality or high density;
(ii) special;
(iii) ordinary or normal or standard;
(iv) for internal use only.

Other gradings – seconds, under- or over-burned etc., relate to shape, character and visual qualities.

Good practice requires the use of high quality bricks in positions of heavy loading and vulnerability, such as below ground and up to two courses above ground.

Special quality bricks are normally called for in conditions of unusual exposure such as the cappings to walls, cills and points of high vulnerability to weathering, such as quoins.

Standard quality bricks are used for external leaf and external facings for weather resistance.

Bricks of internal quality are used for bonding into the inner skins of walls for internal partitions and internal leaves.

Some brick works are able to produce the entire range of these qualities from one clay, but the difference in firing will usually cause the appearance of the bricks to differ, the internal quality bricks usually being low-fired and therefore lighter in colour, and increasing redness and thereafter blueness indicating (crudely) increasing strength etc.

Since these classifications apply to good practice in new brickwork the same principles should be applied where possible in the selection of bricks for reuse and for historic buildings purposes generally. Failure to differentiate between the different qualities of brick will lead to random weathering and unnecessarily high levels of decay in vulnerable positions. Quoins and plinths will fail earlier and a wall will become pockmarked with crumbling brickwork unnecessarily soon, if care is not taken in selection. It is therefore simply inadequate to specify the use of new or second-hand brick without controlling the cleaning, testing and assessment of quality. Specifications should always lay down the quality of brick to be applied in particular areas of work. Where new bricks are being used care should be taken over sulphates in the brick or in the ground water and if the presence of sulphates is suspected low sulphate cement may be applicable with or in preference to lime. Manufacturers can be expected to provide the results of salt analysis across a range of samples. An individual sample is not adequate. The emergence of other salts (efflorescence) is unsightly, but generally not damaging other than in exceptional instances where soluble salts crystallize below the brick surface. In specifying strength and quality bricks of the lowest suitable grade should always be used. Mortar should always be less strong than the brick.

14.4 Mortar characteristics and mixes

Hydraulic lime is prepared from the dry hydrate and unless it is very weakly hydraulic should be used within eight hours of preparation. In essence it is a weak cement which gains further strength by carbonation. The same principles of use apply to lime putty to which a pozzuolan has been added, i.e. PFA, brick dust or a cement (not less than 50% of the putty by volume). Bagged, hydrated, non-hydraulic lime with above 50% by volume ordinary Portland cement (OPC) should be treated as a modified cement mix.

All the foregoing may be used with proportions of sand upwards of 3 to the cementing agent. 'Stronger' mixes than this are unlikely to be considered in historic brickwork, except in special circumstances of industrial, high density and highly loaded brickwork. Strength in this context refers to amount of cementing agent. Increase in its volume by above 25% may in fact produce a weaker mortar.

Non-hydraulic putty should be considered separately from the cementitious mixes, including hydraulic lime. This material depends upon carbonation from the atmosphere for the setting. Water loss precedes setting. Shrinkage occurs with water loss and, therefore, the volume of matrix and volume of filler – sands – have a critical ratio. Equally it is important that sands are graded with a strong preference for the finer elements because this produces finer voids and, therefore, smaller volumes of the cementing matrix in which shrinkage will take place. The strength of the mortar is therefore a product of the quality of the putty and its preparation and the selection of the sands. Sands of uniform size content are to be avoided.

The appropriate ratio of sand to putty can be selected by the simple method of determining the voids left in dry sand. A sample of dry sand is placed in a glass jar with parallel sides filling it about half way. The height is measured. A similar volume of water in a similar jar is measured exactly. This water is used to fill the sand with water until the top surface is exactly reached. The amount of water used is now measured. Its volume is calculated against the amount of sand by simple measurement of reduced depth of water. In the process of filling the sand will have 'bulked' so the total final volume is of no concern. The dry depth of sand is compared with the water used. The amount of water will be found to be in the order of one quarter of the equivalent volume of sand and may be expressed exactly as a percentage of the original height of the sand. If this is 25% it shows that the voids in the dry sand are one quarter of the volume. This may be interpreted as showing that a mix of one part putty to three of sand will contain sufficient putty to fill all the voids. The value of this surprisingly unknown test is that it gives the amounts required for a mix in which the sand particles will remain in physical contact. Therefore, at this ratio (1:3, lime putty:sand) the contraction of the putty on drying will not cause shrinkage in the mortar because the sand grains remain in contact: they have not been bulked out or forced apart by an excessive volume of putty. The point of this test therefore is simply to determine carefully whether the particular sands in use require a cement:sand ratio more or less than one to three, the variation not normally being above 3% either way. Gauge boxes can then be 'fine-tuned'.

Improved workability in mortars may be obtained with the addition of a very small amount of kaolin, a non-expanding clay in preference to organic materials. About 5% of a clay type material passing a 200 mesh sieve is permissible (BS 1199).

As it is the function of a mortar to form the continuous matrix in brickwork the strength of the mortar determines the strength of the brickwork. This does not argue for the use of strong mortars, but rather the other way round. Mortars should be designed to be less strong than the brick, to have good adhesion to it, to have minimal shrinkage and not to be prone to brittle cracking. In many instances a wall of greater strength is produced by a weaker mix because shrinkage cracks occur in cement-rich mortars. The weaker mix is not only more pliable: its performance coefficients are closer to brick itself.

Cements have been weakened traditionally with lime, with inert fillers and by entraining air. Materials such as the white of egg (albumen) have been incorporated in medieval and other historic mortars for this purpose. However, except for specific and well-tried organic air-entraining agents and surfactants organic compounds should not be used in conjunction with mortar mixes. Carbohydrates such as sugars and similar organic materials must be regarded as potential pollutants that must be avoided.

To produce mortars compatible with historic work there are two options: to simulate exactly the original material where it has proved satisfactory or to use a modern equivalent. To do the first, careful analysis is necessary to determine sands and cementing agents. For the latter, careful judgement of the constituents and testing will be necessary to ensure compatibility.

Modern cements are:

1. ordinary Portland cement;
2. rapid hardening Portland cement;
3. Portland blast furnace cement;
4. sulphate resisting Portland cement;
5. other rapid hardening cements;
6. further variants or mixtures made for specific applications, including high alumina cement, white cement and masonry cement.

Rapid hardening and high alumina cements are not applicable to work in historic buildings. Blast furnace and sulphate resisting cements are used only in special circumstances. Masonry cement, which is a mixture of Portland cement with an inert filler (powdered limestone or equivalent up to 40%) and sometimes an air-entraining agent, is designed to have good plasticity and minimal shrinkage. It is, therefore, already a modified cement mix.

All Portland cements may be mixed with hydrated or putty lime, but never with gypsum-based materials.

To improve workability or plasticity a sand carrying a small proportion of a non-expanding clay may be found to be an advantage. Alternatively, as already noted, a small proportion of kaolin (the most readily available pure non-expanding clay) may be incorporated. If artificial plasticizers are used their effect derives from alteration to the surface tension of the water and the consequent incorporation of more air. Overworking will produce an excess and consequently a weakened mortar.

Wherever possible mortars should be coloured by the choice of sands primarily, secondly by the cements and lime used and only if satisfactory tones cannot be achieved should mineral pigments be added. Caution should be exercised over mixes containing more than 3% of mineral additive by weight of cement, although some authorities permit up to 10% by weight other than for carbon black.

14.4.1 Batching and mixing

Good practice requires care in measurement. For proportioning gauge boxes should be used. A careful and experienced operative can achieve reliable mixtures by the shovel but if permitted this method should always derive from initial trials with gauge boxes, i.e. the first mixings should always be based on the number of carefully and consistently estimated shovelfuls used to fill the gauge boxes. Used with less than great care this method is unreliable. The danger is its convenience; it is a short cut.

Sand bulks when damp. Damp sand will be in the order of 140% by volume to 100% volume of dry sand. Specifications will call for dry sand and proportions are given by volume. Hydrated lime will normally be mixed with sand in the dry state prior to the addition of cement.

The standard proportion of lime and cement in relation to dry sand is 1:3 due to the void ratio in sands and this translates frequently into 1:1:6, lime:cement:sand if an equal volume of lime and cement are desired. This commonly used mix is a good compromise between softness and sufficient strength to resist erosion. It is used primarily because of its mathematical simplicity which makes reliable control easier. Weaker mixes provide for voids within the cement matrix and stronger mixes cause the sand particles to be held apart.

The setting action of most mortars has significantly begun after one hour and after two hours is well under way. Beyond this time mortars tend to lose strength because in the knocking up and laying some of the crystalline structure has already formed and is damaged beyond recovery. Mixing should therefore take place at two hourly intervals where mortars contain cement or, in the case of limes, are eminently hydraulic. Non-hydraulic lime mortars may be mixed and retained in the wet and indeed pure lime mortars may not be retained until some days if not weeks have elapsed. Mixes are normally specified by ratio of volume as cement:lime:sand. Thus:

1:3 1:1:5–6 1:2:8–9 1:3:10–12

Where lime and sand have been mixed initially the wet product is referred to as 'coarse stuff' which may be specified in volumes with ratios of 1 lime to anything between 3 and 12 of sand. These mixes may then in turn be specified by ratio of cement to coarse stuff, the weaker mixes accepting a higher proportion of cement. For coarse stuff the specifier may substitute a dry mix of hydrated lime and sand. The objective of this two-part specification is to obtain the right balance of lime in a mix which is then allowed to mature for a long period. In work with pre-industrial historic buildings a mix of 1:3 or 1:4 will be used without further modification.

With such two-part specifications the objective will always be to provide a ratio in which the cement and lime combined represent between up to 25% of the total mortar by volume and the right balance is achieved between cement and lime. Therefore a weak premix of lime and sand – say 1:5 used in a ratio 1:4 cement to coarse stuff or premix – gives a mix 1:4:20 cement:lime:sand, which is a significantly modified cement mortar a little high in sand and therefore below maximum strength. Cement in this instance refers to ordinary Portland cement or its equivalent.

Hydraulic mortars vary according to the bed of limestone used as source material. The Lias lime currently (1996) in production in Britain is derived from Lias beds in south Somerset (Hydraulic Lias Limes Ltd, Sherborne, Dorset). Such mortars will set under water after 50 to 60 hours and will reach a 28 day compressive strength of 1.5 N/mm^2 in a 1:2 (by volume) mix. This strength will increase to 2.5 N/mm^2 within a year.

This material should always be used within two hours of mixing. It must not be 'knocked up' after that time and the minimum of water will be used in the mortar. Good practice in prewetting and maintaining moisture during setting is essential. It may be used alone or with 'coarse stuff' gauged 1:6, matured and then used 1:1 with 'coarse stuff', giving a 1:1:6 mixture.

Good mortar specification is the result of careful judgement and experience. Since sands, plasticizers, limes and cements can all vary very considerably the options are very wide, although all mixes within the range normally acknowledged will produce satisfactory mortars. Experience is gained and certainty in work is best assured by careful experiment. Because mortars derive their strengths and other characteristics over a substantial period the best practice involves either the transfer of experience from one project to the next or initial experimental mortar mixes and trial repairs or walling well ahead of the construction period. Where there is any lack of certainty and adequate time experimental mixes and appropriate use should be tried out three months ahead of the general construction work in conditions equivalent to practical site work. Mixing is now carried out mechanically by methods which simulate turning-in by hand. The effect of this operation is to trap air. With the appropriate additive in place this effect is enhanced and if mixing is carried out for too long a period the mortar will be over-light and significantly weakened. In the mixing of lime alternative methods are recognized as being important. The objective of mixing lime putty is to ensure that every sand particle is coated with the putty and for this purpose the rotary mixer is much less effective than the roller pan type of mixer in which rollers on rotating arms traverse the coarse stuff. This method, which is in effect a form of milling, has the effect of compressing the particles together and forcing the putty against the sand. Traditional hand mixing techniques as normally practised by operatives are inadequate and if hand mixing is employed it must be carried out in conjunction with ramming or pounding. The process is laborious and about an hour's work should be allowed for each mix, followed by a lengthy period of storage.

Two-stage mixing involves the use of a wet semi-plastic material with further additions such as a cement and/or hair or more exotic dry additives such as chopped polypropylene. This second stage mixing is best carried out in a trough which will stand pounding using rake furl and a flat-ended plunger. A mechanical paddle-ended mixer may also be useful at some stages and with some materials. Additional water is added only with caution. Water should not be added as a lubricant because its evaporation from the placed mix causes shrinkage.

Similar equipment with protective covers may be used for hot mixes. It is possible to work with hand or mechanical tools and proper hand and arm protection in a trough over which heavy gauge clear polythene sheeting has been arranged so that there can be no accidental discharge over or on to operatives. An enamelled iron bath set over a gentle fire or a tank into which hot water is circulated can be used to prepare a hot mix using cold slaked lime putty. Tallow or other organic materials can be introduced and dispersed in the normal way. Alternatively powdered quicklime mixed with dry sand may be introduced with care to the mixing water in such a vessel and the heat produced by the exothermic reaction of slaking may then be sufficient for a hot mix to be prepared. Protective clothing for these operations will include overalls, gloves, a face mask and goggles.

The prime objective of a hot mix is to ensure the effective introduction of secondary materials, and the procedure is used for making lime slurries and washes. The effect of the hot caustic slaked lime upon sand is to etch the surfaces of the grains of silica and improve the bond. However a phenomenon sometimes called 'water burn' can result from the use of excess water in the preparation of hot mortars. If this operation is attempted it must be done with the benefit of full protection and by experienced, prudent operatives. Very thorough dry mixing is essential. Eye and skin protection must always be used and ample supplies of cold water must be at hand.

14.5 Damp inhibition and protection

Builders have long battled to keep water out of their buildings and to prevent the migration of water from saturated sections of brickwork into dry walling or into other materials. Four methods of protection have been used – damp-proof or damp-inhibiting layers or coursing, rendering, the construction of separate leaves of brickwork and the application of renders. In addition the construction of brickwork using string courses and ensuring the discharge of rainwater away from the wall have worked on the principle that brickwork should have a greater capacity for evaporation than absorption in its specific circumstance. The first element in good practice is, therefore,

Figure 14.1 Diamond bits for cutting out mortar joints. Industrial diamonds bedded in nickel are used to cut mortar joints cleanly and without damage. An entire brick can be isolated for removal. (Courtesy Progrout Ltd, Plymouth.)

to ensure that brickwork is sufficiently protected for this simple rule to apply. High level brickwork under an eaves is clearly a very different problem from brickwork immediately above wet ground on the weather-face of a wall. An indicator is plant growth and a typical example of failure to ensure that the brickwork is able to meet these criteria may often be seen where a water discharge runs on to brickwork through failure of rainwater goods or an unattended overflow. The saturated brickwork will grow mosses and lichen and if left will be susceptible to frost damage and decay, pointing and the weaker bricks failing first. Adjoining identical brickwork which is protected from saturation will behave normally. In conditions of serious exposure builders have often resorted to rendering which increases the rate of run-off and, therefore, the apparent protection of the wall. Much pebbledash rendering has been applied with this philosophy, sometimes over important and valuable historic brickwork. Strong coatings of this sort will often crack and where the cracks run at or near the horizontal they can prove serious sinks for water coursing down the wall in heavy rainfall. The water builds up behind the semi-pervious render and worsens the protection rather than otherwise. A render to be effective must be free of cracks and be well maintained. It can then be an effective method of reducing the amount of absorption and if painted can be even more efficient. The render does, however, inhibit evaporation, particularly if there is an external coat such as a paint on the outer face. It is then particularly important to ensure that moisture can migrate out of the wall if satisfactory conditions are to be maintained. In particular it becomes important to avoid putting an impervious material on the inner surface of the wall. A mistake is often made by failure to understand the principles involved where parapet walls are rendered externally. Where possible the inner face of an externally rendered wall should be left free of covering to improve evaporation. Good practice is good maintenance plus an eye for weaknesses in detail in the original construction.

Damp-proof courses are commonly applied in the bases of walls under cappings and parapets and in more sophisticated construction in chimneys or within the wall thickness at window and door openings. Before the nineteenth century damp courses in these positions were usually slate laid broken bond, blue brick or perhaps both with the occasional but increasing use of bitumen. From the mid-nineteenth century lead, although perhaps the most efficient form of damp-proof coursing, seems not to have been used to any significant extent in historic buildings, despite its constant use for roofing and ready availability.

Much historic use of damp-proof membranes and courses was not well considered. Bituminous layers can appear where they have no function and where used logically may be bypassed or stopped short. Indeed the practice of damp-proof coursing was generally more an ad hoc matter than a carefully thought-through practice and if considered at all might go no further than the use of denser or near waterproof bricks used perhaps more for their frost resistance and suitability in foundations than for their waterproofing quality.

In historic brickwork the essential principle is that a positive balance in favour of evaporation is the objective. Where the judicious introduction of a semi-permeable layer will sufficiently reduce absorption, this is sufficient.

In brickwork it is generally good practice to include a damp-proofing or damp-inhibiting layer or membrane where water can saturate the topmost or lowest courses of the wall and where the distance between the outer face of brickwork and the inner face is less than 300 mm. This situation usually occurs at a door or window opening which if made in solid work may (structural considerations apart) be chased or cut to allow the insertion of a barrier to water which will increase the length of path. A 100 mm cut into brickwork behind the line of a window frame will effectively increase the path to be travelled by water by 200 mm, provided that the cut and the inserted membrane do not themselves open a track for free liquid. Careful insertion, bonding, sealing, laps and weep holes are, therefore, essential in good historic building detailing where water entry has to be overcome.

A danger in the use of modern materials is the introduction of slip planes or fracture lines not previously considered. Materials such as polyethylene introduced into brick coursing do not have the same bond of adhesion through the mortar as there is between bricks and the mortar bed. A mass of masonry can therefore move on a slip plane created by the introduction of a damp-proof membrane given conditions where the building is shaken by wind or vibration. Such contingencies must be foreseen. Where possible the general standards and codes of practices applicable to modern brickwork should be taken as a guide from which departures are to be made in deference to historic construction although not so slavishly as to perpetuate operations which would now be regarded as bad practice and the cause of building failure. Minor deviations may be acceptable where, for instance, it would be normal building practice to bring a damp-proof membrane to the outer surface of the mortar. It may perhaps be stopped off at pointing depth and concealed in the final pointing, the conservator being satisfied that the inhibition of rising damp over 95% of the wall area is sufficient to achieve his or her objectives.

Water discharge is a frequent cause of failure. In some instances the effects may be obvious. Water being driven the wrong way off a roof or coming from a broken gutter will be seen to be streaking the brickwork below it and be the cause of saturation. Repair or alteration of the water flow are straightforward remedies. Other problems are more subtle. A worn brick cill or a string course eroded sufficiently for water to continue on its way down the wall rather than be thrown clear are likely to be less apparent as defects. They may, however, increase the absorption in localized areas beyond the capacity for evaporation and so cause the build-up of moisture within the wall itself. Careful examination of the functioning of wall details, and sensitive remedy as necessary, are an essential part of good maintenance.

Alien materials in brickwork sometimes give rise to problems. Calcareous stones used as dressings, cills and string courses may be the cause of staining on brickwork which can be removed by the use of hydrochloric acid solutions. The introduction of titanium, stainless steel or similar metallic strips into or under an appropriate course can be used to throw a discharge clear of the wall to avoid recurrence of staining. Other special circumstances may suggest the use of further types of impervious barrier. Termites may be blocked by the introduction of copper sheeting in rebuilding walling and this in turn, in conjunction with the use of proofed timber, may avoid the extensive use of chemicals in poisoning the

Figure 14.2 Mechanical grouting: a non-pressurized auger fitted to a standard electric drill. This tool places grout accurately and precisely very much more quickly than the hawk and trowel. (Courtesy Progrout Ltd, Plymouth.)

ground – a practice of last resort. More commonly vapour barriers may be required. While it is not usual to seek to inhibit the movement of water vapour through or out of brickwork there may be circumstances in which brickwork is to be protected against condensation. It will be good practice to investigate changes of use which produce changes in probable maintained water vapour pressures. Cold solid brickwork enclosing a space which becomes inhabited and is therefore filled with warm moist air continually will accumulate condensation within its structure. It may have insufficient 'breathing' or evaporative surface. It is normally possible to calculate such effects, and where the balance is of moisture entering rather than leaving the wall the application of a material containing a plastic or metallic vapour barrier against the wall surface is to be considered against other historic buildings criteria, such as the nature of the internal surface.

14.6 Pointing

The general principle must be that new mortar is durable and compatible with the host brickwork, being of sufficient depth to remain adequately secure. To achieve this the course must be cut or raked out to a distance at least equal to its face depth. The inserted pointing mortar will then be approximately square in section with parallel faces. It should not lie in a 'V' joint. In placing mortar modern equipment can deliver material more quickly and accurately than hand methods.

14.7 Bonding with new structures

It is sometimes necessary to bond new and existing work. Good practice demands attention to the possible need for movement tolerance and also for the transfer of loads. Physical bonding by toothing, stepping or by the introduction of pins may achieve transverse or vertical restraint leaving the use of mastics or membranes to provide for lateral movement. It will be important to understand the bonding in the historic wall before designing the joint.

Although it may seem obvious an essential component of good practice is to ensure that the structure contains a full adequate bonding and this may not always be the case. All junctions and outer skins secured by relatively few header bricks are cases in point and it may be necessary to introduce stitching members of stainless steel or polyester bonded glass fibre rod to re-establish linkages and ensure the transfer of loads.

Early cavity walls were linked by normal or special bricks to the inner wall and various patent types, cranked and glazed, were introduced. Rat trap bond is successful on the principle of a positive balance in favour of evaporation, but in any such construction dependent upon header bricks for the bond the fracturing of a number of linking bricks seriously threatens the viability of the whole construction. Their integrity is often difficult to verify.

Bonding new work to a damaged structure may be profoundly unsuccessful, damage to the original work and loss of restraint being the prime failures. A new structure should not impose damaging loads on an historic building. With this in mind the ground effects of the new work must be foreseen and separately provided for. A toothed brickwork connection between old and new structures should not be expected to transfer additional loads to a historic building, but it may be used to provide restraint.

Patent stainless steel fixings and ties now available should be employed where movement is to be anticipated.

14.8 Insulation

Very few historic buildings have been insulated to the standards now applicable to well-heated modern buildings. Not only does a change in the internal regime and the possible changes or intensifications of use produce high vapour pressures: regulations, prudence and good practice suggest that where possible improved standards of thermal insulation are achieved in the interests of the user, the owner and the environment in reduction in fuel consumption.

Where historic buildings have been plastered internally there may arise the opportunity to apply a new wall lining carrying insulation and possibly a vapour barrier over the historic plasterwork without necessarily destroying it. The quality of historic finishes may not, however, permit lining out nor will there be a cavity which can receive blown insulation. The most that can be done is to ensure that the wall is kept dry and is therefore less thermally conductive.

14.9 Monitoring and recording

Good practice in assessing the state of a structure extends more widely than the condition of the brick, tile and terracotta to the circumstances of the whole built fabric. While initial measurement of the visible structure is straightforward, extended time may be required to establish the thermal and moisture regime and to determine the nature of movement of cracks or other structural discordances. If mature judgements are to be achieved sufficient time should always be allowed for the analytical investigation of the condition of a historic building. Monitoring may include the use of solid state logging equipment which can record temperature, conductivity (moisture content, salt content indicators) and RH values. This information can be recorded electronically and presented graphically and modern equipment will differentiate between normal values and precautionary or abnormal conditions if required. The use of solid state electronics in modern equipment allows selectivity in records held – only values falling outside acceptable levels being held on store. Warning devices can be used and information may be remotely downloaded so that several sites can be monitored simultaneously. These techniques apply to overall environmental conditions in buildings rather than the specific performance of brickwork but a sampling may include humidity in masonry as a specific or as part of an overall monitoring pattern. Where sophisticated methods are not appropriate visual or hand-metered records of moisture or other conditions should be made over a planned period. Only careful recording can provide a good long-term base for planning a campaign of conservation.

14.10 Purposes and principles

The conservation of brickwork will always have a structural and a cultural objective. There may also be an economic objective in minimizing further investment or achieving maximum utility from a structure, but this should always be a secondary consideration. It is always a prime requisite that the entire operation from conservation decision-making through to the entire range of skills is in the hands of people properly trained and equipped with the facilities and time to monitor the structure.

When the building has been fully understood and the condition of brickwork defined and recorded the statement of objectives can be formulated. Such judgements cannot be made in isolation. The insistence on the retention of an entire brickwork mass might have the effect of preventing the new use of a historic structure which thereby becomes redundant. A balance of objectives must therefore be identified and translated appropriately into a campaign of repair. This should be a conscious, intellectual decision based on careful analysis recorded and presented for the reference of all concerned and for the future.

Minimum intervention is a well recognized principle applicable as much to masonry as to any other structure. It is allied to a conservative approach which avoids unnecessary work and repair. The impact of repairs should always be kept to a minimum and the least harmful or radical solution to a problem is generally the most satisfactory way forward.

A programme of repair and restoration should always aim to retain the time-worn characteristics of brickwork while at the same time retaining the structural integrity. Where restoration or renewal is undertaken time-worn characteristics should not be faked. It should always be possible to introduce sympathetic new work in a manner which is compatible with the existing structure. If salvaged material such as tile, brick or terracotta is reused the nature of the workmanship should make clear the nature of the intervention. Restoration and replacement need not be aggressive and the choice of new materials should be both sympathetic and designed to weather towards the condition of the historic structure. Restoration should not aim to produce an immediately indistinguishable effect but to produce a sympathetic restored structure in which the new components weather down ultimately to a condition compatible with the historic structure but always distinguishable from it. Sometimes it may be necessary to reface an entire surface in which case historical evidence on the obscured surface can be repeated in the new material thereby repeating and continuing to portray evidence which allows the observer to interpret the building. There should be a general presumption against restoration and visible intervention. Where this presumption is overridden by the broader objectives of the conservation of the monument it is important that the qualities of the building in historic terms, including the narrative that arises from their weathering, are respected and not diminished.

Figure 14.3 Identifying replacement materials. A modern brick used in repairs carrying (on the reverse) the maker's name (Redbank), with date and serial stamp from which its precise origin can be determined.

The overall objective dominates the particular but where the particular is a specifically important piece of work its retention will become part of the overall objective. Thus a singular and important element of brickwork might assume a special place in the overall plan of conservation demanding sacrifices of use or special arrangements to ensure its survival.

Special criteria may apply to particular objects. An ornamental glazed tile, a panel of encaustic tile flooring or a fine piece of terracotta may all deserve special protection and even removal as objects of art value, their places being taken by appropriate replicas.

Careful recording requires that evidence is provided of the condition prior and subsequent to the intervention.

There will generally be a presumption against the alteration of details and materials, so that historic brickwork repair anticipates that the structure will subsequently be as unaltered and as authentic as possible. The presumption against the introduction of new materials, new methods, new details, new techniques may, however, necessarily be overcome in order to achieve the prime objective of retention of authentic material. A dripping brick vault may be waterproofed by the superimposition of an asphalt layer at a higher level and the provision of pipework capable of extracting excess moisture from above it. A teetering pier may be secured by visible new brickwork bonded to it by stainless steel stitching. An ancient tiled floor may be protected by the removal of traffic on to a modern walkway. The ramifications of the essential principles translate into a wide variety of possible actions and the problem does not cease at the perimeter of the existing physical structure. An elevation of fine polychrome brickwork may have been conceived as a significant architectural entity and its value extends into the space surrounding it, so that the introduction of other masses, other colours and other patterns impact upon its quality even though they may be physically separated. Simply to extend the polychromy may be the worst possible solution, apparently extending the statement beyond the original intention.

So it is that the ambience of the structure, its visible effect and its structural nature extend beyond its periphery. Brickwork, much as any other powerful form of building, creates about it a resonance to which materials newly introduced respond with sympathy, and react with a satisfactory harmony or with discord. The ultimate aspect of satisfactory conservation is the control of this aesthetic response in the curtilage. The art, the objects and the intentions of conservation in brick, as in other aspects of building, demand clear mindedness and the application of dedicated and balanced skills.

Appendix

Checklists for conservation activities in brickwork

Checklist for support, under-building and cutting out

Avoid sustained heavy percussive activity in favour of cutting, grinding and light percussive methods.

Allow all impact a point of pressure release by creating adjacent void by drilling or grinding.

Apply support pressures evenly and without total reversal of loadings.

Avoid sudden upthrusts which might damage internal bonds in brickwork.

Ensure that temporary support is even and stable.

Control lift using screws and folding wedges, avoiding sudden leverage and reversals of thrust.

Checklist for repointing, pointing and finishing

Ensure that the face material will extend to a depth in the joint at least equal to the width of the joint, with a minimum depth of 20 mm (¾") in fine joints.

Use drilling techniques in cutting out – i.e. routers in preference to hammer and chisel. Where chisels are used, work away from a void with small impacts.

Create a void to be filled with parallel sides the full width being held over the whole depth.

Clean the void of loose and friable materials by washing, air blasting or brushing out.

Wet bricks and brickwork well in advance. Do not allow surplus water to be available to enter the superficial layers of the incoming mortar. Wetting must be such that water is neither drawn out of the mortar into dry masonry nor introduced to the mortar making parts too wet.

Avoid techniques which overwork the mortar, drawing binders to the surface.

Avoid mortar profiles which form thin edges. These will be a source of failure.

Avoid profiles which allow rain to dwell on the detail.

Use equipment (such as low pressure power grouters) which deliver the correct amount of mortar with minimal loss on to adjacent brickwork.

Protect from drying out, from excessive heat, from frost during the whole of the setting period.

Repoint to a colour which will weather towards that of existing work rather than creating an immediate match which will become progressively unsatisfactory.

Provide freshly mixed material at frequent intervals – perhaps 20–30 minutes.
Do not 'knock up' or re-temper old mixes.
Protect from rain and run-off during setting.
If desired hose or dry-sponge the finished surface while still 'green'.
Use appropriate profiles.

Checklist for cleaning

Do not abrade a fired surface which is part of the required finish.
Use materials and pressures which are softer than the brick being cleaned.
Do not use materials or equipment which can leave a staining residue.
Clean off or wash off all agents used.
Undertake careful trials using various methods where applicable.
Dissolve rather than abrade.
Use minimal volumes of liquid – intermittent washing may be appropriate.
Avoid pollutants being sucked into brick.
Avoid wetting brickwork when frost can occur.

Checklist for cutting and abrading bricks

Protect sound material.
Support soft brick edges during abrasion with wooden boxes or supports.
Use techniques that provide continuous cutting rather than substantial impact.
Avoid shock, physical or thermal.
Use mechanical tools such as grinders in a fixed bench rather than hand-held.
Use a circular motion when hand-rubbing on York stone or carborundum bed.
Prepare accurate templates.

Checklist for mortar preparation

Ensure that the mortar design suits the application.
Check that all dry materials are fresh, clean and free of contaminants.
Check that putties and coarse stuff have been retained wet.
Mix powders thoroughly before wetting and ensure that components are batched, weighed or measured carefully.
Use minimal water, in successive additions.
Avoid synthetic additives for colour or workability.
Prepare sufficient only for a work period shorter than the time taken to start to 'go off'.
Use mixing technique which applies pressure to the mortar – paddles or augurs in preference to lifting and dropping, dense rubber in preference to steel.

Checklist for mortar design

Ensure that the design will produce a mortar less hard and more permeable than the brick with which it is used.
Include a balance of particle sizes with appropriate angularity.
Provide an appropriate setting agent(s) – hydraulic lime, cement(s), pozzuolan or carbonating lime. Where previously used, provide gypsum mortar.
Adjust colour by the choice of materials in preference to additives.

Procedural checklist

All conservation activities should be carried out against a clear discipline to ensure that all (and only) appropriate steps are taken.

This checklist sets out a procedure widely applicable and not confined but adapted to brickwork. Short cuts are always applicable, provided they are taken knowingly, recognizing the processes or steps being by-passed. In the sequencing of choice the checklist shows which operations are being passed over.

Preliminary assessment should determine:

- nature of construction;
- type of failure(s);
- possibility of latent faults;
- stability of rate of change;

and may include determination of structure and thickness, consistency of brick type, nature of mortar and renders, provision of water inhibitor or barrier, possible aquifers and/or discontinuities, location of faults or failures, effects of adjoining structures and support, failures of associated materials and/or services, rates of change, movement or penetration.

Need for temporary restraint indicates:

- immediacy;
- options in methods of restraint;
- availability of resources;
- cost consequences and choices;
- side-effects of restraining actions;
- notifications;
- preparatory or warning arrangements;
- other structures or services required or affected;

and must take account of speed and type of known and potential movements and loads, assessing brick and mortar strengths, bonding and construction, repair and support available in terms of type and cost of equipment, the transfer of loads, changes induced by alterations in static conditions, humidity and activities in associated structures.

Recording, ensures permanent knowledge of:

- the shape and nature of the structure;
- the fault(s);
- measurement and sampling for analytical purposes, followed by relevant archiving.

and will include descriptions, historic evaluations as well as dimensions, to differentiate types of brick, their origins, possible discontinuities, alignments of visible defects, interactions with other material services etc. Damp and salt impregnation measurements may be added and decrepitation, plant growth or other visible symptoms may require delineation.

Detailed investigation demonstrates:

- nature of construction and type of material;
- origins of materials and sub or supporting structures;
- history of construction;
- history of fault(s) or failure(s);
- possibility of contributory unknown and unobserved factors.

which may involve physical removal of parts of the structure to provide evidence of types of material, penetration by foreign materials, decay or oxidation, and voids. Laboratory analysis may be required for evaluation of water content, salts, growths, mortar type or even historic evidence of origins of brickwork. Non-destructive investigations may be used to determine voids, reinforcement and internal construction. Thermography, ultrasonics, magnetometry and ground radar may in some circumstances have a part to play.

Investigations may have to take place over an extended period before helpful results are available.

Strain gauges, flat jacks and other testing devices may be required if structural information is needed.

The programme for intervention determines:

- whether there is need or justification for action;
- the ethics of intervention;
- availability of resources and funds;
- test results required;
- options open and choices to be made;
- sequence of operations;
- involvement of and notification to other parties;
- continuing investigation of unknown or indeterminate factors;
- priorities;
- contractual actions;
- contingency planning;

Implementation of programme to finalize conservation, repair or restoration by:

- builder's work
- protective or prevention activities;
- provision of information;

in particular pre-programming of interventions will yield savings and improve quality, eliminating unnecessary delays (e.g. while special brick or terracotta is manufactured), allowing other useful work such as glazing or pointing to proceed while scaffolding is in place. The integration of all work processes cannot be left to commercial operators, each of whom is concerned only with a specific activity even at the expense of another operation. Careful contingency planning will ensure the availability of materials, equipment or services which might otherwise cause serious delay if the need had not been foreseen.

Contingency planning must extend to financial and regulatory matters which can halt operations completely.

Ultimate recording should include:

- intervention(s) or other action, taken;
- historical information obtained and the provision of explanatory material;
- further actions required with reference to timing and methods.

Final recording will add to initial surveys an account of work done. For this purpose the initial analyses may be rationalized for the superimposition of final information.

Photography in the course of work is a valuable contribution. Internal construction of brick walls can be shown to supplement 'before and after' information.

The placing of records in secure storage is vital. In important cases duplication of some other form (e.g. electronic or film) of retention is necessary.

Instructions on future management of the brickwork will be placed in the hands of those responsible for the building as a part of overall guidance.

In all these matters brickwork is only one component of the wider conservation activity.

Bibliography

Bibliography

In addition to the books listed below, important material appears in journals and occasional publications. The range is very wide. Some key references are given under case study subject headings. The location of the publisher is quoted if the source is outside the UK.

Addleson L. *Building Failures: Diagnosis, Remedy and Prevention*, 3rd edition. 1992.

Ashurst J. & N. *Practical Building*. Vols 2 and 3. 1988.

Barley M. W. *The English Farmhouse and Cottage*. 1961.

Barley M. W. *Houses and History*. 1986.

Beswick M. *Brickmaking in Sussex*. 1993.

Brick Development Association. *Bricks, Their Properties and Use*. 1974.

Brodribb G. *Roman Brick and Tile*. 1987.

Brunskill R. W. *Brick Building in Britain*. 1990.

Brunskill R. W. *Illustrated Handbook of Vernacular Architecture*. 1970, 1978, 1987.

Brunskill R. W. and Clifton-Taylor A. *English Brickwork*. 1977.

Clifton-Taylor A. *The Pattern of English Building*. 1952.

Cox A. *Brickmaking: a History and Gazetteer*. 1979.

Davey N. *Building in Britain*. 1964.

A History of Building Materials. Phoenix, 1961.

Dobson E. AZ. *Rudimentary Treatise on the Manufacture of Bricks and Tiles*. 1850 (reprinted with an introduction by F. Celoria, 1971).

Fawcett J. *Historic Floors: Their History and Conservation*. 1998.

Franke L. and Schuman I. *Causes and Mechanisms of the Decay of Historic Brick Buildings in Northern Germany*. NATO-CCMS Pilot Strudy. 1992. Berlin.

Hammond M. D. P. *Bricks and Brickmaking*. 1981.

Handisyde C. C. and Haseltine B. *Bricks and Brickwork*. 1975.

Historic Scotland. *Preparation and Use of Lime Mortars*. 1995.

Horie C. V. *Materials for Conservation*. 1987.

Jenkins D. *Architectural Brickwork*. 1990.

Lizzi F. *The Static Restoration of Monuments*. Genoa. 1981.

Lloyd N. *A History of English Brickwork ... from Medieval Times to the End of the Georgian Period*. 1925 (reissued 1983).

London M. *Masonry: How to Care for Old and Historic Brick and Stone*. Washington, DC. 1986.

Mack R. C. and Askins J. S. *The Repointing of Historic Masonry Buildings*. Sermac, IL. 1971.

National Materials Advisory Board (USA) *Conservation of Historic Stone Buildings and Monuments*. Washington, DC. 1982.

Salzman L. F. *Building in England down to 1540.* 1952.

Stratton M. *The Terracotta Revival.* 1993.

Teutonico J. M. *A Laboratory Manual for Architectural Conservators.* Rome. 1988.

Teutonico J. M. *Architectural Ceramics.* 1996.

Torraca, Giorgio. *Porous Building Materials,* 3rd edition. ICCROM, Rome. 1988.

Wright J. *Brick Building in England from the Middle Ages to 1550.* 1972.

Worrall W. E. *Raw Materials.* The Institute of Ceramics. 1964.

In addition to the books listed above, the following journals and occasional publications are devoted to the subject, or specific aspects of it.

NATO CCMS. *Proceedings of Expert Meetings,* Nos 1–7. Berlin. 1988–1994.

British Brick Society. *Information.* London. 1973 (compilation volume 1, 1973–1981).

Further reference to case studies may be found in the following books and journals.

Ashburnham Brickworks (Chapter 2) Harmer *et al. Brick and Tile Making at Ashburnham.* 1991; also in Sussex Industrial History, 1977 and 1981.

Hampton Court (Chapter 13) Hyett J. (Feilden and Mawson) *Transactions of the Association for Studies in the Conservation of Hstoric Buildings.* 1992. No. 17. pp. 1–14.

Laughton Place (Chapter 13) Warren J. and Haslam C. *Journal of the Ancient Monuments Society.* 1986.

Farrant *et al.* Sussex Archaeological Collections. 1991.

San Marco (Chapter 13) Dr Ettore Vio, *Proceedings of the Nono Centenario di San Marco* (Italian). Venice. 1996.

The Smeaton Project (Chapter 12) Teutonico J. M. *et al.* The Smeaton Project: factors affecting the properties of lime-based mortars. *Bulletin of the Association for Preservation Technology,* 25: 3,4. 1994.

also *Lime News* 1994 and *Eurolime* 1994 and revised, forthcoming, English Heritage.

Ashurst J. *et al.* Development of lime mortar formulations for use in historic buildings. A report on the Smeaton Project. *Proceedings of the Seventh International Conference on Durability of Building Materials and Components.* ed. C. Sjöström. Stockholm. 1996.

Traditional buildings in Baghdad (Chapter 13) Warren J. and Fethi I. *Traditional Houses in Baghdad.* 1983.

Aga Khan Foundation. Proceedings of Symposium. Harvard. 1982.

Amanat al-Asima. *Conservation in Baghdad.* 1984.

The following British Standards are relevant to maintenance of brickwork (British Standards Institution, London).

Overridingly: *Guide to the Principles of the Conservation of Historic Buildings.* (1998).

BS 4729 1971 Special bricks.

BS 890 1972 Building limes.

BS 102 1973 Damp-proofing.

BS 5262 1976 External rendering.

BS 1199 & 1200 1976 Building sands.

BS 402 1979 Tiles for roofing.

BS 5837 1980 Trees near buildings.

BS 72 1981 Scaffolding.

BS 6270 Pt 1. 1982 Cleaning, surface repair, pointing.

BS 6270 1982 Abrasive cleaning.

BS 6263 1982 Tile floors.

BS 6180 1982 Protective barriers.

BS 3892 Pt 2. 1984 Pulverised Fuel Ash.

BS 6477 1984 Water repellents.

BS 5628 Pt 3. 1985 Use of masonry.

BS 8210 1986 Maintenance management.

BS 8210 1986 Cleaning.

BS 6463 1987 Analysis and tests for limes.

BS 5492 1997 Plastering

Index